NEUROMETHODS

Series Editor
Wolfgang Walz
University of Saskatchewan
Saskatoon, SK, Canada

For further volumes:
http://www.springer.com/series/7657

Neuromethods publishes cutting-edge methods and protocols in all areas of neuroscience as well as translational neurological and mental research. Each volume in the series offers tested laboratory protocols, step-by-step methods for reproducible lab experiments and addresses methodological controversies and pitfalls in order to aid neuroscientists in experimentation. *Neuromethods* focuses on traditional and emerging topics with wide-ranging implications to brain function, such as electrophysiology, neuroimaging, behavioral analysis, genomics, neurodegeneration, translational research and clinical trials. *Neuromethods* provides investigators and trainees with highly useful compendiums of key strategies and approaches for successful research in animal and human brain function including translational "bench to bedside" approaches to mental and neurological diseases.

Neuromuscular Assessments of Form and Function

Edited by

Philip J. Atherton

MRC-Versus Arthritis Centre for Musculoskeletal Ageing Research and NIHR Nottingham BRC, Centre of Metabolism, Ageing and Physiology (COMAP), School of Medicine, University of Nottingham, Derby, UK

Daniel J. Wilkinson

MRC-Versus Arthritis Centre for Musculoskeletal Ageing Research and NIHR Nottingham BRC, Centre of Metabolism, Ageing and Physiology (COMAP), School of Medicine, University of Nottingham, Derby, UK

Humana Press

Editors
Philip J. Atherton
MRC-Versus Arthritis Centre for Musculoskeletal Ageing Research and NIHR Nottingham BRC
Centre of Metabolism, Ageing and Physiology (COMAP), School of Medicine
University of Nottingham
Derby, UK

Daniel J. Wilkinson
MRC-Versus Arthritis Centre for Musculoskeletal Ageing Research and NIHR Nottingham BRC
Centre of Metabolism, Ageing and Physiology (COMAP), School of Medicine
University of Nottingham
Derby, UK

ISSN 0893-2336 ISSN 1940-6045 (electronic)
Neuromethods
ISBN 978-1-0716-3314-4 ISBN 978-1-0716-3315-1 (eBook)
https://doi.org/10.1007/978-1-0716-3315-1

This Humana imprint is published by the registered company Springer Science+Business Media, LLC, part of Springer Nature.
The registered company address is: 1 New York Plaza, New York, NY 10004, U.S.A.

Preface to the Series

Experimental life sciences have two basic foundations: concepts and tools. The Neuromethods series focuses on the tools and techniques unique to the investigation of the nervous system and excitable cells. It will not, however, shortchange the concept side of things as care has been taken to integrate these tools within the context of the concepts and questions under investigation. In this way, the series is unique in that it not only collects protocols but also includes theoretical background information and critiques which led to the methods and their development. Thus, it gives the reader a better understanding of the origin of the techniques and their potential future development. The Neuromethods publishing program strikes a balance between recent and exciting developments like those concerning new animal models of disease, imaging, in vivo methods, and more established techniques, including, for example, immunocytochemistry and electrophysiological technologies. New trainees in neurosciences still need a sound footing in these older methods in order to apply a critical approach to their results.

Under the guidance of its founders, Alan Boulton and Glen Baker, the Neuromethods series has been a success since its first volume published through Humana Press in 1985. The series continues to flourish through many changes over the years. It is now published under the umbrella of Springer Protocols. While methods involving brain research have changed a lot since the series started, the publishing environment and technology have changed even more radically. Neuromethods has the distinct layout and style of the Springer Protocols program, designed specifically for readability and ease of reference in a laboratory setting.

The careful application of methods is potentially the most important step in the process of scientific inquiry. In the past, new methodologies led the way in developing new disciplines in the biological and medical sciences. For example, Physiology emerged out of Anatomy in the nineteenth century by harnessing new methods based on the newly discovered phenomenon of electricity. Nowadays, the relationships between disciplines and methods are more complex. Methods are now widely shared between disciplines and research areas. New developments in electronic publishing make it possible for scientists that encounter new methods to quickly find sources of information electronically. The design of individual volumes and chapters in this series takes this new access technology into account. Springer Protocols makes it possible to download single protocols separately. In addition, Springer makes its print-on-demand technology available globally. A print copy can therefore be acquired quickly and for a competitive price anywhere in the world.

Saskatoon, SK, Canada *Wolfgang Walz*

Preface

Neuromuscular physiology has a strong relevance in clinical and basic sciences. In a clinical sense, skeletal muscles and their innervating nerves play a role in genetic muscle (such as muscular dystrophies) and/or neuromuscular diseases (such as amyotrophic lateral sclerosis). Equally, the neuromuscular system is subject to age-related declines—or sarcopenia—which remains a significant problem of the twenty-first century due to the burgeoning increase in lifespan owing to modern medicine and nutrition. Similarly, both infectious (including SARS-CoV2/COVID-19) and myriad non-infectious diseases are associated with neuromuscular declines—including rheumatic diseases, organ failures, diabetes, and cancers to name but a few. Finally, a basic understanding of neuromuscular biology remains underpinning to the deterministic basis of neuromuscular dysfunction, e.g., signal transduction, mitochondrial biology, and molecular motors powering contraction. It is this widespread relevance of neuromuscular biology that underpins the vast amount of research across aforementioned fields. At the heart of bioscience discovery are cutting-edge methodologies aimed to maximize biological insights in health, aging, and disease. It is to this end that the purpose of this book is to collate important methods spanning imaging techniques, metabolic tracing, and deep muscle phenotyping.

For readers interested in means to quantify attributes of muscle mass, we include chapters relating to the imaging of skeletal muscle mass, i.e., by magnetic resonance imaging (MRI) and dual-energy x-ray absorptiometry (DXA), computed tomography (CT), and ultrasound.

For readers interested in muscle functional aspects and nerve-muscle physiology, we include chapters covering measures of neuromuscular function such as power/strength/force. We also include both surface electromyography (EMG) and intramuscular (iEMG). On a molecular level, study of the neuromuscular junction in muscle biopsy tissue is also covered with a focus on microscopy and immunohistochemistry. Ex vivo fiber function and isolation is also included for readers with an interest in the study of fiber mechanics in vitro.

From a metabolic perspective, we include chapters on stable isotope tracing of muscle proteostasis (i.e., protein turnover) and also the quantification of blood flow within muscle tissue using contrast-enhanced ultrasound approaches. Magnetic resonance imaging (MRI)-based energetics (phosphocreatine re-synthesis) techniques will showcase in vivo imaging of dynamic metabolism in humans. Recent advances in our understanding of muscle endocrine signaling will be covered in the form of the measurement of so-called "myokines."

In being a unique aspect of skeletal muscles in the context of resident stem cells, a chapter focusing on measures of skeletal muscle satellite cell physiology and function is included. Finally, in recognizing the contribution of model organisms in the field, approaches using the nematode *C. elegans* are included in relation to the genetic study of neuromuscular disorders.

As editors, we hope this book will represent a one-stop shop for those interested in either singular or multiple methods to study skeletal muscle neurophysiology.

Derby, UK — *Philip J. Atherton*
Derby, UK — *Daniel J. Wilkinson*

Contents

Contributors

JESPER L. ANDERSEN • *Institute of Sports Medicine Copenhagen, Department of Orthopaedic Surgery, Copenhagen University Hospital – Bispebjerg and Frederiksberg, Copenhagen, Denmark; Center for Healthy Aging, Department of Clinical Medicine, University of Copenhagen, Copenhagen, Denmark*

MATTHEW S. BROOK • *MRC-Versus Arthritis Centre for Musculoskeletal Ageing Research and NIHR Nottingham BRC, Centre of Metabolism, Ageing and Physiology (COMAP), School of Medicine, University of Nottingham, Derby, UK; School of Life Sciences, University of Nottingham, Nottingham, UK*

COLLEEN S. DEANE • *Department of Sport and Health Sciences, College of Life and Environmental Sciences, University of Exeter, Exeter, UK; Living Systems Institute, University of Exeter, Exeter, UK*

JASON M. DEFREITAS • *Health and Human Performance, Applied Neuromuscular Physiology Laboratory, Oklahoma State University, Stillwater, OK, USA*

TIMOTHY ETHERIDGE • *Department of Sport and Health Sciences, College of Life and Environmental Sciences, University of Exeter, Exeter, UK*

KATHERINE L. FORD • *Human Nutrition Research Unit, Department of Agricultural, Food & Nutritional Science, University of Alberta, Edmonton, AB, Canada*

MARTINO V. FRANCHI • *Department of Biomedical Sciences, University of Padova, Padova, Italy*

SUSAN T. FRANCIS • *Sir Peter Mansfield Imaging Centre, MRC-Versus Arthritis Centre for Musculoskeletal Ageing Research, National Institute for Health Research Nottingham Biomedical Research Centre, Queen's Medical Centre, University of Nottingham, Nottingham, UK*

M. CRISTINA GONZALEZ • *Postgraduate Program in Health and Behavior, Catholic University of Pelotas, Pelotas, Brazil; Pennington Biomedical Research Center, Louisiana State University, Baton Rouge, LA, USA*

PAUL L. GREENHAFF • *School of Life Sciences, MRC-Versus Arthritis Centre for Musculoskeletal Ageing Research, National Institute for Health Research Nottingham Biomedical Research Centre, Queen's Medical Centre, University of Nottingham, Nottingham, UK*

SAMANTHA HUGHES • *Department Environment and Health, Vrije Universiteit Amsterdam, Amsterdam, The Netherlands*

ELEANOR J. JONES • *Centre of Metabolism, Ageing and Physiology (COMAP), MRC-Versus Arthritis Centre of Excellence for Musculoskeletal Ageing Research, Nottingham NIHR Biomedical Research Centre, School of Medicine, University of Nottingham, Derby, UK*

ANA TERESA LIMON-MIRO • *Department of Medicine, University of Alberta, Edmonton, AB, Canada*

ABIGAIL L. MACKEY • *Institute of Sports Medicine Copenhagen, Department of Orthopaedic Surgery, Copenhagen University Hospital – Bispebjerg and Frederiksberg, Copenhagen, Denmark; Center for Healthy Aging, Department of Clinical Medicine, University of Copenhagen, Copenhagen, Denmark*

EDUARDO MARTINEZ-VALDES • *Centre of Precision Rehabilitation for Spinal Pain, School of Sport, Exercise and Rehabilitation Sciences, University of Birmingham, Birmingham, UK; MRC-Arthritis Research UK Centre for Musculoskeletal Ageing Research, University of Birmingham, Birmingham, UK*

JORDAN J. MCGING • *School of Medicine, University of Nottingham, Nottingham, UK*

GORDON W. MORAN • *School of Medicine, MRC-Versus Arthritis Centre for Musculoskeletal Ageing Research, National Institute for Health Research Nottingham Biomedical Research Centre, Queen's Medical Centre, University of Nottingham, Nottingham, UK*

ROBERT H. MORRIS • *School of Science and Technology, Nottingham Trent University, Clifton Campus, Clifton Lane, Nottingham, UK*

MARCO V. NARICI • *CIR-MYO Myology Centre, Department of Biomedical Sciences, University of Padova, Padova, Italy*

FRANCESCO NEGRO • *Department of Clinical and Experimental Sciences, Università degli Studi di Brescia, Brescia, Italy*

ANDY NOLAN • *Research Institute of Sport and Exercise Sciences, Liverpool John Moore's University, Liverpool, UK*

CAMILA L. P. OLIVEIRA • *Human Nutrition Research Unit, Department of Agricultural, Food, and Nutritional Science, University of Alberta, Edmonton, AB, Canada*

ANA P. PAGANO • *Human Nutrition Research Unit, Department of Agricultural, Food, and Nutritional Science, University of Alberta, Edmonton, AB, Canada*

BETHAN E. PHILLIPS • *Centre of Metabolism, Ageing and Physiology (COMAP), MRC-Versus Arthritis Centre of Excellence for Musculoskeletal Ageing Research, Nottingham NIHR Biomedical Research Centre, School of Medicine, University of Nottingham, Derby, UK*

MATHEW PIASECKI • *Centre of Metabolism, Ageing & Physiology (COMAP), MRC-Versus Arthritis Centre for Musculoskeletal Ageing Research, National Institute for Health Research (NIHR) Nottingham Biomedical Research Centre, School of Medicine, University of Nottingham, Nottingham, UK*

CARLA M. PRADO • *Human Nutrition Research Unit, Department of Agricultural, Food & Nutritional Science, University of Alberta, Edmonton, AB, Canada*

CARLO REGGIANI • *University of Padova, Department of Biomedical Sciences, Padova, Italy; ZRS Koper, Science and Research Center, Koper, Slovenia*

MICHAEL D. ROBERTS • *School of Kinesiology, Molecular and Applied Sciences Laboratory, Auburn University, Auburn, AL, USA*

CRAIG SALE • *Department of Sport and Exercise Sciences, Manchester Metropolitan University Institute of Sport, Manchester, UK*

SÉBASTIEN SERRES • *School of Life Sciences, MRC-Versus Arthritis Centre for Musculoskeletal Ageing Research, National Institute for Health Research Nottingham Biomedical Research Centre, Queen's Medical Centre, University of Nottingham, Nottingham, UK*

ADAM SHARPLES • *Institute for Physical Performance, Norwegian School of Sport Sciences, Olso, Norway*

BRUNA RAMOS DA SILVA • *Human Nutrition Research Unit, Department of Agricultural, Food & Nutritional Science, University of Alberta, Edmonton, AB, Canada; Department of Health Sciences, Ribeirão Preto Medical School, University of Sao Paulo, Ribeirão Preto, SP, Brazil*

KEN SMITH • *MRC-Versus Arthritis Centre for Musculoskeletal Ageing Research and NIHR Nottingham BRC, Centre of Metabolism, Ageing and Physiology (COMAP), School of Medicine, University of Nottingham, Derby, UK; School of Life Sciences, University of Nottingham, Nottingham, UK*

CASPER SOENDENBROE • *Institute of Sports Medicine Copenhagen, Department of Orthopaedic Surgery, Copenhagen University Hospital – Bispebjerg and Frederiksberg, Copenhagen, Denmark; Center for Healthy Aging, Department of Clinical Medicine, University of Copenhagen, Copenhagen, Denmark*

DANIEL W. STASHUK • *Department of Systems Design Engineering, University of Waterloo, Waterloo, ON, Canada*

CLAIRE STEWART • *Research Institute of Sport and Exercise Sciences, Liverpool John Moore's University, Liverpool, UK*

NATHANIEL SZEWCZYK • *Department of Biomedical Sciences, Heritage College of Osteopathic Medicine, Ohio University, Athens, OH, USA*

MARK VIGGARS • *Research Institute of Sport and Exercise Sciences, Liverpool John Moore's University, Liverpool, UK*

DANIEL J. WILKINSON • *MRC-Versus Arthritis Centre for Musculoskeletal Ageing Research and NIHR Nottingham BRC, Centre of Metabolism, Ageing and Physiology (COMAP), School of Medicine, University of Nottingham, Derby, UK; School of Life Sciences, University of Nottingham, Nottingham, UK*

CRAIG R. G. WILLIS • *Department of Sport and Health Sciences, College of Life and Environmental Sciences, University of Exeter, Exeter, UK; School of Chemistry and Biosciences, University of Bradford, Bradford, UK; Ohio Musculoskeletal and Neurological Institute, Ohio University, Athens, OH, USA*

Chapter 1

Estimation of Lean Soft Tissue by Dual-Energy X-Ray Absorptiometry as a Surrogate for Muscle Mass in Health, Obesity, and Sarcopenia

Camila L. P. Oliveira, Ana P. Pagano, M. Cristina Gonzalez, and Carla M. Prado

Abstract

Dual-energy X-ray absorptiometry (DXA) is a three-compartment, molecular-level method of body composition assessment widely available in clinical settings. It measures lean soft tissue (LST), fat mass, and bone mineral content and is currently a surrogate method to assess muscle mass. As DXA does not directly measures skeletal muscle, appendicular LST (ALST) is used for its estimation, as 75% of whole-body skeletal muscle is located in the arms and legs, and ~74% of the ALST is comprised of skeletal muscle. In fact, The European Working Group on Sarcopenia in Older People consensus on the definition and diagnosis of sarcopenia supported the use of this technique to confirm sarcopenia diagnosis. However, some question the ability of DXA to reflect functional muscle mass as DXA measures LST, and, therefore, skeletal muscle mass and composition are not being directly assessed.

DXA shows very high precision and accuracy; its measurements are considered quick, safe, noninvasive, and observer-independent; emits very low radiation doses; and exhibits minimal individual burden. On the other hand, DXA is costly, is not portable, and needs a trained technician and cross-calibration studies for direct comparison of results among different makes and models. Furthermore, pregnant women should not be exposed to DXA scans.

Among all commercial manufacturers of DXA, the General Electric (GE) Healthcare, Ltd. (Madison, WI, USA) is the most popular one. For this reason, this chapter is focused on the model Lunar iDXA and the enCORE software version 18 (GE Healthcare, Ltd., Madison, WI, USA). Although general concepts in this chapter can be generalized to different manufactured instruments, operational descriptions are unique to the model hereby presented.

Key words Dual-energy X-ray absorptiometry, Body composition, Lean soft tissue, Fat mass, Bone mineral content

1 Introduction

Dual-energy X-ray absorptiometry (DXA) is one of the most commonly used methods for body composition assessment [1]. It is a three-compartment, molecular-level method of body composition assessment, which includes regional (arms, legs, and trunk) or

Philip J. Atherton and Daniel J. Wilkinson (eds.), *Neuromuscular Assessments of Form and Function*, Neuromethods, vol. 204,
https://doi.org/10.1007/978-1-0716-3315-1_1,

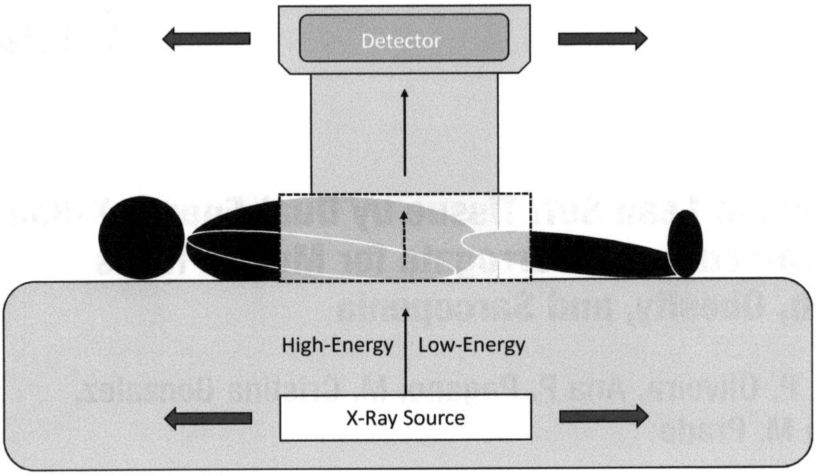

Fig. 1 Fundamental principles of dual-energy X-ray absorptiometry

whole-body lean soft tissue (LST), fat mass (FM), and bone mineral content (BMC) [2–4]. In addition to its initial purpose, DXA is currently a surrogate method to assess *muscle mass*. Although it does not directly measure this body compartment, muscle mass is part of the LST, which consists of total body water, protein, carbohydrates, nonfat lipids, and soft tissue minerals [1, 2]. More specifically the appendicular skeletal muscle mass, more correctly termed appendicular LST (i.e., the sum of the LST from upper and lower limbs, ALST), can be used as an estimation of skeletal muscle. The rationale is that 75% of the whole-body skeletal muscle is located in the arms and legs and approximately 74% of the ALST comprises skeletal muscle [5]. In addition to skeletal muscle, the ALST also includes water and fibrotic and connective tissues [6].

The physical principles of DXA are based on the attenuation of X-rays of high- and low-photon energy levels that are passed through the human body, allowing the distinction of tissues of higher and lower densities (i.e., bone and soft tissue, respectively) (Fig. 1) [1, 4]. The discrimination of soft tissue between FM and LST is based on attenuation values previously obtained from measurements of pure fat and bone-free soft tissue [7]. The X-ray beams are emitted from below the scanner table and are detected by a photon detector located above the individual [1]. The distribution and detection of X-rays depend on whether the scanner uses a pencil beam or fan beam geometry [8]. The pencil beam uses a single beam of X-ray and a single detector to generate the image, resulting in a measurement time of approximately 15 min [8]. In the early 1990s, the fan beam geometry was introduced, in which X-rays are distributed across a wider area of the body and captured by several detectors, improving the scanning time (approximately 5 min) as well as image resolution and quality [8].

A literature review conducted by Toombs et al. [9] revealed that in vivo precision of DXA is very high with coefficients of variation ranging from 0.4% to 1.3% for LST, 0.8% to 2.7% for FM, and 0.6% to 4% for %FM, making it a useful tool for tracking changes in body composition over time. Compared to the four-compartment model, the accuracy of DXA for measurement of % FM in weight-stable healthy adults is also high with a mean difference between techniques ranging from −5.3% to 2.9% [9]. Several factors can affect DXA's precision and accuracy, such as hydration status, body thickness and size, food intake, level of training of the technician, software packages, and machine calibration procedures [1, 2]. This technique's measurements are considered quick, safe, noninvasive, observer-independent, and this technique emits very low radiation doses (1–7 μSv) compared to other imaging techniques such as computed tomography (CT) (range between 2.1 and 31 mSv or higher, depending on imaging purpose) with a minimal individual burden [1, 2, 10, 11]. On the other hand, DXA is costly, is not portable, and needs a trained technician and cross-calibration studies for direct comparison of results among different makes and models [9, 12]. Furthermore, The International Society of Clinical Densitometry recommends that pregnant women should not be exposed to DXA scans [13]. The authors justify this recommendation based on the lack of valuable diagnostic information this technique can provide during pregnancy, precluding radiation exposure justification [13].

DXA was first created in the 1980s to assess BMC and bone mineral density (BMD) for the detection of osteopenia/porosis and still represents the gold-standard method for this purpose [1]. For this reason, DXA is widely available in clinical settings. The American Society for Parenteral and Enteral Nutrition Clinical Guidelines recognized DXA as a valid technique for the assessment of FM in patients with a variety of disease states, but with an unknown validity for LST assessment [14]. Notably, the number of studies in clinical populations is limited [14].

In regard to aging, DXA has been recommended in the diagnosis of sarcopenia due to the measurement of ALST. The European Working Group on Sarcopenia in Older People consensus on the definition and diagnosis of sarcopenia supported the use of this technique in both clinical and research settings to confirm sarcopenia diagnosis and have proposed cutoff points for low muscle quantity based on ALST and ALST index (i.e., ALST/height^2) [15]. An ALST <20 kg for men and <15 kg for women has been established based on data from $n = 26{,}625$ older adults [16], and an ALST index (i.e., ALST/height^2) <7.0 kg/m^2 for men and <6.0 kg/m^2 for women has been established based on $n = 682$ Australians aged 20 to 39 years [17]. In addition to the ALST index, the absolute ALST can also be adjusted for weight (i.e., ALST/weight) and body mass index (i.e., ALST/body mass

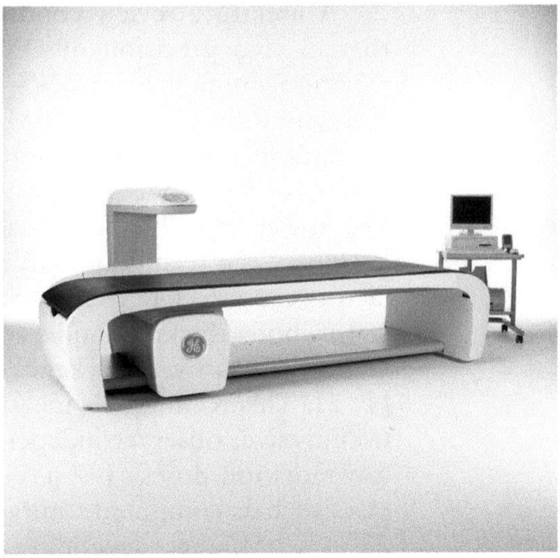

Fig. 2 Lunar iDXA (GE Healthcare, Ltd., Madison, WI, USA). (Used with permission of GE Healthcare, Ltd. (Madison, WI, USA))

index) if data are available for a relevant normative population [18]. Although the absolute and adjusted ALST values obtained from DXA are valuable and currently recommended for sarcopenia diagnosis, these are not widely accepted. Many cutoffs to define low levels of ALST have been proposed. Additionally, some question the ability of DXA to reflect functional muscle mass as skeletal muscle is not being directly assessed. This has led to disagreement regarding DXA use in the definition of sarcopenia. For example, the Sarcopenia Definition and Outcomes Consortium recently suggested that LST should not be included in the definition of sarcopenia [19]. However, evidence suggests the latter can explain the poor association of DXA-derived ALST with physical performance and functional outcomes when compared to techniques that directly measure total body muscle mass (i.e., functional muscle mass), such as the D_3-creatine dilution [6].

There are several commercial manufacturers of DXA, and the ones leading the market are General Electric (GE) Healthcare, Ltd. (Madison, WI, USA), Hologic, Inc. (Bedford, MA, USA), and Cooper Surgical, Inc. (Trumbull, CT, USA). Among these, GE Healthcare, Ltd. (Madison, WI, USA) is the most popular one. For this reason, the next sections of this chapter focused on the model Lunar iDXA and the enCORE software version 18 (GE Healthcare, Ltd., Madison, WI, USA) (Fig. 2). All information contained herein was used with permission of GE Healthcare, Ltd.

2 Materials

The Lunar iDXA (GE Healthcare, Ltd., Madison, WI, USA) must be placed in a clean, ventilated area with a minimum clearance of 24 cm at the head and foot ends of the scanner table, at least 15 cm for the arm side, 45 cm for the operator side, and at least 15 cm between the console table and the wall. The recommended room dimensions for the scanner are 4.0 m × 3.35 m (13.2 ft. × 11 ft); however, if the computer and peripherals are powered by an isolation transformer, the minimum room dimensions are 3.2 m × 3.35 m (10.5 ft. × 11 ft). The humidity for the scanner area must be kept between 20 and 80% and the temperature between 65 and 81 °F (18 and 27 °C).

2.1 Maintenance

2.1.1 Daily Quality Assurance

Quality assurance is a procedure that should be performed every morning before measurements using a calibration block.

2.1.2 Weekly Cleaning

The scanner table environment should be cleaned weekly using a vacuum cleaner and nonabrasive cleaners. The surface can be dusted regularly to remove dirt.

2.1.3 Annually - GE-Authorized Maintenance

An annual preventive maintenance is recommended by a GE-authorized service engineer.

2.2 Measurement

The following are recommended prior to a DXA measurement for body composition assessment [1]:

1. Overnight fasting period of 12 h.
2. Minimal clothing, such as a jogging suit. Alternatively, a hospital gown can be provided.
3. Empty bladder before the scan.
4. All metal items, such as clothing with zippers, snaps, buckles, buttons, and belts, as well as any jewelry must be removed.
5. Exercise must be avoided for approximately 12 h prior to scanning.
6. GE Healthcare, Ltd. (Madison, WI, USA) recommends that the following population groups should not be exposed to a DXA test:
   (a) Subjects who have ingested or have been injected with radionuclides or radiopaque agents in the past 3 to 5 days.
   (b) Pregnant women.

3 Methods

The following sections are worded per the user manual, used with permission as mentioned above.

3.1 Maintenance

3.1.1 Daily Quality Assurance

The quality assurance test verifies the functionality, accuracy, and precision of the machine. It should be performed each morning before any measurement or at least once a week if the scanner is not being used. If the room temperature changes ±5 °C, an additional quality assurance test is recommended during the day. For a quality assurance test of the Lunar iDXA enCORE version 18 (GE Healthcare, Ltd., Madison, WI, USA), the following steps should be completed:

1. Select *QA* from the common toolbar.
2. Select *Start*. A message instructs you to position the calibration block.
3. Put the calibration block on the pad so that the laser light rests in the center of the crosshair label on the calibration block, and the brass is on the bottom.
4. Select *OK*, and follow the screen prompts to complete the quality assurance procedure. If the test did not pass, reposition the calibration block and repeat the procedure. If the procedure fails a second time, GE Support should be called for assistance.
5. To print the results, select *Report*. Keep the printout for 1 year.
6. To trend the results, select *Settings*, choose *Table Columns* or *Graph Y-Axis*, and select the result that you would like to trend. Select *OK*. Failures are represented by red dots on the graph.

3.2 Measurement

To measure body composition using the Lunar iDXA enCORE version 18 (GE Healthcare, Ltd., Madison, WI, USA), the following steps should be completed:

1. Record information for a new subject or select an existing record from the database.
2. Position the subject onto the scanner table:
   (a) Position the subject's body in the center of the scanner table. If a subject is wider than the scan area, the subject can be positioned for a half-body scan, in which all of the right side of the body, head, and spine are included in the scan window, Fig. 3 (*see* Section 3.3, **step 3**).
   (b) The subject's head must be ~3 cm below the horizontal line on the table pad (Fig. 3a).

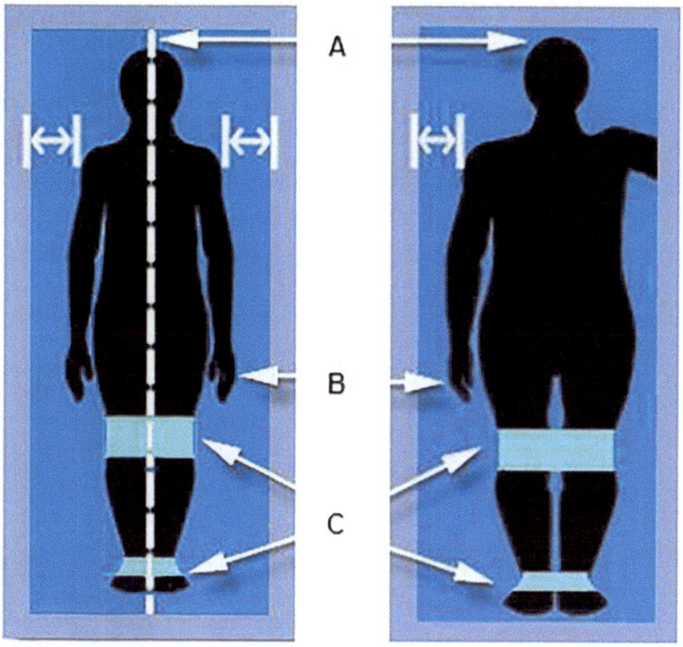

Fig. 3 Subject positioning on the scanner table. (Used with permission of GE Healthcare, Ltd. (Madison, WI, USA))

(c) The subject's arms should be within the scan area, hands turned on the sides with thumbs up, palms facing legs, and arms alongside the body. Hands should not touch the legs. A ~1 cm air gap should be left between the arms and torso (Fig. 3b).

(d) Use the Velcro straps on the subject's knees and ankles to prevent movement during measurement (Fig. 3c).

3. Select *Position* from the New Measurement toolbar.
4. Remove the subject's shoes.
5. To start the measurement, select *Start* from the *New Measurement* toolbar.
6. Monitor the image to make sure it is correct. If the image is not correct, select *Abort* and reposition the subject. To complete another measurement for the subject, select *Set Up* from the *New Measurement* toolbar.
7. When the measurement is complete, select *Home* to move the scanner arm to the home position.
8. Select *Analyze* to proceed to analysis or *Close* to exit the measurement screen.

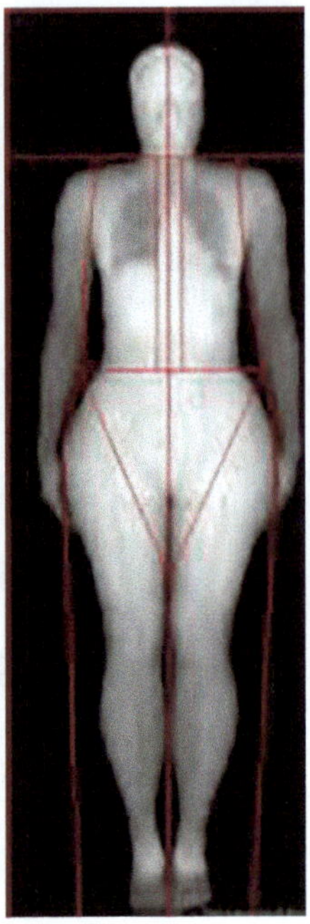

Fig. 4 Body composition analysis landmarks. (Used with permission of GE Healthcare, Ltd. (Madison, WI, USA))

3.3 Analysis

The following analysis procedures are applicable to the Lunar iDXA enCORE version 18 (GE Healthcare, Ltd., Madison, WI, USA):

1. Total body:
   (a) If an image is not already open, select an image file for analysis. Composition results are shown on the screen when you select the *Composition* tab in the *Analyze* window.
   (b) Use the cut positions to define the tissue regions (Fig. 4).
   (c) To print a report, select the *Composition* report option in the *Reports* dialog box.
2. Regional options:
   (a) Right/left: the option to generate left and right results is located in *Tools* → *User Options* → *Analyze tab* → *Total Body Analysis Options*. Enable the option to calculate *Left/Right* results.
   (b) Additional options such as the *Total Body Less Head* and upper and lower arm and leg are available in the features *Adults TBLH* and *Small Body Comp ROIs*, respectively.
3. Half-body scan (Fig. 5):
   (a) Click *ROIs*, and then click *Estimate* to estimate one side of the body from the other.

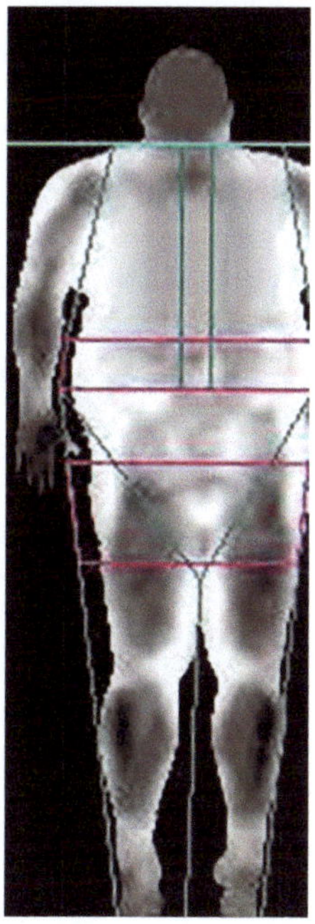

Fig. 5 Half-body scan. (Used with permission of GE Healthcare, Ltd. (Madison, WI, USA))

(b) If the body region is derived from an estimate from one half of the body, *(e)* appears next to the Region column in Densitometry and Composition results tables and next to the Measured Date column in Trend tables.

4. Android and gynoid fat (Fig. 6):

(a) Android: lower boundary at pelvis cut. Upper boundary above the pelvis cut by 20% of the distance between the pelvis and neck cuts. Lateral boundaries are the arm cuts.

(b) Gynoid: upper boundary below the pelvis cut line by 1.5 times the height of the android region. Gynoid height is equal to 2 times the height of the android region. Lateral boundaries are the outer leg cuts.

(c) The android/gynoid (A/G) ratio is the ratio between the %fat of the android (central) and %fat of the gynoid (hip and thigh) regions. The Enhanced Composition report page displays the android/gynoid ratio.

5. Visceral adipose tissue (VAT) and subcutaneous adipose tissue (SAT):

The software provides estimates of VAT and SAT, which would be more correctly termed visceral FM and subcutaneous FM. Adipose tissue consists of all connective tissue formed by adipocytes, collagenous and elastic fibers, fibroblasts, and capillaries [2]. This body compartment consists of approximately 60 to 85% of FM, 2 to 10% of protein, and 10 to 20% of water [1].

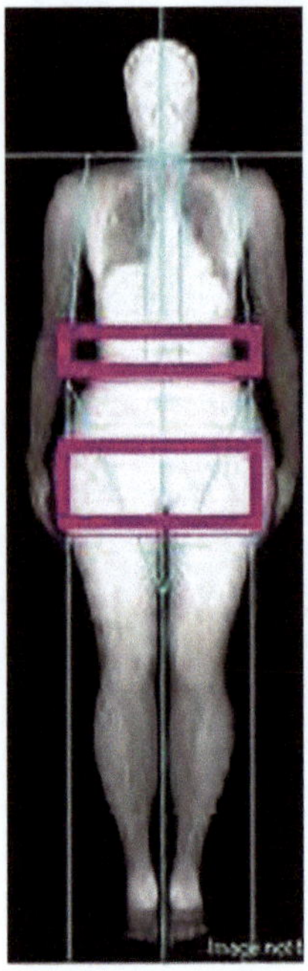

Fig. 6 Android and gynoid fat. (Used with permission of GE Healthcare, Ltd. (Madison, WI, USA))

(a) The CoreScan software option estimates the VAT and SAT within the android region (i.e., defined as the lower boundary at pelvis cut, upper boundary above the pelvis cut by 20% of the distance between the pelvis and neck cuts, and lateral boundaries are the arm cuts). The content that is estimated is the VAT mass, area, and volume and the SAT mass, area, and volume.

(b) The results have been previously validated by Kaul et al. [20]. The values are displayed on the *CoreScan* tab on the *Analyze* screen.

(c) CoreScan results are found on the *Total Body Composition Ancillary Report*, and Composer field codes are available.

6. Composition report options:

(a) In addition to the main Composition report, enCORE also features a Composition Ancillary Report, which includes:

(i) Fat mass ratios of trunk/total, legs/total, and (arms + legs)/trunk.

(ii) Resting metabolic rate: calculated from anthropometric values using equations from Harris and Benedict [21] or Mifflin et al. [22]. The software also has the option of manually entering additional equations.

(iii) Relative skeletal muscle index (RSMI): corresponds to the ALST index (i.e., the sum of the LST from the upper and lower limbs divided by height^2) and is calculated using the Baumgartner equation [23]: RSMI or ALST index = (LST of arms [kg] + LST of legs [kg]) / (height [m^2]).

(iv) Total body, intracellular, and extracellular water: these values are not measured by DXA but can be assessed by another body composition technique and manually entered into the software.

(b) The above metabolic information is displayed on the Metabolic tab on the Analyze screen for total body exams. You can configure the display of this information in *Tools → User Options → Results Display → Metabolic Options.*

7. Sarcopenia software feature:

This feature allows for the categorization of subjects into being diagnosed with sarcopenia or not. This can be done using one of the six official definitions of sarcopenia [15, 16, 23–26] or entering a new definition criterion manually.

(a) The Sarcopenia software feature can be enabled by selecting *Tools → User Options → Analyze tab → Total Body Analysis Options.*

(b) Select the *Change* button in the Sarcopenia Definition group to display the Sarcopenia Definition dialog.

(c) Configure the sarcopenia definition according to site needs by selecting a default sarcopenia definition and selecting *OK*. The software gives the option to select one of the six official definitions of sarcopenia [15, 16, 23–26]. There is also the option of entering a new definition criterion.

(d) The Sarcopenia Calculator is available for total body analysis when the subject meets the minimum sarcopenia definition age, which will depend on the definition previously chosen [15, 16, 23–26].

(e) Sarcopenia report is available on the Reports Center dialog to show either current exam results (non-trending) or trend results based on the *Report Center Trend* checkbox.

4 Notes

DXA assumes a constant hydration of fat-free mass (i.e., the sum of LST and BMC) of approximately 73% [27]. For this reason, factors such as hydration alterations (i.e., variations >5%), digestive tract content, and previous exercise practice can alter the subject's hydration status and hence affect body composition measurement causing an overestimation of the LST compartment [27, 28].

The most common DXA scans have a weight capacity between 114 and 204 kg and a scan area between 193 to 197.5 cm (length) and 57.6 to 66 cm (width) [29]. The Lunar iDXA (GE Healthcare, Ltd., Madison, WI, USA) has a weight capacity of 227 kg and a maximum scanning area for total body measurements of 196.8 cm (length) and 66 cm (width). Although the weight capacity and dimensions of most DXA scans can accommodate most of the population, there are subjects who exceed these limits. Alternative techniques have been developed to assess their body composition, such as the sum of two separate scans (head separated from trunk and limbs), adopting a knee-bent position, or performing half-body scans [29]. Half-body scans using the Lunar iDXA (GE Healthcare, Ltd., Madison, WI, USA) accurately predicted whole-body FM (kg and %) and LST in individuals with obesity [30]. This result can be explained by the symmetry between the left and right sides of the body, which has been previously reported as 0.03 ± 0.08 kg for BMC ($r^2 = 0.89$), -0.04 ± 0.62 kg for FM ($r^2 = 0.99$), and -0.04 ± 0.86 kg for LST ($r^2 = 0.97$) [31]. In addition to limitations regarding the weight capacity and scan area dimensions, tissue thickness > 25 cm can influence X-ray penetration into tissues, which causes attenuation errors, resulting in an underestimation of FM [32]. The standard body composition analysis from the Lunar iDXA enCORE version 18 (GE Healthcare, Ltd., Madison, WI, USA) reports %FM as a percentage of LST and FM (i.e., excludes BMC - tissue %FM) and as a percentage of total body mass (i.e., includes BMC - region %FM). Depending on the researcher's and clinician's goals, either can be used; however, if the intention is to report total body FM, the region %FM should be used.

The use of different DXA scanners for repeated measures is associated with precision errors. These errors may occur due to differences in calibration between and within manufacturers, within equipment of same make and model, as well as according to the software in use [33]. Therefore, the International Society for Clinical Densitometry recommends the use of the same scanner for repeated measures [12]. When multiple scans are obtained from different machines, cross-calibration is needed to reduce precision errors [33], allowing a direct comparison of results from different equipment [34]. The cross-calibration can be performed using in vitro and/or in vivo scans. To perform in vitro cross-calibration, an adequate phantom should be used to mimic in vivo scans [12]. The International Society for Clinical Densitometry recommends the use of an adequate phantom for cross-calibration of different scanners from the same make and model; however, in vivo cross-calibration is indicated for equipment from different manufacturers [12]. Cross-calibration equations have been published and should be considered when comparing different equipment [33]. According to the International Society for Clinical Densitometry, the minimum acceptable precision is 2% for LST, 2% for FM%, 3% for FM, and 1% for BMD [12]. Additional information on cross-calibration can be found elsewhere [12, 33].

Acknowledgments

We would like to thank Dan Hill (Aymes Medical, Newmarket, ON, Canada) for sharing his expertise and facilitating access to some of the information hereby presented.

References

1. Blew RM, Sardinha LB, Milliken LA (2020) Body composition laboratory methods. In: Lohman TG, Milliken LA (eds) ACSM's body composition assessment. Human Kinetics, Champaign, pp 31–58
2. Prado CM, Heymsfield SB (2014) Lean tissue imaging: a new era for nutritional assessment and intervention. JPEN J Parenter Enteral Nutr 38(8):940–953
3. Fosbol MO, Zerahn B (2015) Contemporary methods of body composition measurement. Clin Physiol Funct Imaging 35(2):81–97
4. Siervo M, Jebb SA (2010) Body composition assessment: theory into practice: introduction of multicompartment models. IEEE Eng Med Biol Mag 29(1):48–59
5. Kim J, Wang Z, Heymsfield SB, Baumgartner RN, Gallagher D (2002) Total-body skeletal muscle mass: estimation by a new dual-energy X-ray absorptiometry method. Am J Clin Nutr 76(2):378–383
6. Cawthon PM, Orwoll ES, Peters KE, Ensrud KE, Cauley JA, Kado DM et al (2019) Strong relation between muscle mass determined by D3-creatine dilution, physical performance, and incidence of falls and mobility limitations in a prospective cohort of older men. J Gerontol A Biol Sci Med Sci 74(6):844–852
7. Roubenoff R, Kehayias JJ, Dawson-Hughes B, Heymsfield SB (1993) Use of dual-energy x-ray absorptiometry in body-composition studies: not yet a "gold standard". Am J Clin Nutr 58(5):589–591
8. Blake GM, Parker JC, Buxton FMA, Fogelman I (1993) Dual X-ray absorptiometry: a comparison between fan beam and pencil beam scans. Br J Radiol 66(790):902–906
9. Toombs RJ, Ducher G, Shepherd JA, De Souza MJ (2012) The impact of recent technological advances on the trueness and precision of DXA to assess body composition. Obesity (Silver Spring, Md) 20(1):30–39
10. Rehani MM, Berry M (2000) Radiation doses in computed tomography. The increasing doses of radiation need to be controlled. BMJ 320(7235):593–594
11. Smith-Bindman R, Lipson J, Marcus R, Kim K-P, Mahesh M, Gould R et al (2009) Radiation dose associated with common computed tomography examinations and the associated lifetime attributable risk of cancer. Arch Intern Med 169(22):2078–2086
12. Hangartner TN, Warner S, Braillon P, Jankowski L, Shepherd J (2013) The Official Positions of the International Society for Clinical Densitometry: acquisition of dual-energy X-ray absorptiometry body composition and considerations regarding analysis and repeatability of measures. J Clin Densitom 16(4): 520–536
13. Kendler DL, Borges JL, Fielding RA, Itabashi A, Krueger D, Mulligan K et al (2013) The Official Positions of the International Society for Clinical Densitometry: indications of use and reporting of DXA for body composition. J Clin Densitom 16(4):496–507
14. Sheean P, Gonzalez MC, Prado CM, McKeever L, Hall AM, Braunschweig CA (2020) American Society for Parenteral and Enteral Nutrition Clinical Guidelines: the validity of body composition assessment in clinical populations. JPEN J Parenter Enteral Nutr 44(1):12–43
15. Cruz-Jentoft AJ, Bahat G, Bauer J, Boirie Y, Bruyère O, Cederholm T et al (2019) Sarcopenia: revised European consensus on definition and diagnosis. Age Ageing 48(1):16–31
16. Studenski SA, Peters KW, Alley DE, Cawthon PM, McLean RR, Harris TB et al (2014) The FNIH sarcopenia project: rationale, study

description, conference recommendations, and final estimates. J Gerontol A Biol Sci Med Sci 69(5):547–558

17. Gould H, Brennan SL, Kotowicz MA, Nicholson GC, Pasco JA (2014) Total and appendicular lean mass reference ranges for Australian men and women: the Geelong osteoporosis study. Calcif Tissue Int 94(4):363–372
18. Kim KM, Jang HC, Lim S (2016) Differences among skeletal muscle mass indices derived from height-, weight-, and body mass index-adjusted models in assessing sarcopenia. Korean J Intern Med 31(4):643–650
19. Bhasin S, Travison TG, Manini TM, Patel S, Pencina KM, Fielding RA et al (2020) Sarcopenia definition: the position statements of the sarcopenia definition and outcomes consortium. J Am Geriatr Soc 68(7):1410–1418
20. Kaul S, Rothney MP, Peters DM, Wacker WK, Davis CE, Shapiro MD et al (2012) Dual-energy X-ray absorptiometry for quantification of visceral fat. Obesity (Silver Spring, Md), 20(6):1313–1318
21. Harris JA, Benedict FG (1918) A biometric study of human basal metabolism. Proc Natl Acad Sci U S A 4(12):370–373
22. Mifflin MD, St Jeor ST, Hill LA, Scott BJ, Daugherty SA, Koh YO (1990) A new predictive equation for resting energy expenditure in healthy individuals. Am J Clin Nutr 51(2): 241–247
23. Baumgartner RN, Koehler KM, Gallagher D, Romero L, Heymsfield SB, Ross RR et al (1998) Epidemiology of sarcopenia among the elderly in New Mexico. Am J Epidemiol 147(8):755–763
24. Cruz-Jentoft AJ, Baeyens JP, Bauer JM, Boirie Y, Cederholm T, Landi F et al (2010) Sarcopenia: European consensus on definition and diagnosis: report of the European Working Group on Sarcopenia in Older People. Age Ageing 39(4):412–423
25. Chen LK, Liu LK, Woo J, Assantachai P, Auyeung TW, Bahyah KS et al (2014) Sarcopenia in Asia: consensus report of the Asian working Group for Sarcopenia. J Am Med Dir Assoc 15(2):95–101
26. Fielding RA, Vellas B, Evans WJ, Bhasin S, Morley JE, Newman AB et al (2011) Sarcopenia: an undiagnosed condition in older adults. Current consensus definition: prevalence, etiology, and consequences. International working group on sarcopenia. J Am Med Dir Assoc 12(4):249–256
27. Messina C, Albano D, Gitto S, Tofanelli L, Bazzocchi A, Ulivieri FM et al (2020) Body composition with dual energy X-ray absorptiometry: from basics to new tools. Quant Imaging Med Surg 10(8):1687–1698
28. Lohman TG, Chen Z (2005) Dual-energy X-ray absorptiometry. In: Heymsfield SB, Lohman TG, Wang Z, Going SB (eds) Human body composition, 2nd edn. Human Kinetics, Champaign, pp 63–77
29. Silva AM, Heymsfield SB, Sardinha LB (2013) Assessing body composition in taller or broader individuals using dual-energy X-ray absorptiometry: a systematic review. Eur J Clin Nutr 67(10):1012–1021
30. Rothney MP, Brychta RJ, Schaefer EV, Chen KY, Skarulis MC (2009) Body composition measured by dual-energy X-ray absorptiometry half-body scans in obese adults. Obesity (Silver Spring, Md) 17(6):1281–1286
31. Tataranni PA, Ravussin E (1995) Use of dual-energy X-ray absorptiometry in obese individuals. Am J Clin Nutr 62(4):730–734
32. Johnson Stoklossa CA, Forhan M, Padwal RS, Gonzalez MC, Prado CM (2016) Practical considerations for body composition assessment of adults with class II/III obesity using bioelectrical impedance analysis or dual-energy X-ray absorptiometry. Curr Obes Rep 5(4): 389–396
33. Jankowski LG, Warner S, Gaither K, Lenchik L, Fan B, Lu Y et al (2019) Cross-calibration, least significant change and quality assurance in multiple dual-energy x-ray absorptiometry scanner environments: 2019 ISCD official position. J Clin Densitom 22(4):472–483
34. Reinhardt M, Piaggi P, DeMers B, Trinidad C, Krakoff J (2017) Cross calibration of two dual-energy X-ray densitometers and comparison of visceral adipose tissue measurements by iDXA and MRI. Obesity (Silver Spring, Md) 25(2): 332–337

Chapter 2

Analysis of Skeletal Muscle Mass from Preexisting Computerized Tomography (CT) Scans

Katherine L. Ford, Bruna Ramos da Silva, Ana Teresa Limon-Miro, and Carla M. Prado

Abstract

Computerized tomography (CT) is a gold standard imaging technique for assessing body composition at the tissue-organ level, particularly for the quantitative and qualitative analyses of skeletal muscle. A single slice measurement at the third lumbar vertebra strongly correlates with whole-body values. Thus, regional skeletal muscle cross-sectional area can be used as such or extrapolated using predictive equations to estimate whole-body skeletal muscle mass. CT images also allow the assessment of skeletal muscle "quality," reflective of myosteatosis (i.e., fat infiltration into muscle). Several software packages and protocols for assessing body composition via CT image are available. In this chapter, the protocol for analyzing a CT image at the third lumbar vertebra using sliceOmatic (Tomovision, Magog, Canada) software will be described.

Key words Computerized tomography, Skeletal muscle, Body composition, Adipose tissue, Protocol, Muscle assessment, Muscle attenuation, Intermuscular adipose tissue, Myosteatosis

1 Introduction

Computerized tomography (CT) is a nuclear medicine imaging technique that produces a reconstructed image of a specific section of the body based on the measurement of repeated radiologic projections at different positions and angles [1]. The varying attenuations from the X-ray beams as they pass through the body are reconstructed to form a grayscale two-dimensional image of a "slice" of the body [1]. The cross-sectional image is composed of pixels of different attenuation values that can be used to explore further the quantity and "quality" of the tissues included in that particular body section (adipose, skeletal muscle, bone, visceral organs, and brain) [2]. The resulting CT images are the gold standard for body composition assessment at the tissue-organ level, particularly for the quantitative analysis of skeletal muscle and adipose tissue areas [2–5]. However, CT is regularly used for

Philip J. Atherton and Daniel J. Wilkinson (eds.), *Neuromuscular Assessments of Form and Function*, Neuromethods, vol. 204,
https://doi.org/10.1007/978-1-0716-3315-1_2,

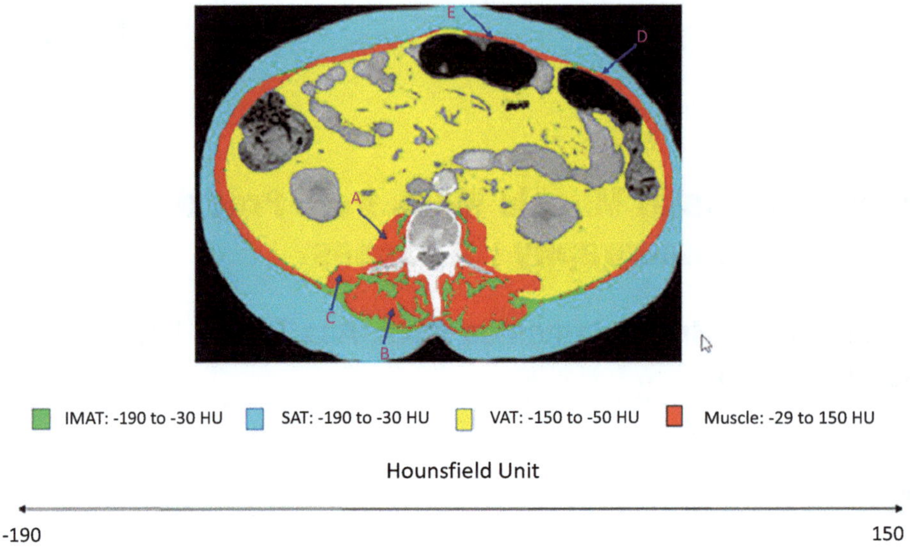

Fig. 1 Body composition profile on a computerized tomography image at the third lumbar (L3) vertebra. Tissues are colored based on their respective Hounsfield units (HU). Abbreviations: IMAT, intermuscular adipose tissue; SAT, subcutaneous adipose tissue; VAT, visceral adipose tissue; Muscle, skeletal muscle. The arrows indicate different muscle groups: (**a**) psoas major; (**b**) erector spinae; (**c**) quadratus lumborum; (**d**) abdominal obliques; (**e**) rectus abdominus

screening and monitoring certain health conditions (e.g., cancer, liver, and pulmonary diseases) rather than for body composition assessment [6].

Body/tissue composition is assessed by quantifying those represented within given CT image(s). The tissues are identified based on their attenuation which is represented through pixels (typically 1 mm × 1 mm) in the image by a numerical value known as a CT number which is expressed in Hounsfield units (HU) [1, 2, 4]. Specific HU ranges represent the different types of tissues and are typically represented by unique colors (e.g., skeletal muscle: red) as shown in Fig. 1. Hounsfield units are primarily determined based on tissue density in relation to air (least dense) and water (most dense) at standard temperature and pressure with air being equal to −1000 HU and water equal to 0 HU [4]. Tissues with a lower HU will appear darker on the image while denser tissues (higher HU) will appear gray or white. Notably, HU ranges for tissues are not consistent throughout the literature and may vary based on the protocol used [7]. In addition to the quantity of muscle tissue (a focus of this chapter), CT images can also assess the integrity (quality) of skeletal muscle. The varying attenuation (represented by HU) allows for the distinction between skeletal muscle and intermuscular adipose tissue (IMAT), with lower attenuation indicating increased intermuscular lipid accumulation [2, 4, 8]. The IMAT identified on CT imaging encompasses intramyocellular and extramyocellular lipid content;

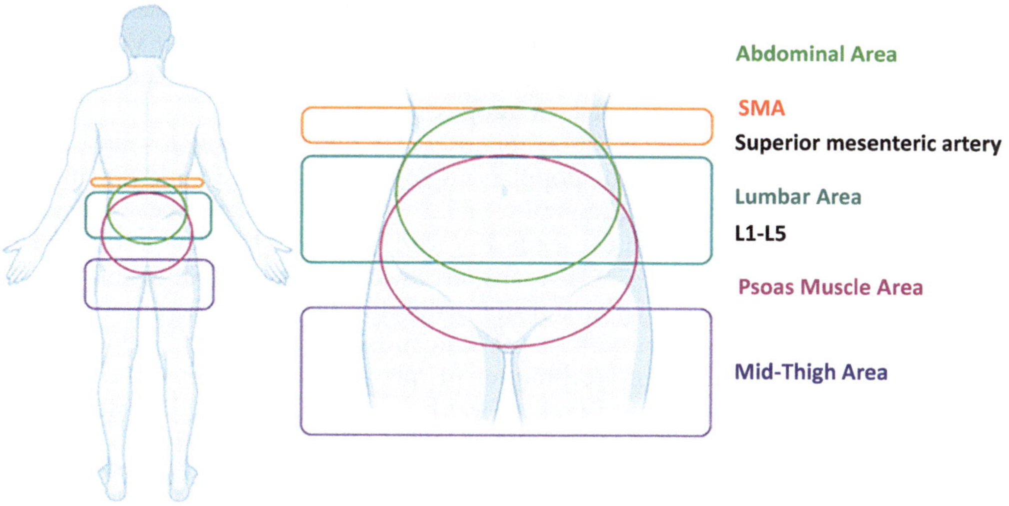

Fig. 2 Muscle groups in the lumbar area

exclusive identification of intramyocellular lipids is possible through a muscle biopsy or magnetic resonance imaging with a magnetic resonance spectroscopy technique applied [4, 9].

For body composition assessment, the cross-sectional area (cm^2) of skeletal muscle at the third lumbar (L3) vertebra correlates best with whole-body skeletal muscle and adipose tissues [4, 7, 10, 11]. Selecting the area of interest (herein, the L3 image) is known as landmarking and is the first step in CT image analysis for body composition assessment. The skeletal muscle index can be determined using skeletal muscle area standardized by height, assessed at the level of the L3 vertebra [12]. Alternative landmarks and approaches have been used for the analysis of body composition using CT images such as the mid-thigh cross-sectional area, thoracic, cervical, single muscle approaches (e.g., psoas), and the superior mesenteric artery level (Fig. 2) [7, 12–15]. Moreover, an agreement between sex-specific L3 cutoffs for low muscle mass and those derived from the tenth thoracic vertebra to the fifth lumbar vertebra has been confirmed in healthy kidney donors [14], and it has been suggested that when the L3 image is unavailable, the following images should be considered in order of L2, L4, L5, L1, T12, T11, and T10 [14]. Various publications have reported limitations to these alternative landmarks when compared to the L3 vertebra reference including a low proportion of total trunk muscles per scan, measurements with a high level of error, and discrepancies among observers [7, 13, 16, 17]. For these reasons, alternative anatomical landmarks have not been recommended by any guideline or expert group for skeletal muscle mass assessment [16]. Although software-specific standards for body composition analysis by CT scans exist, assessing the L3 vertebra using the Edmonton Protocol [18] is one of the most common approaches within the literature and will be discussed. All information contained herein was used with permission of Voronoi Health Analytics (Coquitlam, Canada) and TomoVision, a division of Virtual Magic, Inc. (Magog, Canada).

Body composition assessment by CT is perhaps most appropriate in patient populations (e.g., aortic valve replacement, cancer, critically ill, respiratory failure, trauma) where preexisting scans are available in the medical record [2, 19–24]. When used under these conditions, CT scans are widely available at a minimal cost [5]. Indeed, it is noteworthy that the high dose of ionizing radiation required for a CT scan curbs the use of this body composition assessment technique in prospective studies when requiring repeated measures of healthy individuals [2, 4]. Exposure to ionizing radiation depends on the type of CT scan and is 31 milli-Sieverts (mSv) for a CT scan of the abdomen and pelvis [25]. Instead, studies can exploit scheduled reassessments when patients are expected to receive a follow-up CT image for clinical reasons within the course of their disease trajectory [26]. Altogether, given the widespread opportunity for those without CT scanning facilities to access preexisting clinical images for repurposing to the study of skeletal muscle-related readouts, this chapter will focus on post-image acquisition analyses of CT scans.

Various software and protocols are available for body composition analysis using CT imaging; this chapter will focus on the Edmonton Protocol, housed within the sliceOmatic (TomoVision, version 5.0) software. Computed tomography image analysis involves the following three steps and will be reviewed in detail in the methods section: (1) landmarking to isolate an L3 CT image; (2) CT image analysis using the sliceOmatic region growing/painting mode; and (3) retrieving region growing analysis output from the CT images.

Best practices for CT image analysis include training to meet precision standards against an experienced trained observer (training for the analysis of images of varying quality); using a single observer in any study (person analyzing the images should remain consistent throughout the study); monitoring data quality; and consulting with a radiologist for questionable images.

2 Materials

1. A computer with sliceOmatic (TomoVision, version 5.0) software. A paid subscription is required to save images.
2. A trained technician with basic knowledge of radiology and anatomy to landmark and segment the images.
3. Access to CT images of interest.

3 Methods

The following methods are based on the software manufacturer's recommendations, training and the authors' combined experience [18]. Some sections are worded per the user manual, used with permission as mentioned above.

3.1 CT Image Analysis

In accordance with the institutional ethics board approval and local health authority processes, obtain CT images from medical records of interest. Using the Alberta Protocol shortcut to access the sliceOmatic (TomoVision, version 5.0) software, read and landmark the obtained CT. When the image of interest (e.g., L3) are obtained, drag and drop the specific DICOM (Digital Imaging and Communications in Medicine) files into sliceOmatic. TAG images should not be imported in this step as this can cause the software to close unexpectedly. Alternatively, navigate the files on the left-hand side of the window to select the study folder. Select the images of interest by clicking the folder name, clicking the icon, or using "Shift" and click to select multiple images or files. Selected files will be outlined in green. Click and move the slider under the preview window to scroll through the image files. Once satisfied with the selected images, click "Read 'x' image files" to view images in the software and click "Exit." The previous steps may be repeated to analyze additional images later.

3.2 Landmarking

Landmarking is the process of selecting the anatomical location of interest for analysis, in this case the third lumbar vertebra. Once images are imported, sliceOmatic will display two image windows. The program is best used when maximized to fit the computer screen size. Click on the image window on the right so that it is outlined in green. Under "Tools", select "Colour Scheme," and then select "Frame Selection" under the subheadings "All Modes" and "Tool 2D." Under "Mode," select "Region Growing." Navigate the "Frame Selection" display on the right-hand side of the screen and click on any axial frame. Select the left-hand image window to transfer the green selection border and renavigate to the "Frame Selection" display to select any scout, sagittal, or coronal image frame displaying the L3 vertebra. Place the cursor over the L3 vertebra in the left window and press the "." button to display the closest axial version of the scout image. The axial slice that is closest to the position of the cursor on the scout image will populate. To view alternate axial images, use the "Page Up" and/or "Page Down" keys. Ensure that the image on the right screen is that of interest as this image that will be segmented for body composition analysis. When only axial images are available, navigate "Frame Selection" in the right-hand column, and click "One" to increase the size of the viewing window. Press the space bar to view the axial images, and select the one displaying the L3 vertebra as described below.

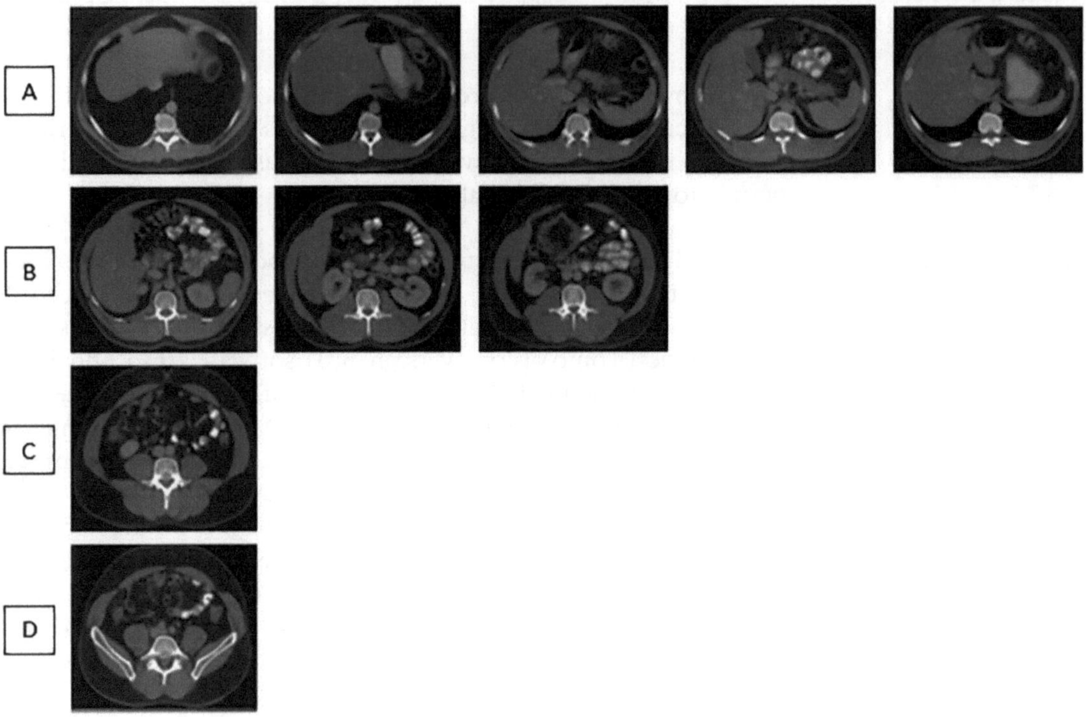

Fig. 3 Axial images of thoracic and lumbar vertebrae without contrast. (**a**) Thoracic vertebrae (T1–T12). Articular facets for ribs, long sharp spinous, and the lungs start to disappear. (**b**) Lumbar vertebrae (L1–L3 from left to right). No articular facets for ribs, short blunt spinous process, and no ribs are visible. (**c**) L4 vertebra. The iliac crest may be visible. (**d**) L5 vertebra. The iliac crest is the main landmark

To select the L3 image from a full-body (or larger) series, hover the mouse over "Frame scale" on the right-hand side of the screen, and use the scroller on the mouse to move through the images. Lumbar vertebrae are distinct from thoracic vertebrae because the former does not have rib attachments while the latter do. Example images of thoracic and lumbar vertebrae are shown in Fig. 3. Scroll down through the images to identify the thoracic vertebrae which have a rib attachment and posterior transverse processes. Scroll past the thoracic vertebrae to identify the first lumbar (L1) vertebra, the first image without rib attachment and with horizontal transverse processes. Using the transverse processes, count down through the images until an image displaying the L3 vertebra is obtained. Typically, three to five slices (images) per vertebra, occasionally up to 11 slices, are available. Notably, each time the transverse process disappears and reappears, a different (and lower) vertebra is being viewed. If the iliac crest (top of the pelvis) or short transverse processes are visible, the L4 or L5 vertebra is being viewed, and the L3 was passed (Fig. 3d). In approximately 10% of the population, an L6 vertebra is visible. Once the L3 vertebra has been identified, select the image within the L3 vertebra frames that best displays the even and thickest transverse processes on either

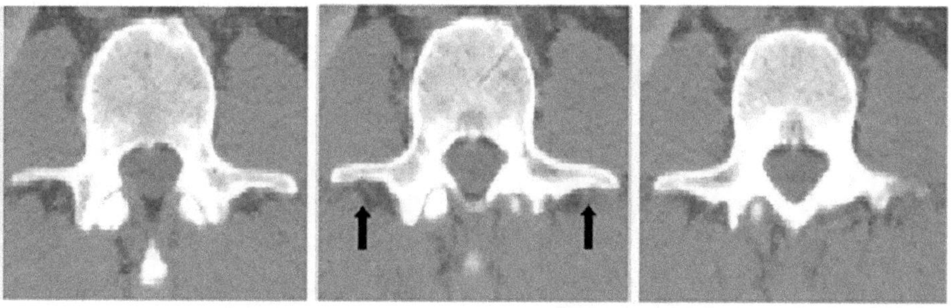

Fig. 4 Three frames showing the transverse processes at the third lumbar (L3) vertebra level. The arrows indicate the thickest and most consistent transverse process within the series of images

side of the spine, as shown in Fig. 4. The middle image within the series of L3 vertebra frames is generally most acceptable. To identify the image, the middle of the L3 is tagged with a dot of color. Once satisfied that the most appropriate L3 vertebra frame has been selected, click "File" and "Save Tag." Under "Mode," select "DB File Management," and select "All/Close Selected/Update DB." Repeat the above-described processes for each subject folder.

3.3 Manual Segmentation

Once the image of interest (L3) is identified, the image is ready for analysis. Segmentation is the process of tagging a grayscale image to produce a color-coded image representing skeletal muscle, IMAT, visceral adipose tissue (VAT), and subcutaneous adipose tissue (SAT), as shown in Fig. 1. These tissues are identified based on defined lower and upper threshold limits of HU. The HU limits and TAG color can be manually adjusted in the "Region Growing" mode, although the colors associated with each tissue (Fig. 1) are those most used. An image must be loaded into sliceOmatic before manually adjusting the HU limits by clicking the TAG color of choice and setting the appropriate HU ranges (both lower and upper). This process is continued for each tissue that needs identifying. Once the TAG colors have been selected with desired HU ranges applied, the process of coloring the image can begin. Traditionally, tissues are tagged in the following order to ensure accuracy: skeletal muscle, IMAT, VAT, and SAT.

The most accurate method of identifying skeletal muscle is through the "Paint" sub-mode. To "paint" an image, select the brush size (the larger, the better) and TAG color based on the tissue that will be evaluated. Left-click and drag the mouse to move the paintbrush over the appropriate areas of the image to manually identify the desired tissue. For example, select TAG 1 (red color) to evaluate skeletal muscle. Note that fascia bundles that connect muscle should not be analyzed as skeletal muscle, as shown in Fig. 5a. The lumbar artery and fascia under the spinalis muscle are not considered part of the muscle area and should not be colored. Muscle wasting near the end of muscles is also not colored.

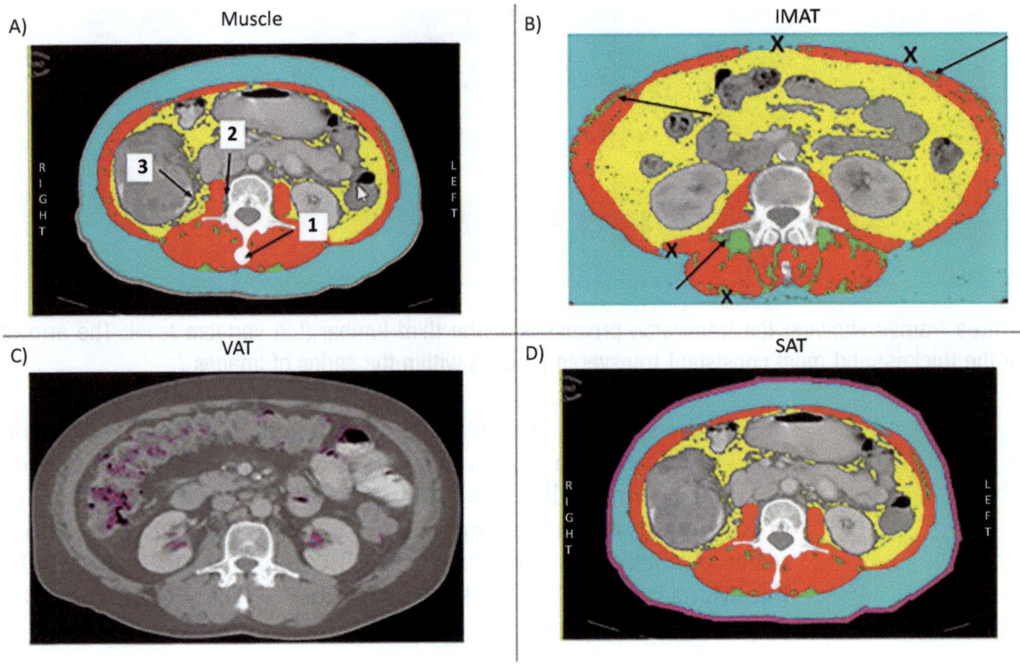

Fig. 5 Potential problematic areas when tagging different tissues. Abbreviations: IMAT, intermuscular adipose tissue; VAT, visceral adipose tissue; SAT, subcutaneous adipose tissue. Skeletal muscle tissue is represented by red color, IMAT is represented by green color, VAT is represented by yellow color, and SAT is represented by cyan color. (**a**) (1) The area underneath the neural spine (if present) should remain untagged; (2) represents a tendon and should not be analyzed as muscle; (3) non-muscle structures touching the psoas or other muscles should not be tagged. (**b**) The arrows indicate areas to examine for the presence of IMAT. The X's indicate areas that should not be tagged. (**c**) Adipose tissues inside the organs (e.g., liver, kidney, and intestine) are not considered VAT. Purple color indicates tissue that should not be included as VAT despite having the same density. (**d**) Purple color indicates the skin. The skin is not analyzed as SAT

Contrarily, abdominal muscles are colored, although it is important to note the ligaments and the linea alba as these are not colored. Next, evaluate the IMAT by selecting TAG 2 (green color) and dragging the paintbrush (left-click and hold on the mouse) over the uncolored areas within the skeletal muscle. The empty spaces between the iliocostalis lumborum muscle, longissimus lumbalis muscle, and fascia are all colored as IMAT, although muscle wasting around the edges of muscles is not tagged. Use the TAG lock feature, as described in the following paragraph, to lock the tagged IMAT. Continue this process for the remaining tissues by selecting TAG 5 (yellow color) for VAT and TAG 7 (cyan color) to evaluate the SAT. The "Grow 2D" feature can be useful for tagging the abdominal cavity as VAT. The intestine and visceral organs are not colored, nor should the skin be tagged as SAT (Fig. 5). Use the TAG lock feature prior to selecting TAG 7 for the SAT. Once satisfied, navigate "File" and select "Save TAG."

Coloring tools within sliceOmatic that have been mentioned and may be useful are outlined below. To fill an area of the image, trace the edges of a large muscle area by painting a line. Once a large tissue area is enclosed, select "Grow 2D" and the smallest brush size to flood a large area of pixels with the selected TAG color. This mode is most effective for VAT and SAT analyses, not the analysis of skeletal muscle or images with minimal VAT and SAT. If the limits of a tissue are hard to visualize, use the "Colour Scheme" tool to change the contrast of black and white to make the edges of a tissue clearer. Increasing the darkness of the image will improve muscle visualization, whereas lightening an image will allow for better visualization of adipose tissue. To erase any errors, right-click and hold the mouse while hovering over the area(s) containing pixels colored in error. Alternatively, select the "none" TAG and left-click and hold while hovering the mouse over the colored pixels to be deleted. To undo previous brush strokes, use the "Ctrl Z" feature on the keyboard. Once satisfied with a color, use "TAG lock" to protect that color and lock it in place. This feature ensures that no other color can override the tagging that has been done previously. The TAG lock feature is especially important when coloring tissues that have overlapping HU (e.g., the three types of adipose tissue). Lastly, "Scale" will increase the size of the image, making it easier to complete the coloring with accuracy.

3.4 Determining Body Composition Measures

Once the images are analyzed, sliceOmatic will provide the TAG surface/volume measurement for each TAG (color) used. Navigate "Tools," select "TAG Surface/Volume," and in the right-hand column, select "DB Surface" to obtain measurement information. The software will prompt the files to be saved. Next, navigate "Mode," and select "DB File Management" and "All/Close Selected/ Update DB". Outside of the software, open the saved file in a spreadsheet. The mean GLI (gray level image) for a TAG is reported in HU and found in row 3 of the spreadsheet. The GLI value represents the physical properties of the scanned tissue (it is a marker of muscle radiodensity). Variables of interest are given in cross-sectional areas in cm^2, which is a direct unit of measurement from CT, and are found in row one of the spreadsheet. Repeat this process in its entirety for each segmented image, and transfer the data to a master spreadsheet.

4 Notes

Data on the cross-sectional area can then be used to stratify people from least to most muscular or to compare muscularity with previously published population-specific cut points used for the identification of low muscle mass [20, 21, 27–29]. Regression equations are also available for the prediction of whole-body skeletal muscle and adipose tissue based on values retrieved from the L3 vertebra

[30, 31]. Data containing the mean value of the pixels in HU is available for each TAG and depicts the mean radiodensity of each tissue. Reflective of attenuation, the mean HU is indicative of fatty infiltration of muscle, and, thus, muscle quality or presence of myosteatosis can be inferred [28, 32].

Body composition assessment from analysis of CT images can be conducted regardless of contrast agent (e.g., barium, iodine dyes) used during the imaging process [5]. Notably, contrast does alter the radiodensity of tissues and thus will equate to different HU compared to an image that did not contain contrast; thus, the use of contrast is an important consideration when analyzing and comparing results of body composition assessment by CT. If available, we recommend analyzing images without contrast.

To determine the type of series that is being viewed in sliceOmatic, navigate the "Tools" tab, and select "2D Overlay." In the bottom right-hand corner of the screen, select "Tech." Technical information for the image will appear [18]. Frame thickness and spacing are important variables that can be found within the technical information and can vary depending on the CT equipment and the technician performing the scan. Frame thickness should be recorded in a separate spreadsheet and will be needed for body composition analysis. Frame thickness can also affect the technician's ability to select the appropriate L3 slice. As mentioned in the methods, the L3 slice with the thickest and most even transverse processes should be selected for segmentation [18].

Image quality can be affected by artifacts within the body, challenging the accuracy of body composition assessment, and in some cases the image may need to be discarded from analysis [33]. For example, cavity fillings, medical implants, and titanium rods can alter the quality of a CT image. If the patient is wearing a watch during the scan, streaks are visible across the image, challenging the accuracy of the segmentation process and, subsequently, body composition assessment. A buildup of fluid (ascites) will cause poor contrast as the fluid has a very similar density to the muscle [18]. The "Colour Scheme" tool within sliceOmatic allows for contrast adjustment, which can be especially useful for analyzing low-quality images. For example, images can be viewed in grayscale, mixed, tinted, or over (brightest option). Use the F1 through F4 keys to toggle through the gray, mixed, tint, or over options, respectively. Gray (F1) is best used for visualizing the boundaries between muscle and organs. Mixed (F2) is ideal during the segmentation (coloring) process. Tint (F3) is ideal when reviewing the image for errors or pixels that are not appropriately colored. Lastly, the over option (F4) is ideal when using the images for presentations or when verifying that no pixels of the skin have been tagged as SAT [18].

Anatomical knowledge is required for CT image analysis. Notably, CT images are inverted whereby the right side of the body appears on the left side of the image and vice versa. The option to identify individual muscle groups in sliceOmatic is possible by setting different TAG colors for specific muscles (e.g., psoas major, erector spinae, quadratus lumborum, abdominal obliques, and rectus abdominus), as shown in Fig. 1 [18].

Beyond the cost, expertise, and ionizing radiation exposure, the size of the person can also present as a limitation to CT analysis for body composition assessment [2]. The CT scanner field of view may not accommodate all persons living with excess body weight, resulting in CT images that are missing sections of tissues (e.g., SAT, skeletal muscle) [2]. The image should not be used for body composition analysis if both sides of the obliques/transversus abdominus, erector spinae, or rectus abdominus are cut off from the image [18]. If the person's body is pressed against the scanner, the CT image will produce an altered image containing photon starvation artifacts and may not be usable in analysis [33]. Additionally, CT images may not capture regional change in skeletal muscle and adipose tissue that results from weight change [4]. Whole-body CT scans are not common due to the ionizing radiation, and weight change may be region-specific and thus not captured within in the landmark image of interest [4]. As discussed, the whole-body composition can be estimated by a single cross-sectional CT image of the abdomen [10, 34]. If weight change has occurred, the data should be interpreted with caution.

Analyzing CT images for the purpose of body composition assessment will take an experienced technician approximately 20 min to segment one image, whereas someone who has less experience will require about 45 min per image. This timeframe is in addition to the time needed for the landmarking process. As mentioned, it is important that the technician used for one study remain consistent throughout the analysis of images for that study to ensure consistency in technique across images. For a visual explanation of landmarking using sliceOmatic, we recommend viewing our YouTube video or the training modules available elsewhere [35, 36]. For detailed anatomical photos to assist with landmarking and segmenting, we recommend referring to the *Pocket Atlas of Sectional Anatomy Vol. II* [37].

In terms of organization, we recommend that the person analyzing the images create a folder for all tagged L3 images and include a subfolder for each participant. The subfolder should contain two files: the TAG and DICOM files. The TAG file is the final segmented file, whereas the DICOM file is needed for any changes to be made; the TAG file cannot be uploaded into sliceOmatic for adjustments. Within a master spreadsheet containing data from all segmented images, we recommend that notes are recorded if the following are present within the images: SAT cutoff and location; presence of a hernia (whether minor or major); ostomy (regardless of location); abnormal anatomy; pannus (skinfold); and presence of artifacts.

Manual segmentation was discussed herein although the future of CT image analysis includes automatic segmentation for the purpose of body composition assessment and is now available [38]. Automatic segmentation of skeletal muscle, VAT, and SAT is taken from an axial CT image accurately, in comparison to manual segmentation using Automatic Body composition Analyser using Computed tomography image Segmentation™ (ABACS™) software (Voronoi Health Analytics) [33, 38–40]. This program has been internally [38] and externally [33] validated in comparison to the manual segmentation of images. External validation in a cohort of patients with nonmetastatic colorectal ($n = 3102$) and breast cancer ($n = 2888$) found that automatic segmentation by ABACS™ underestimates skeletal muscle by a mean of -2.35 cm^2 [33].

New technological developments, recently released by the company, include the ABACS™ plug-in, which is an artificial intelligence program integrated within the Data Analysis Facilitation Suite (DAFS) 3.0 platform and automatically segments landmark images of interest [39]. The entire body composition analysis workflow, including data curation, vertebral slice landmarking, tissue segmentation, manual editing/corrections, and final tissue statistics computation, can be performed using DAFS [41]. DAFS includes two key automation engines, the ABACS™ 3D which provides a volumetric segmentation of the entire CT scan containing any field of view within the thoracolumbar region, and the Automated Vertebral Annotator engine that enables automated landmarking of the thoracic to sacrum vertebral locations [41]. Such advances in software technology and the use of artificial intelligence in body composition analysis make for a promising future within the field, ideally with the integration of body composition into clinical care.

References

1. Goldman LW (2007) Principles of CT and CT technology. J Nucl Med Technol 35:115–128. https://doi.org/10.2967/jnmt.107.042978
2. Prado CMM, Heymsfield SB (2014) Lean tissue imaging: a new era for nutritional assessment and intervention. J Parenter Enter Nutr 38:940–953. https://doi.org/10.1177/0148607114550189
3. Wang ZM, Pierson RN, Heymsfield SB (1992) The five-level model: a new approach to organizing body-composition research. Am J Clin Nutr 56:19–28. https://doi.org/10.1093/ajcn/56.1.19
4. Ross R, Janssen I (2005) Computed tomography and magnetic resonance imaging. In: Heymsfield SB, Lohman T, Wang Z, Going S (eds) Human body composition, 2nd edn. Human Kinetics, Champaign, IL, pp 89–108
5. Gomez-Perez SL, Haus JM, Sheean P, Patel B, Mar W, Chaudhry V, McKeever L, Braunschweig C (2016) Measuring abdominal circumference and skeletal muscle from a single cross-sectional computed tomography image. J Parenter Enter Nutr 40:308–318. https://doi.org/10.1177/0148607115604149
6. Kang HR, Cho JY, Lee SH, Lee YJ, Park JS, Cho YJ, Il YH, Lee KW, Lee JH, Lee CT (2019) Role of low-dose computerized tomography in lung cancer screening among

never-smokers. J Thorac Oncol 14:436–444. https://doi.org/10.1016/j.jtho.2018.11.002

7. Faron A, Luetkens JA, Schmeel FC, Kuetting DLR, Thomas D, Sprinkart AM (2019) Quantification of fat and skeletal muscle tissue at abdominal computed tomography: associations between single-slice measurements and total compartment volumes. Abdom Radiol 44:1907–1916. https://doi.org/10.1007/s00261-019-01912-9
8. Goodpaster BH, Thaete FL, Simoneau J, Kelley DE (1997) Subcutaneous abdominal fat and thigh muscle composition predict insulin sensitivity independently of visceral fat. Diabetes 46:1579–1585
9. Shen W, Mao X, Wang Z, Punyanitya M, Heymsfield SB, Shungu DC (2003) Measurement of intramyocellular lipid levels with 2-D magnetic resonance spectroscopic imaging at 1.5 T. Acta Diabetol 40:1–7. https://doi.org/10.1007/s00592-003-0026-x
10. Shen W, Punyanitya M, Wang ZM, Gallagher D, St.-Onge MP, Albu J, Heymsfield SB, Heshka S (2004) Total body skeletal muscle and adipose tissue volumes: estimation from a single abdominal cross-sectional image. J Appl Physiol 97:2333–2338. https://doi.org/10.1152/japplphysiol.00744.2004
11. Irlbeck T, Massaro JM, Bamberg F, O'Donnell CJ, Hoffmann U, Fox CS (2010) Association between single-slice measurements of visceral and abdominal subcutaneous adipose tissue with volumetric measurements: the Framingham heart study. Int J Obes 34:781–787. https://doi.org/10.1038/ijo.2009.279
12. Faron A, Sprinkart AM, Kuetting DLR, Feisst A, Isaak A, Endler C, Chang J, Nowak S, Block W, Thomas D, Attenberger U, Luetkens JA (2020) Body composition analysis using CT and MRI: intra-individual intermodal comparison of muscle mass and myosteatosis. Sci Rep 10:1–10. https://doi.org/10.1038/s41598-020-68797-3
13. Marquis K, Debigaré R, Lacasse Y, Leblanc P, Jobin J, Carrier G, Maltais F (2002) Midthigh muscle cross-sectional area is a better predictor of mortality than body mass index in patients with chronic obstructive pulmonary disease. Am J Respir Crit Care Med 166:809–813. https://doi.org/10.1164/rccm.2107031
14. Derstine BA, Holcombe SA, Ross BE, Wang NC, Su GL, Wang SC (2018) Skeletal muscle cutoff values for sarcopenia diagnosis using T10 to L5 measurements in a healthy US population. Sci Rep 8:1–8. https://doi.org/10.1038/s41598-018-29825-5
15. Bril SI, Wendrich AW, Swartz JE, Wegner I, Pameijer F, Smid EJ, Bol GH, Pothen AJ, de Bree R (2019) Interobserver agreement of skeletal muscle mass measurement on head and neck CT imaging at the level of the third cervical vertebra. Eur Arch Otorhinolaryngol 276:1175–1182. https://doi.org/10.1007/s00405-019-05307-w
16. Baracos VE (2017) Psoas as a sentinel muscle for sarcopenia: a flawed premise. J Cachexia Sarcopenia Muscle 8:527–528. https://doi.org/10.1002/jcsm.12221
17. Rutten IJG, Ubachs J, Kruitwagen RFPM, Beets-Tan RGH, Olde Damink SWM, Van Gorp T (2017) Psoas muscle area is not representative of total skeletal muscle area in the assessment of sarcopenia in ovarian cancer. J Cachexia Sarcopenia Muscle 8:630–638. https://doi.org/10.1002/jcsm.12180
18. TomoVision (2017) sliceOmatic Alberta protocol user's manual version 5.0 rev: 11
19. Prado CMM (2013) Body composition in chemotherapy: the promising role of CT scans. Curr Opin Clin Nutr Metab Care 16:525–533. https://doi.org/10.1097/MCO.0b013e328363bcfb
20. Martin L, Birdsell L, MacDonald N, Reiman T, Clandinin MT, McCargar LJ, Murphy R, Ghosh S, Sawyer MB, Baracos VE (2013) Cancer cachexia in the age of obesity: skeletal muscle depletion is a powerful prognostic factor, independent of body mass index. J Clin Oncol 31:1539–1547. https://doi.org/10.1200/JCO.2012.45.2722
21. Prado CM, Lieffers JR, McCargar LJ, Reiman T, Sawyer MB, Martin L, Baracos VE (2008) Prevalence and clinical implications of sarcopenic obesity in patients with solid tumours of the respiratory and gastrointestinal tracts: a population-based study. Lancet Oncol 9:629–635. https://doi.org/10.1016/S1470-2045(08)70153-0
22. Sheean P, Peterson S, Gomez Perez S, Troy K, Patel A, Sclamberg J, Ajanaku F, Braunschweig C (2014) The prevalence of sarcopenia in patients with respiratory failure classified as normally nourished using subjective global assessment and computed tomography. J Parenter Enter Nutr 38:873–879. https://doi.org/10.1177/0148607113500308
23. Moisey LL, Mourtzakis M, Cotton BA, Premji T, Heyland DK, Wade CE, Bulger E, Kozar RA (2013) Skeletal muscle predicts ventilator-free days, ICU-free days, and mortality in elderly ICU patients. Crit Care 17. https://doi.org/10.1186/cc12901
24. Foldyna B, Troschel FM, Addison D, Fintelmann FJ, Elmariah S, Furman D, Eslami P, Ghoshhajra B, Lu MT, Murthy VL, Hoffmann U, Shah R (2018) Computed tomography-based fat and muscle

characteristics are associated with mortality after transcatheter aortic valve replacement. J Cardiovasc Comput Tomogr 12:223–228. https://doi.org/10.1016/j.jcct.2018.03.007

25. Smith-Bindman R, Lipson J, Marcus R, Kim K, Mahesh M, Gould R, de Gonzalez A, Phil D, Miglioreti D (2009) Radiation dose associated with common computed tomography examinations and the associated lifetime attributable risk of cancer. Arch Intern Med 169:2078–2086
26. Solheim TS, Laird BJA, Balstad TR, Bye A, Stene G, Baracos V, Strasser F, Griffiths G, Maddocks M, Fallon M, Kaasa S, Fearon K (2018) Cancer cachexia: rationale for the MENAC (Multimodal-Exercise, Nutrition and Anti-inflammatory medication for Cachexia) trial. BMJ Support Palliat Care 8: 258–265. https://doi.org/10.1136/bmjspcare-2017-001440
27. Caan BJ, Meyerhardt JA, Kroenke CH, Alexeeff S, Xiao J, Weltzien E, Feliciano EC, Castillo AL, Quesenberry CP, Kwan ML, Prado CM (2017) Explaining the obesity paradox: the association between body composition and colorectal cancer survival (c-scans study). Cancer Epidemiol Biomark Prev 26: 1008–1015. https://doi.org/10.1158/1055-9965.EPI-17-0200
28. Van Der Werf A, Langius JAE, De Van Der Schueren MAE, Nurmohamed SA, Van Der Pant KAMI, Blauwhoff-Buskermolen S, Wierdsma NJ (2018) Percentiles for skeletal muscle index, area and radiation attenuation based on computed tomography imaging in a healthy Caucasian population. Eur J Clin Nutr 72:288–296. https://doi.org/10.1038/s41430-017-0034-5
29. Caan BJ, Cespedes Feliciano EM, Prado CM, Alexeeff S, Kroenke CH, Bradshaw P, Quesenberry CP, Weltzien EK, Castillo AL, Olobatuyi TA, Chen WY (2018) Association of muscle and adiposity measured by computed tomography with survival in patients with nonmetastatic breast cancer. JAMA Oncol 4:798–804. https://doi.org/10.1001/jamaoncol.2018.0137
30. Mourtzakis M, Prado CMM, Lieffers JR, Reiman T, McCargar LJ, Baracos VE (2008) A practical and precise approach to quantification of body composition in cancer patients using computed tomography images acquired during routine care. Appl Physiol Nutr Metab 33:997–1006. https://doi.org/10.1139/H08-075
31. Holt DQ, Strauss BJG, Lau KK, Moore GT (2016) Body composition analysis using abdominal scans from routine clinical care in patients with Crohn's disease. Scand J Gastroenterol 51:842–847. https://doi.org/10.3109/00365521.2016.1161069
32. Goodpaster BH, Kelley DE, Thaete FL, He J, Ross R (2000) Skeletal muscle attenuation determined by computed tomography is associated with skeletal muscle lipid content. J Appl Physiol 89:104–110. https://doi.org/10.1152/jappl.2000.89.1.104
33. Cespedes Feliciano EM, Popuri K, Cobzas D, Baracos VE, Beg MF, Khan AD, Ma C, Chow V, Prado CM, Xiao J, Liu V, Chen WY, Meyerhardt J, Albers KB, Caan BJ (2020) Evaluation of automated computed tomography segmentation to assess body composition and mortality associations in cancer patients. J Cachexia Sarcopenia Muscle 11:1258–1269. https://doi.org/10.1002/jcsm.12573
34. Shen W, Punyanitya M, Wang ZM, Gallagher D, St-Onge M-P, Albu J, Heymsfield SB, Heshka S (2004) Visceral adipose tissue: relations between single-slice areas and total volume2. Am J Clin Nutr 80:271–278
35. UofANutrition - Dr. Carla Prado (2015) sliceOmatic for body composition analysis. In: YouTube. https://youtu.be/KJrsQ_dg5mM
36. Abbott Nutrition Health Institute (2020) Body composition & health, modules 1 & 2. https://anhi.org/education/course-catalog/body-composition-with-carla-prado
37. Moeller T, Reif E (2013) Pocket atlas of sectional anatomy, vol II: Thorax, heart, abdomen and pelvis, 4th edn. Thieme Medical Publishers, Inc., New York
38. Popuri K, Cobzas D, Esfandiari N, Baracos V, Jägersand M (2016) Body composition assessment in axial CT images using FEM-based automatic segmentation of skeletal muscle. IEEE Trans Med Imaging 35:512–520. https://doi.org/10.1109/TMI.2015.2479252
39. Voronoi Health Analytics (2021) DAFS 3.0. https://www.voronoihealthanalytics.com/abacs3d. Accessed 15 Apr 2021
40. Voronoi Health Analytics ABACS-sliceOmatic Tutorial. In: video. https://drive.google.com/file/d/1DXAY8vC17EgDaynvBD0Pjx47lT2k1eHq/view
41. Voronoi Health Analytics (2021) General user guide

Chapter 3

Imaging Skeletal Muscle by Magnetic Resonance Imaging (MRI)

Robert H. Morris and Craig Sale

Abstract

Skeletal muscle comprises a significant proportion of whole-body mass, and its maintenance is critical to the quality of life, particularly with aging. It is therefore important to determine changes in skeletal muscle mass and quality, accurately and reliably. Magnetic resonance imaging (MRI) is a commonly used noninvasive imaging method to quantify whole-body skeletal muscle mass, but it can also be used to examine muscle quality. In this chapter we outline the main materials and methods required to determine skeletal muscle mass using MRI. We initially provide a brief overview of the underlying principles of MRI before moving on to key methodological considerations in determining skeletal muscle volume, muscle to fat ratio and the tracking of individual skeletal muscle fibers. We will finally describe approaches to automatic segmentation and processing before finishing with some future considerations and potential directions.

Key words Magnetic resonance imaging, Skeletal muscle, Muscle mass, Muscle quality, Muscle fiber imaging

1 Introduction

Skeletal muscle is the largest, non-adipose tissue in the body comprising up to 55% of an individual's body mass and being critical for locomotion, structural support, and metabolism. The maintenance of muscle mass is important to health across the lifespan, and the loss (or atrophy) of skeletal muscle is a characteristic of aging [1] and age-related diseases [2–4]. Loss of muscle mass is also associated with reductions in physical activity, such as with bed rest, immobilization, or reduced ambulation [5]. Atrophy of skeletal muscle, therefore, has significant effects on an individual's morbidity [6] and mortality [7].

Given the implications of skeletal muscle atrophy, methods to accurately and reliably determine changes in skeletal muscle mass and quality are required to monitor condition or disease progression and to determine the efficacy of any interventional strategies (e.g., nutrition, exercise, electrical stimulation) designed to

Philip J. Atherton and Daniel J. Wilkinson (eds.), *Neuromuscular Assessments of Form and Function*, Neuromethods, vol. 204, https://doi.org/10.1007/978-1-0716-3315-1_3,

maintain or increase skeletal muscle mass. This was somewhat difficult to do directly in vivo, until the early 1980s when several key imaging modalities (magnetic resonance imaging, computed tomography, and photon absorptiometry) were developed that could be employed for this purpose. While a detailed review of these developments is beyond the scope of this chapter, a good description can be found in Heymsfield et al. [8], and the assessment of muscle mass by computed tomography and dual-energy X-ray absorptiometry (DXA) is covered by Prado et al.'s chapters of this book.

Magnetic resonance imaging (MRI) is a commonly used noninvasive imaging method to measure whole-body skeletal muscle mass, but it can also be used to examine muscle quality (e.g., via examination of the degree of skeletal muscle fat infiltration). Early on, Foster et al. [9] reported the application of pulse sequences that could determine differences between adipose and skeletal muscle tissues. By the mid-1990s, MRI studies were being published regarding the quantification of adipose tissue and skeletal muscle masses in humans in response to dietary and exercise interventions [10, 11]. MRI is now broadly accepted as a benchmark measurement technique of skeletal muscle primarily because of its accuracy and reproducibility [12]. Mitsiopoulos et al. [13] demonstrated the validity of adipose tissue-free skeletal muscle area determinations by MRI (and computed tomography) against cadaver estimates. Adipose tissue-free skeletal muscle area measured by MRI (39 ± 22 cm^2) was not significantly different from cadaver measurements (40 ± 23 cm^2); MRI values were also strongly correlated with corresponding cadaver values. The intra-observer correlation for duplicate measurements of adipose tissue-free skeletal muscle areas by MRI was 0.99. As with any measurement method, there are also disadvantages to the use of MRI methods for the determination of skeletal muscle mass; the main issues being the level of resources required, both in terms of cost and time. To a lesser extent, it is also important to consider exclusion criteria for studies using MRI: although many implants are compatible with MRI, participants with some implants (e.g., cochlear implants) are not normally able to be scanned safely, and while no ill effects are observed in participants with most modern orthopedic implants, image quality surrounding the implants is often poor owing to interference with the applied magnetic fields. This chapter will focus on outlining the main materials and methods required to determine skeletal muscle mass using MRI; this will also include some lessons learned.

1.1 A Primer on MRI Principles

MRI is possible with a wide range of different nuclei. The most encountered in imaging of the human body is the proton since they are abundant—with two being found in every water molecule. Readers with a strong interest in a more complete understanding of the underlying physics of the technique are directed to some excellent works in this area [14, 15]. An overview of the principles is presented here to the level that is required to fully appreciate the methods presented in this chapter (*see* **Note 1**).

There are three main parts of an MRI scanner that are needed to generate the signals required to produce an image.

The first is a strong and homogenous magnetic field, which is stated in Tesla (T). The most common fields in practice are 1.5 T and 3 T. As the field increases, the amount of signal that can be collected in a given amount of time also increases; yet this comes with a trade-off of cost (a 3 T system is approximately three times the price of a 1.5 T system). This magnetic field is normally generated by a hollow cylindrical superconducting magnet inside which the participant is placed by means of an electromechanically controlled table. The participant is positioned (usually with the assistance of a laser emitted by the MRI) such that the body part of interest is within the isocenter (center from all directions) of the system, as this is where the field is the most homogenous and thus the images are of the highest quality.

This magnetic field has two effects on nuclei, such as protons. The first is that they will align themselves along the direction of the applied magnetic field. The second is that they become responsive to the application of a second rapidly changing field. For each type of nuclei, there is a specific frequency at which this field needs to oscillate based on the magnetic field strength. This frequency is determined by multiplying the magnetic field by a constant called the gyromagnetic ratio (which is different for different nuclei). In the case of protons, this constant is approximately 42.6 MHz/T giving a frequency of 63.9 MHz for 1.5 T magnets. This frequency is called the Larmor frequency, and because the values are normally in MHz, the field is most commonly referred to as the radiofrequency (RF) field.

Once a participant is centered in the magnet, a short burst of this radiofrequency field can be used to rotate a small number of the protons within the body away from their equilibrium state aligned with the large magnetic field. Once this burst (normally called an RF pulse) ceases, a behavior called precession occurs. The protons that were rotated by the RF pulse begin to spin around the main magnetic field, at the Larmor frequency, slowly returning to their equilibrium alignment. A graphic of this motion can be seen in Fig. 1. The time that this takes to occur is dependent upon how readily the nuclei can return their energy to their environment, which happens with a time constant known as T_1 and is the first of several characteristic measurements that can be made on materials within the MRI scanner (also shown in Fig. 1).

A carefully positioned metal coil surrounds the part of the body being imaged, and the rotation of the protons is equivalent to the spinning magnet in a dynamo, which generates a very small, but measurable, voltage detectable by sensitive amplifiers in the MRI system. The time it takes for this signal to disappear is mostly

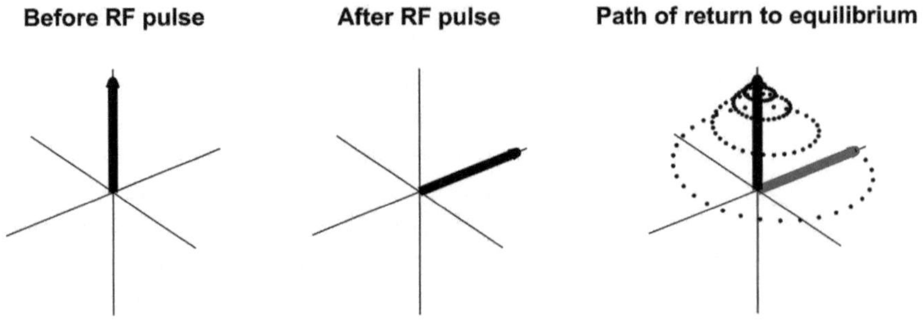

Fig. 1 Left to right: all the protons are aligned in the same direction (their equilibrium state). An RF pulse is applied to tip them by 90°. While processing, they return to equilibrium. The path of the tip of the arrow is traced out with dots separated by the same amount of time. The dots here represent five times the T_1 time

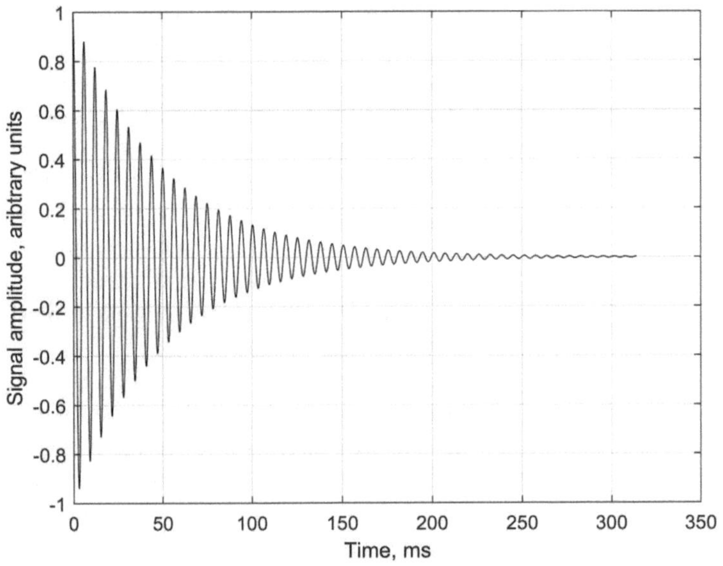

Fig. 2 An example free induction decay signal. The signal drops off exponentially with a characteristic time constant T_2*, which in this case is equal to 50 ms

dependent upon the chemical structure and motion of the molecule that it is a part of. The time constant with which this happens is known as T_2* and is another key measurement that can be made. This signal, known as a free induction decay (FID), can be seen in Fig. 2. It is not, however, the image that we set out to acquire and is instead representative of all tissues within the center of the magnet. To produce an image, we require the third part of the system: a gradient coil set.

The magnetic field determines both the frequency at which RF pulses must be applied and that at which precession occurs. If comparatively small but linear variations are introduced in the magnetic field at points in space throughout the volume of interest inside the bore of the magnet, it can be determined from where physically in space the signal originated owing to its frequency. It is possible to introduce these linear variations, or gradients in the magnetic field, in the three orthogonal directions, such that images can be created in three dimensions (*see* **Note 2**). Just like the RF pulse, these gradients are turned on and off in pulse sequences, to obtain different contrasts and observe particular features. In the simplest of these pulse sequences, FID signals are collected repeatedly following the application of field gradients in three dimensions. By changing the strength and duration of these gradient pulses, signals which correspond to points across the region of interest can be collected, one at a time. This is a slow and rarely used method, since the next signal cannot be collected until the system has returned to equilibrium, with the protons aligned with the magnetic field (possibly taking several seconds).

Fortunately, there are numerous methods that are significantly faster, permitting tens or even hundreds of these points to be collected from a single excitation pulse. These faster methods take advantage of a phenomenon known as an echo; in almost all samples, the time it takes for the FID to decay (with T_2*) is significantly less than the time it takes for the signal to return to equilibrium (with T_1).

There are two methods for generating an echo: the first is to form a so-called "spin echo." Sometime after the FID has disappeared, an RF pulse, which is twice the amplitude of the excitation, is applied (known as a refocusing pulse). This causes the direction of precession to be inverted, and in an equal amount of time later, the protons will come back together giving a signal with the appearance of two FIDs mirrored back-to-back. This can be repeated several times, but each echo will be formed by less protons coming back together as they move into different environments and have distinct Larmor frequencies. The time between echoes is determined by the time between refocusing pulses which is known as the echo time (T_E). Such a pulse sequence and corresponding signal can be seen in Fig. 3.

This process is often represented by the analogy of a running race: sometime after the runners leave the starting blocks, a whistle is blown, and the runners turn around (the same action as the refocusing pulse); the fastest runners will have traveled the furthest and thus are more distant from the starting line than the slowest runners, who have less distance to cover to return. However, if the runners' speeds remain the same as on the outward leg, they will all return to the starting point at the same time (*see* **Note 3**). Now, instead of waiting several T_1 times before collecting the next imaging point, it is only necessary to wait an "echo time," since each echo can represent the signal at a different gradient strength location. In fact, significant speed gains can be made by collecting an echo, while one gradient is left on (referred to as a readout gradient), which leads to every point *in* the echo corresponding to the signals that were acquired in the single-

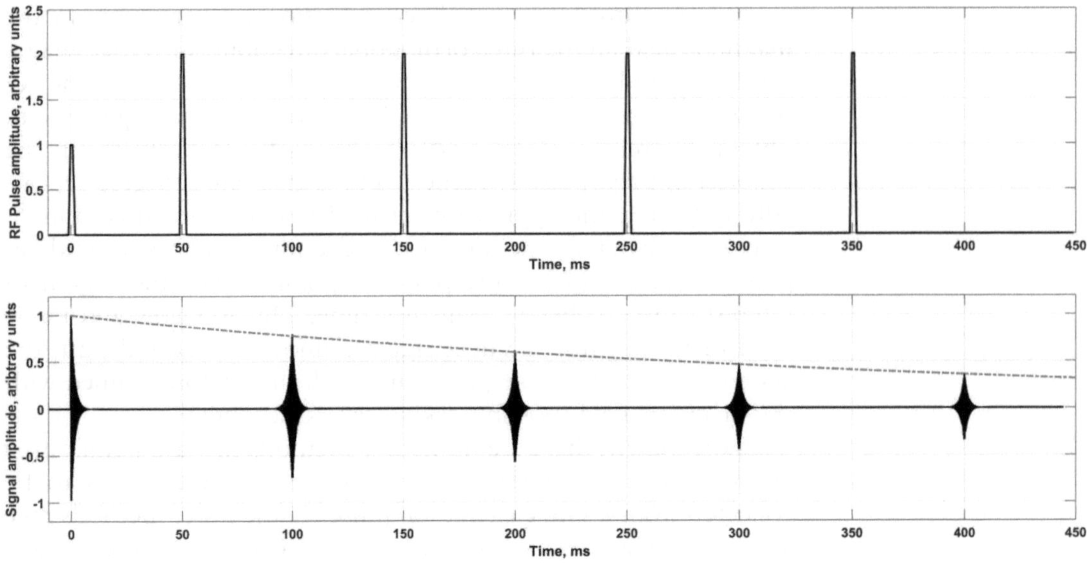

Fig. 3 Spin echo pulse sequence. The top plot shows blocks representing the RF pulses (the first pulse is the excitation pulse which has the effect of rotating the protons by 90°, and the subsequent pulses are refocusing pulses, which rotate the protons by 180° and are thus twice the amplitude). The lower plot is the resulting signal showing the FID from the excitation pulse followed by a series of echoes generated by the refocusing pulses. The dashed line represents the decay in echo amplitude caused by protons moving into different environments

point imaging. With methods such as this, it is possible to collect whole images in a matter of milliseconds, which is invaluable for time-resolved studies.

The second method for echo formation is to apply a gradient pulse to the excited system and then some time (the echo time) later apply a gradient that has the opposite effect, such that the protons come back together in much the same way. In practice, this is much more sensitive to local differences in the body, e.g., between tissue types (such as fat and muscle fibers), and can lead to more image artifacts than are present in spin echo-based imaging techniques.

There are several techniques that build upon these basic principles to provide a wide range of possible contrasts between different tissues. The most important of these and that which is used extensively when imaging the human body is a family of techniques known as fat suppression. These techniques use one of the several methods to eliminate the signal from fat (actually from triglycerides), either by their characteristic relaxation times or owing to the unique offset in their Larmor frequency, caused by the presence of many protons connected to different nuclei. This is clinically useful in removing the highly dominant fatty tissues revealing smaller changes due to underlying pathologies, such as edema. It also has value in scientific circles as it allows for quantified measurements of fat and muscle content by subtraction of images with and without fat suppression, for example. One of the techniques that is most often applied for such measurements is the DIXON sequence [16], which takes advantage of the fact that the different precessional frequencies of triglycerides and non-triglycerides will cause them to come repeatedly in and out of phase with each other in a known time interval. By setting the echo time so that even echoes occur when water and fat are in phase and that odd echoes occur when they are out of phase, it is possible to get images purely from fat and purely from water by addition and subtraction.

The following sections will consider applications of these techniques along with starting points for their implementation. Each section begins with a rationale for the measurement and a review of the literature covering key studies in the area and their findings, followed by a short commentary on important considerations when conducting such an experiment based upon personal experiences.

1.2 Image Processing and Software

As the quality of MRI images has improved, offering ever higher resolutions and increased signal to noise ratios, the burden on manual processing has concomitantly increased. However, the availability and usability of semiautomated or fully automated processing tools have also grown. The number of publications for new methods of automatic MRI data processing has been at an average of 150 since 2015 (Web of Knowledge citation report), although this has not translated into as many useful tools for the non-imaging expert (*see* **Note 4**). There are, however, several tools that are readily available and will be discussed throughout the following sections. For all applications discussed in this chapter, automated tools are of primary benefit to undertake segmentation (the process of identifying and grouping similar tissues). For volume measurements, this volume of the segmentation directly correlates with muscle volume. With muscle to fat ratio calculation, this segmentation allows for a rapid calculation of the ratio in individual muscles. In most cases, the time to process the data outweighs the time to collect the data, often by several orders of magnitude. Any steps that can, therefore, be removed brings about significant benefits.

There are three terms that are used to describe the tools that will be covered briefly in this section. They are defined here to assist the reader in identifying the tools that may be useful [17, 18]:

- Fully automated methods take in the raw images and provide the final desired result with no further user input. For example, a 3D dataset of the thigh may be added to the tool, and the volume of adipose tissue will be returned. These are rarely available as a standard tool but are being developed for individual applications and often for a single study. In addition, there is no opportunity to modify segmentations where they are clearly incorrect.

- Semiautomated methods require the operator to undertake some of the operations. This could be to draw an outline around individual muscles within a group at a small number of slices or to adjust a threshold value so that processing can take place. These may be available as part of available tools and offer far greater flexibility. They cannot, however, work without human interaction.
- Supervised methods are also available that are capable of automated segmentation, but with the option for the operator to intervene if fully automated processing fails. These approaches may offer representative slices for checking by the operator or adjustable parameters to provide good segmentation.

Care should be taken when designing an experimental protocol to ensure that the whole workflow is planned out and tested before starting the study. The temptation to collect data and assume that processing will take place as predicted is high but often leads to extensive data processing which could have been avoided by a simple change of MRI parameters. For example, for compartmental VAT estimation, the focus is typically on achieving high contrast between adipose and muscle tissues. However, automated processing tools can successfully resolve relatively small differences between these tissue types, since even with relatively low contrast, they are well differentiated. Excellent adipose to muscle contrast does not necessarily, however, delineate well between individual muscles, potentially leading to many (hundreds or more) hours of manual segmentation to delineate muscle boundaries, which could be avoided by improved muscle to muscle contrast. Collection on a single trial dataset followed by complete processing would have highlighted this issue from the beginning and saved a significant amount of time in the overall study.

Useful tools for processing MRI images are described in each of the sections which have been chosen from the wide range of tools that are available for their (relative) ease of use for the nonexpert imager and processor.

1.3 Measurement of Skeletal Muscle Volume

Measurement of the volume of muscles, without external manipulation, is one of the primary benefits of the MRI technique. With modern processing methods, which are discussed in a later section, it is possible to routinely estimate volumes from three-dimensional (3D) data rather than relying upon estimates based on thickness at individual locations. Such processing also yields useful information regarding the shape of the muscle rather than simply its volume, which can have significant influence on function [19]. The bulk of studies in the literature focuses on the validation of 2D estimates of muscle volume by comparing the two approaches of full muscle measurements versus key slice muscle body thickness measurements, often using the Anatomical Cross-Sectional Area or ACSA.

For example, Morse et al. [20] demonstrated that the ACSA, at 60% length from the distal end, correlates to within 10% with the gold standard of manually calculated volumes from slice-by-slice segmented T_1-weighted MRIs for healthy young adult males. A slightly modified approach achieving improved correlation to within 8% RMS differences was reported by Mersmann et al. [21] for healthy males and females, despite different muscle geometries. Their approach of estimating the volume of the *m. quadriceps vastii* from measurements of its length, maximum ACSA, and shape factor was compared to manually calculated volumes from slice-by-slice segmented T_1-weighted MRIs. Empirically determined shape factors for each of the vastus muscles yield good results (*see* **Note 5**).

Where full muscle volume measurements are used (e.g., [22, 23]), a similar approach is employed but using contiguous slices which cover the whole length of the muscle. The sum of areas determined on each slice is then multiplied by the slice thickness to determine the volume. In this study, volume estimation is undertaken on two scan sessions, a week apart, and by two observers. Confidence over 0.99 was found between observers and scan sessions demonstrating the benefit of whole muscle measurements where study responsiveness is likely to be small. Similar studies have also been conducted on other muscle groups, for example, *m. triceps brachii* [24], spinopelvic muscles [25], and shoulder [26], with similar findings.

Despite a growing trend toward automated segmentation of whole muscle volumes with image processing, the ACSA approaches reduce scan time and thus remain very popular. The collection of a single image in each orthogonal plane is sufficient to estimate volume using such methods and can be achieved in less than 1% of the total scan time required for full volume analysis. Where responsiveness is expected to exceed 10%, these approaches offer rapid imaging prospects for longitudinal studies, allowing for higher numbers of participants within a given timeframe and imaging cost.

Imaging for all studies discussed in this section is T_1-weighted. Using a gradient echo pulse sequence (*see* **Note 6**), common imaging parameters (*see* **Note 7**) are:

- Short echo time between 10 ms and 16 ms
- Gradient echo sequence repetition times around 100 ms
- Spin echo sequence repetition times between 500 ms and 700 ms
- Matrix sizes between 256 × 256 and 300 × 300
- Resolutions between 0.5 mm and 1 mm isotropic
- Slice thicknesses between 4 mm and 6 mm (based on required accuracy)

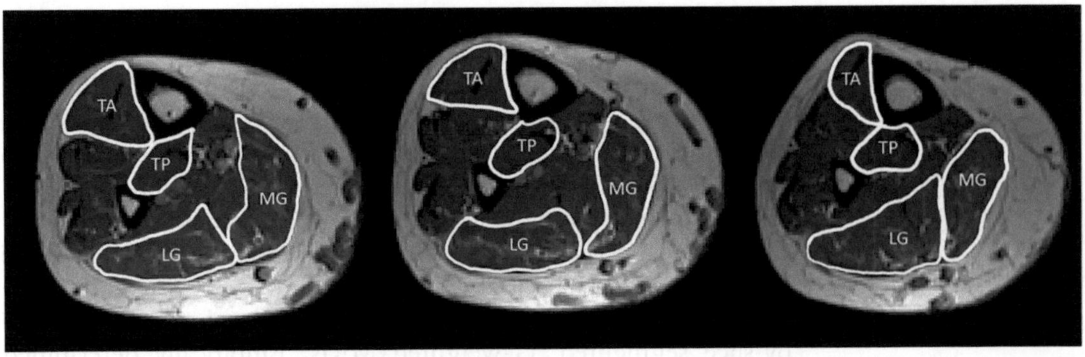

Fig. 4 Images showing the impact of various positions of the foot on imaging of the m. tibialis anterior and posterior (TA and TP) and the m. lateral and medial gastrocnemius (LG and MG). ACSA are outlined for each muscle. The left and center images are collected using different support techniques resulting in a difference in foot position of less than 10° to the leg. The right image shows the foot in dorsiflex. Significant differences in the shape of TA and LG are apparent owing to their functions in foot positioning. The resulting cross-sectional areas are correspondingly different leading to limited analytical value if these represented different timepoints in a study. While the differences between the dorsiflexion and relaxed positions will not be surprising, the difference in the cross-sectional area of several muscles is significant even for minor changes in positioning

Since the volume of a muscle is not necessarily constant in different positions [27], consideration must be given to participant positioning before any scanning is undertaken. This is of particular importance in studies where different participant groups are to be imaged. For example, in studies including active and sedentary populations, it is worth, at the study design stage, dedicating time to ensure that both populations can position the joints in a suitable position to achieve consistent measurements of relaxed muscles. Large and highly defined leg muscles, such as those of the athlete, require significantly different cushioning and support to achieve a natural 180° positioning of the knee. Furthermore, ensuring accurate measurements of muscle volume changes over time require significant planning to accomplish adequate repeat positioning. Considering briefly by way of example, imaging of the hamstring muscles, this problem is highlighted: whilst knee angle is the obvious focus for ensuring a relaxed state measurement, anterior pelvic tilting, owing to the use of a different number of pillows to support the head and torso, will change the stretch of the muscle leading to variable results. Wherever possible, the use of indexable patient support systems, which are commonly used in radiotherapy planning and treatment, minimizes this source of possible measurement error. Where such equipment is not available, a rigorous approach to consistent patient positioning and accurate notes detailing the padding and supports used are essential to minimize errors (*see* **Note 8**). An example of the pitfalls of positioning can be seen in Fig. 4.

1.3.1 Suggested Scan Card

There is significant variation in the protocols used between scanner vendors and centers resulting from hours of invested time optimizing each parameter to local systems. Presented below is a useful starting point for the scientist conducting their first measurements in each of the areas described in this chapter. Some minor changes to parameters may lead to better images on the specific system used, but these should serve as a good starting point.

Scout image—three-slice gradient echo

Sequence	Siemens: FLASH GE: SPGR Phillips: T1-FFE Canon: RF-spoiled FE
Plane	Axial
T_E	3 ms
T_R	8 ms
Flip angle	20°
FOV	500 mm × 500 mm
Slice thickness	6 mm
Matrix size	256 × 192

Localizer image—3 × 3 slice gradient echo

Sequence	Siemens: FLASH GE: SPGR Phillips: T1-FFE Canon: RF-spoiled FE
Plane	Axial, sagittal, coronal
T_E	3 ms
T_R	8 ms
Flip angle	20°
FOV	500 mm × 500 mm
Slice thickness	6 mm
Matrix size	256 × 192

T_1-weighted image

Sequence	Siemens: FLASH GE: SPGR Phillips: T1-FFE Canon: RF-spoiled FE
Plane	Axial (perpendicular to muscle fiber direction)

(continued)

T_E	12 ms
T_R	100 ms
FOV	256 mm × 256 mm
Slice thickness	5 mm
Matrix size	256 × 256
Averages	1

1.3.2 Software Tools for Volume Estimation

There are several tools available to aid in segmentation for muscle volume estimation. At the time of writing, there are no freeware tools available that are capable of fully automated processing of muscles from anywhere in the body. Discussed here are four tools that are reported in multiple publications and are regularly used by the authors alongside several which directly compare their features (e.g., [28–30]). We do not consider the myriad of approaches that are covered in the literature, where advanced skills and knowledge are required for implementation, as these are outside the scope of this chapter.

ImageJ [31] *and Fiji* are exceptionally popular (e.g., [32–34]). *Fiji* is simply a distribution of *ImageJ* bundled with lots of useful plug-ins. It is included here not because it is particularly powerful, but because it is a tool with which many readers will already be familiar and is suitable for use in the segmentation process. There are two main tools used within ImageJ or Fiji. The first is VOLUMEST, which is a generic volume estimation plug-in that prompts the user to draw around the tissue to be measured in each of the slices it can be seen on. It will then interpolate between slices to estimate the volume of the region of interest. Although very slow, this method is present in almost all studies comparing fully automated segmentation methods as the gold standard for volume estimation. The second is bundled with *Fiji*; the segmentation editor can also be used with other 3D software tools and allows for measurements of multiple segments (i.e., multiple muscles) in parallel, in that all regions can be traced on the same slice. Slices like the last can be skipped, and the tool will automatically interpolate them, which can result in a significant time saving for long muscles. It should be noted that not all DICOM exports correctly report the slice thickness in the way that it is expected by *ImageJ*, and thus volumes must be sanity checked when using a new scan protocol or system. *Fiji* is available here: https://imagej.net/Fiji.

3D Slicer is a powerful and open-source freeware application for 3D reconstruction (e.g., [30, 35, 36]), which makes it highly useful for volumetric measurements. DICOM files can be imported directly and processed with supervised and semiautomated approaches. As previously discussed, additional effort in optimizing scan parameters to achieve a high contrast and low noise dataset

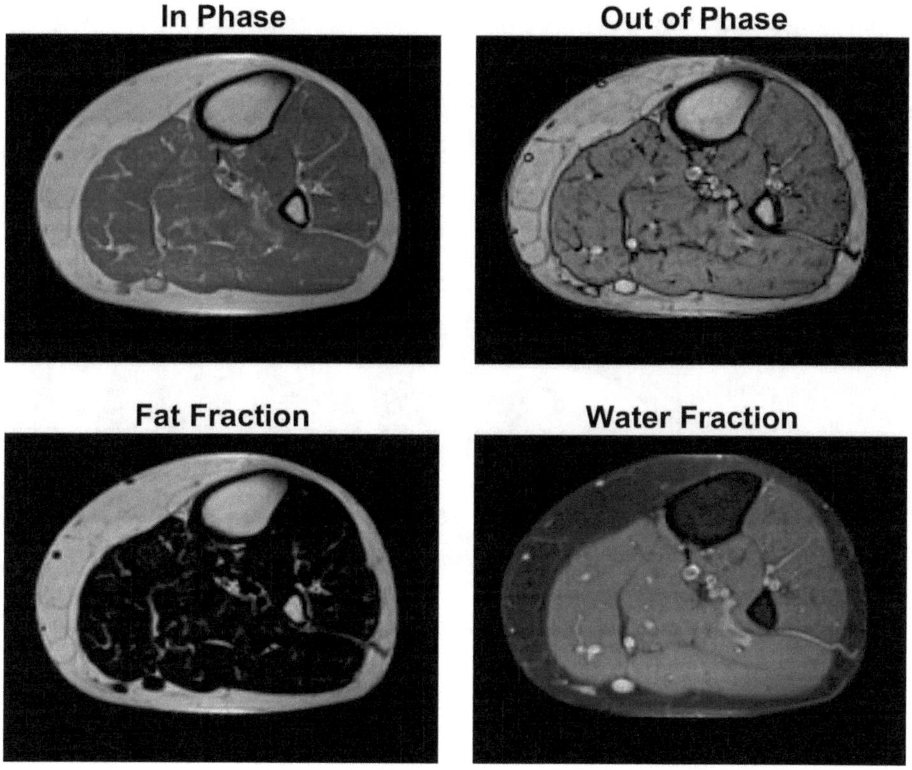

Fig. 5 DIXON images of the lower leg of a healthy volunteer which have been corrected for field inhomogeneities using a line of pixels through the fat on the left-hand side of the image. The in-phase image is much like the T_1-weighted image shown in this figure. The out of phase image is visually very different. The addition and subtraction of these two images reveals the fat fraction and water fraction in each pixel. These images can be used quantitatively to determine overall and compartmental fat to muscle ratio. The outline of different tissue types on the out of phase image is an artifact caused by equal quantities of fat and water in the same pixel which cancel each other out, leaving approximately zero signal. This image is very useful for segmentation, which is discussed later in this section

pays dividends in reducing the human effort involved in segmentation. The out of phase DIXON image is often used for segmentation, making use of the dark borders around tissues as seen in Fig. 5; thus, muscles are often well delineated. This makes it possible to perform semiautomated segmentation by identifying a so-called seed in each of the muscles of interest, which are grown up to a border in three dimensions, thus identifying each structure.

This is only as effective as the delineation between the muscles in the groups and may require several iterations with additional manual outlines to be successful (*see* **Note 9**). This software also has a mode of operation where outlines are drawn on several noncontiguous slices, and semiautomated processing traces contours between the slices to determine the volume. Manual paint-like tools can correct any errors in the predicted slices before calculating the volume of each muscle or muscle group. An example of

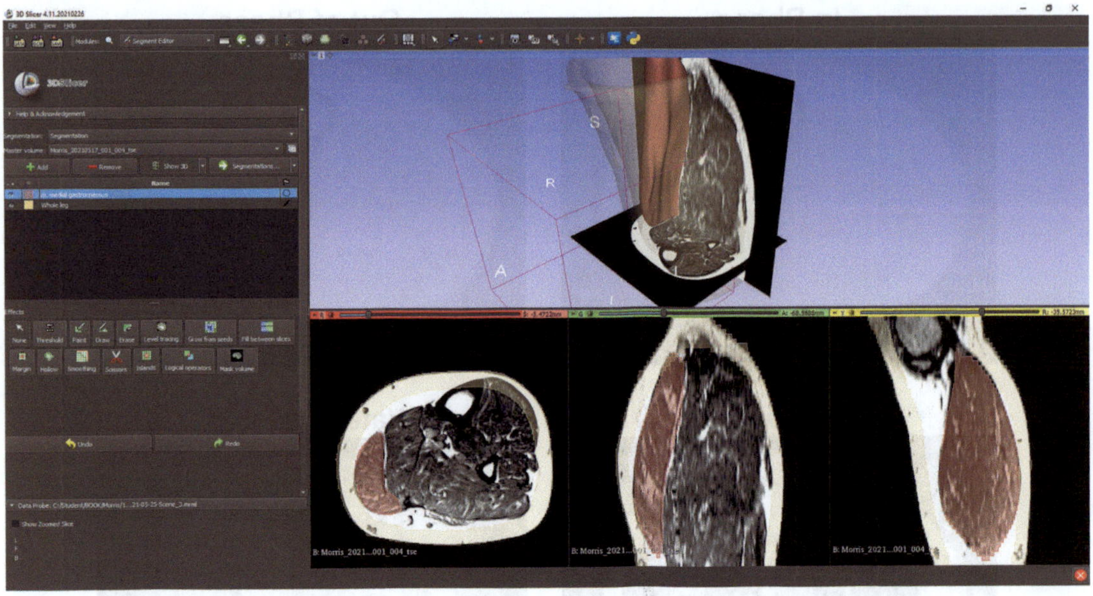

Fig. 6 An example of a 3D segmentation of m. medial gastrocnemius and the whole leg using 3D Slicer

segmentation showing m. medial gastrocnemius and a whole leg outline is shown in Fig. 6. It is available here: https://www.slicer.org/.

OsiriX is another tool that is often cited in the literature (e.g., [37–39]). Only available for Apple Mac systems, it is primarily intended as a DICOM viewer but has powerful 3D visualization and segmentation capabilities, thanks to a growing number of available plug-ins. Of these, the Segmentation tool of Brian Jensen (http://campar.in.tum.de/Main/BrianJensen) offers a nice interface, which also has region-growing algorithms working outwards from a manually planted "seed" up to a boundary, minimizing human input. A previously available segmentation plug-in called iSix has been offline now for some time and so is not discussed here, although it appears in several published works. The software is free for trial but must be purchased for custom plug-in use. Full details are available at https://www.osirix-viewer.com/osirix/osirix-md/.

In addition to the free and freemium options, there is a commercial application of note: *SliceOmatic* has very advanced segmentation algorithms and is well cited (e.g., [40–43]) for accuracy in both compositional analysis and segmentation. A limited demonstration version is available for evaluation here: https://www.tomovision.com/products/sliceomatic.html.

1.4 Measurement of Muscle to Fat Ratio

The muscle to fat ratio is an important measurement in understanding the health of an individual and is often poorly correlated with body mass index or waist size [44]. In many studies, a holistic view of adipose tissue distribution throughout the body is undertaken using T_1-weighted whole-body MRI, which is important in determining the overall health of the individual and understanding risks to metabolic health. This same technique is now also used routinely on individual muscles for the calculation of the compartmental volume of adipose tissue (compartmental VAT). The validity of this approach using MRI was demonstrated by Mitsiopoulos et al. [13] in comparison with cadaveric dissection with correlations better than 99%. Such measurements of compartmental VAT, also referred to as intramuscular adipose tissue or IMAT, allow for reliable prediction of reduced mobility, as demonstrated by measurements at the mid-thigh in older individuals [45]. In combination with the fact that infiltration of intramuscular adipose tissue increases with age, this measurement is a key predictor of mobility, or loss thereof, in studies concerned with an aging population.

A useful review is presented by Borga et al. [46] that covers the whole-body compositional analysis (*see* **Note 10**) to which the interested reader is directed for such measurements. Of primary interest here, however, is the compartmental VAT measurements, which are typically achieved using one of two families of methods.

The first employs some form of image analysis on a standard T_1-weighted image. Since adipose tissues are very bright on T_1 images (see Fig. 4), through a histogram of pixel intensities or by manual regions of interest in adipose and muscle tissues (a useful comparison of these methods is presented by Ogawa et al. [47]), a threshold is determined above which pixels that represent AT can be labeled. The number of pixels labeled in this fashion is then counted and multiplied by the pixel surface area to reveal the cross-sectional AT area. This can be repeated on several slices if needed and multiplied by slice thickness to determine the volume, although, much like the ACSA approach in volume measurements, any slice of a muscle near the maximum thickness is representative of the overall quantity. An example of this processing using a histogram approach can be seen in Fig. 7.

The second approach is to utilize the DIXON sequence introduced in the basic principles section, which produces two images representing pixels containing triglycerides and pixels containing other proton-containing molecules. These images are typically referred to as fat and water images, respectively. A simple ratio of these two images reveals the overall muscle to fat ratio. This is normally undertaken in a manual region of interest to determine the intramuscular infiltration. For example, Kim et al. [48] produced manual regions of interest on standard T_1-weighted images, which were then replicated on DIXON images to measure fat fractions of the upper thigh. An example is shown in Fig. 5.

Advances in the DIXON technique in the last decade have resulted in multiple echo DIXON sequences that utilize more than one each of in and out of phase echoes to provide quantitative estimates of the fat within a single voxel (the volume equivalent of a

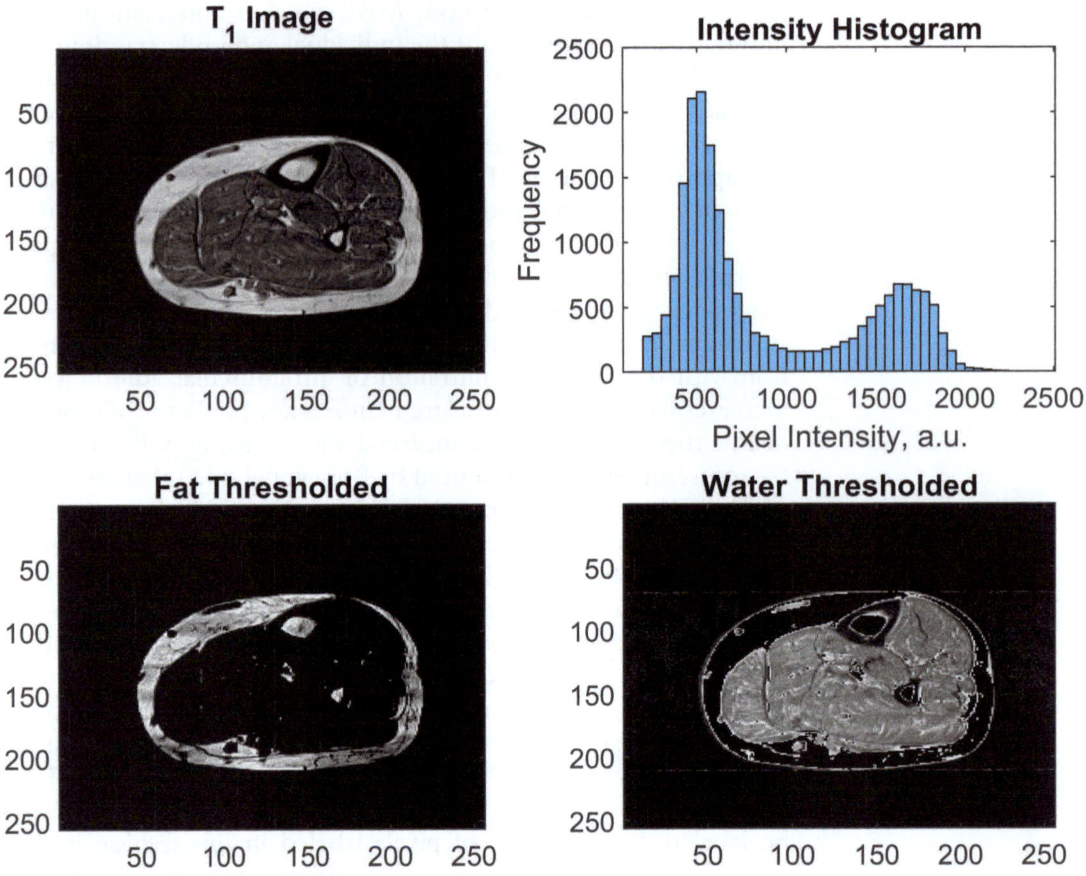

Fig. 7 An example of fat to muscle ratio image processing using T_1. The T_1 image as collected can be seen in the top left showing bright signal from the adipose tissue and reduced intensity from other tissue types. A histogram of this image reveals a bimodal distribution which can be seen top right. The lower peak represents non-adipose tissue. This can be used to determine the fat to muscle ratio over the whole image by taking the integral of the two peaks. However, thresholding and using the images can reveal information about the location and thus the nature of the distribution of the adipose tissues which can be seen in the bottom images. Thresholding has been undertaken using a cutoff value of 1000 estimated using the histogram

pixel). Of particular note is the mDIXON sequence available on Phillips scanners [49], which utilizes three of each echo to produce a color-coded map representing the percentage of fat within each voxel. Originally produced for assessment of fatty liver disease (*see* **Note 11**), this has been successfully demonstrated as a useful technique to measure accurate fat contents in muscles, for example, in paraspinal muscles [50], with interobserver reproducibility exceeding 98%.

Both approaches lend themselves to relatively straightforward imaging protocols that are available on all MRI systems. The T_1-weighted imaging with thresholding approach can be undertaken with any T_1-weighted imaging protocol, giving good differentiation between fat and muscle. This can be achieved using a turbo spin echo sequence with the following:

- A repetition time between 400 ms and 600 ms
- An echo time of 16 ms or lower
- A resolution around 0.7 mm, with a 250 mm to 350 mm field of view, both isotropic
- A slice thickness between 2 mm and 5 mm (depending upon the size of the muscle and the accuracy needed).
- No averaging

DIXON sequences have similar resolutions and slice parameters. Gradient or spin echo DIXON sequences can be used without preference with the following example sequence parameters:

- Repetition times between 10 ms and 20 ms for gradient echo.
- Repetition times between 4000 ms and 5000 ms for spin echo sequences.
- Echo times should be kept as short as can be reliably achieved on free echo time sequences or according to field-dependent values otherwise.

Where single limbs are scanned, phase encoding should always be set to anterior to posterior (or vice versa) to avoid ghosting of the unscanned limb (*see* **Note 12**).

The T_1 thresholding approach is straightforward to implement. As the pixel result is either fat or not fat, accuracy in the result is the price for this simplicity. Minor variations in the thresholding value also radically change which population each pixel belongs to, this impacts the intramuscular fat, and care must, therefore, be taken when using such a method. In contrast, the DIXON methods offer highly accurate results; they are well correlated with other measures and provide good predictions of performance and mobility, although this does come at the expense of additional effort in setup and processing. The DIXON technique relies upon knowing the difference in precessional frequency between triglycerides and other proton-containing molecules, which requires a homogenous field that can only be assured in the center of the bore. Accordingly, care should be taken to avoid imaging too close to the end of the bore. Well-shimmed magnets offer the best imaging results, although there are strategies available for the correction of homogeneity and RF nonuniformity. The simplest to implement is based on the assumption that the water and fat images experience this variation equally and that by identifying 50 or so high fat content pixels along the image, a linear correction can be applied to both the water and fat images before further processing (*see* **Note 13**). This and other useful corrections are discussed in more detail by Leinhard et al. [51].

1.4.1 Suggested Scan Card

There is a significant variation in the protocols used between scanner vendors and centers resulting from hours of invested time optimizing each parameter to local systems. Presented below is a useful starting point for the scientist conducting their first measurements in each of the areas described in this chapter. Some minor changes to parameters may lead to better images on the specific system used, but these should serve as a good starting point.

Scout image—three-slice gradient echo

Sequence	Siemens: FLASH GE: SPGR Phillips: T1-FFE Canon: RF-spoiled FE
Plane	Axial
T_E	3 ms
T_R	8 ms
Flip angle	20°
FOV	500 mm × 500 mm
Slice thickness	6 mm
Matrix size	256 × 192

Localizer image— 3 × 3 slice gradient echo

Sequence	Siemens: FLASH GE: SPGR Phillips: T1-FFE Canon: RF-spoiled FE
Plane	Axial, sagittal, coronal
T_E	3 ms
T_R	8 ms
Flip angle	20°
FOV	500 mm × 500 mm
Slice thickness	6 mm
Matrix size	256 × 192

T_1-weighted image

Sequence	Siemens: TSE GE: FSE Phillips: TSE Canon: FSE
Plane	Axial (perpendicular to muscle fiber direction)
T_E	12 ms

(continued)

T_R	500 ms
FOV	256 mm × 256 mm
Slice thickness	3 mm
Matrix size	256 × 256
Averages	1

DIXON

Sequence	Siemens: FLASH GE: Flex Phillips: mDIXON Canon: WFOP
Plane	Axial (perpendicular to muscle fiber direction)
T_E	16 ms
T_R	500 ms
FOV	256 mm × 256 mm
Slice thickness	3 mm
Matrix size	256 × 256
Averages	1

1.4.2 Software Tools for Muscle to Fat Ratio

There are two principal applications which are available for use in the automated determination of muscle to fat ratio.

FATCALC [46, 52] is a fully automated fat volume estimation macro for ImageJ [31]. Designed to process two-point (standard) DIXON images, the plug-in automatically segments images and determines the visceral adipose tissue and subcutaneous adipose tissue. This tool is particularly useful for large cohort studies, as the processing can be fully unsupervised. A calibration factor is available to ensure that the results are consistent with manual segmentations based upon different scan sequence inputs. It is also possible to combine the macro with a 3D region of interest to undertake VAT measurement in a single muscle. It is available from SourceForge here: https://sourceforge.net/projects/fatcalc-imagej/files/.

AMRA® Researcher is often used in studies of adipose tissue volumes (e.g., [53–55]). This software is not free but is mentioned here, since it is used extensively in the literature. It is capable of fully automated processing of MRI images to reveal visceral adipose tissue, muscle fat infiltration, subcutaneous adipose tissue, and lean thigh muscle volume. Reports in the literature suggest a good correlation with manual segmentation techniques. Like FATCALC, this software is capable of individual muscle-level processing and includes tools for segmentation in the application. Full details of licensing are available here: https://www.amramedical.com/academic-research.

1.5 Measurement of Skeletal Muscle Fibers

Fiber tractography is an advanced MRI technique that is often used in the brain to determine interconnected regions. It works by measuring how quickly protons diffuse in different directions. In a bulk sample, this will occur similarly in all directions as there are no restrictions. In many samples, such as those that are highly ordered (e.g., those that are fibrous), it is much easier to move in some directions than others, and thus there is the appearance of anisotropy. A good analogy is to think of water molecules in narrow straws. If the molecules move along the length of the straws, they can do so unimpeded; however, should they try to move across the straw, they will quickly encounter a boundary. Thus, in each time period, they will statistically have moved further along the straw than across it. By measuring the mean free path (a measure of the distance moved in a second) in different directions, it is possible to determine the so-called diffusion tensor, which is a measure of the magnitude and direction of the diffusion in 3D space. It is then possible to estimate which of these tensors join, and thus the path of the straws can be traced out. In the body there are many tissues that have this same effect on the water molecules present, such as axons in the brain or muscle fibers, throughout the body. The MRI pulse sequence used to determine these tensors is known as diffusion tensor imaging or DTI. A single coefficient defines how anisotropic the diffusion is, which is known as fractional anisotropy. On a local level, this coefficient can take on values between 0 (completely isotropic) and 1 (diffusion in a single direction) and is an indicator of how fibrous a sample is. Where fibers cross perfectly perpendicular, this coefficient increases to $\sqrt{2}$ and takes on values in between this and 1 for different angles of crossing. Table 1 shows a graphical representation of diffusion tensor and fractional anisotropy value for a range of samples.

There is some conflict in the early literature regarding the ability to undertake fiber tractography on muscles. For example, Finch et al. [56] showed that the diffusion was mostly isotropic, rendering it impossible to undertake tractography. This was fortunately revisited in 1976 by Cleveland et al. [57], who reported fractional anisotropies in the order of 0.4, which are suitable for tractography. The primary application of tractography in the imaging of muscles is in measuring their structural properties, such as pennation angle, which is a good predictor of maximum force generation. Early work was undertaken in animal models [58, 59] demonstrating the ability to measure pennation angle with an r^2 against dissection of 0.89 and to determine fiber lengths in line with anatomical expectations. A review by Damon et al. [60] covers in detail the best approaches to undertaking diffusion tensor-based measurements of muscles and is a useful resource for the reader interested in undertaking such measurements. To summarize, carefully following the experimental setup requirements for muscle DTI measurements are essential for obtaining data of sufficient

Table 1
A comparison of fractional anisotropies and diffusion tensor shapes for various sample configurations

Sample	Diffusion tensor	Fractional anisotropy value
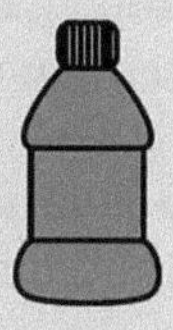		0
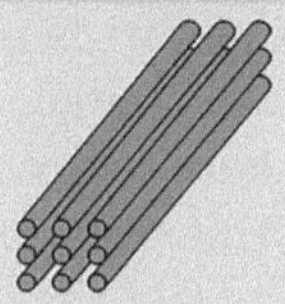		1
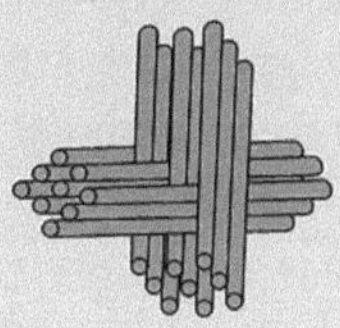		$\sqrt{2}$

quality. This includes, but is not limited to, appropriate immobilization of the muscle to eliminate motion artifacts during the lengthy imaging protocols, using rapid imaging protocols to minimize the chance of motion during measurements and appropriate imaging parameters. Data processing is not trivial for DTI measurements and requires co-registration of images and computationally intensive mathematical translations, as well as an iterative search process to determine the paths of the muscle fibers (*see* **Note 14**).

In addition to informing traditional measurements for the characterization of muscle architecture and predictions of performance, muscle fiber tractography has been used to determine the strain tensor during muscular contractions [61]. The fiber orientation was measured during separate isometric contractions of the muscle, which were then used to determine the strain tensor in different compartments to gain an understanding of muscle dynamics. This has been used in several studies since to investigate different muscles, such as the m. gastrocnemius [62], where it was shown that muscle loading does not necessarily result in a reduction in force production and postulates a link to muscle architecture and muscle length differences.

In addition to complete fiber tractography, the measurements required to undertake computationally intensive tracking have also been shown to have merit. For example, fractional anisotropy correlates over 80% with the maximal muscle power of the *m. soleus* [63]. Similarly, variations in the fractional anisotropy are a good predictor of muscle health, particularly in relation to the detection of low-level muscle damage, such as grade 1 strains, which are not well visualized on traditional T_2-weighted MRI images [64]. At present, however, it has been shown that these approaches are less useful than a radiologist's visual assessment at predicting a return to play after an injury [65], which is one of the potential applications for such diagnostic measures.

While most MRI systems will have rudimentary diffusion tensor imaging sequences, typically within advanced brain imaging protocols, a significant investment in time and effort is needed to undertake fiber tractography on muscles. Some initial sequence parameters, as used in the studies referenced herein, are provided as a starting point, but it should be noted that this technique is not straightforward to implement. Issues include the fact that the T_2 in most brain tissues is very long (around 100 ms) in comparison to those in muscle (around 30 ms); thus, the default sequence parameters, when combined with the necessary gradient strengths, typically result in very dark (and largely unusable) images. Where it is available, the use of stimulated echo (STE) diffusion preparation should be used (as per Damon et al. [66]) to allow for sufficiently short effective echo times (less than 50 ms). On newer and research-grade systems, a new family of DTI sequences, often called HARDI (high angular resolution diffusion imaging), has been anecdotally found to make setup less challenging, since the determination of fractional anisotropies and diffusion tensors is more robust, although the effective echo time must still be kept lower than that which is typical for neurological imaging. A b-value of 400 s/mm^2 to 500 s/mm^2 with between 6 and 15 gradient directions is suggested in most studies in this area, although robust tensor measurements have been shown for values up to 725 s/mm^2 [60]. For efficient acquisition, an EPI-based DTI sequence should be used.

Other imaging parameters are:

- Matrix sizes between 128 × 128 and 256 × 256
- Resolutions between 1 mm and 3 mm isotropic
- Echo times between 36 ms and 42 ms
- Repetition times between 6300 ms and 7500 ms
- A train length of between 40 and 64
- SPAIR fat suppression
- Average of 8–12 (at 1.5 T to have sufficient SNR for efficient tractography, less will be required at 3 T)

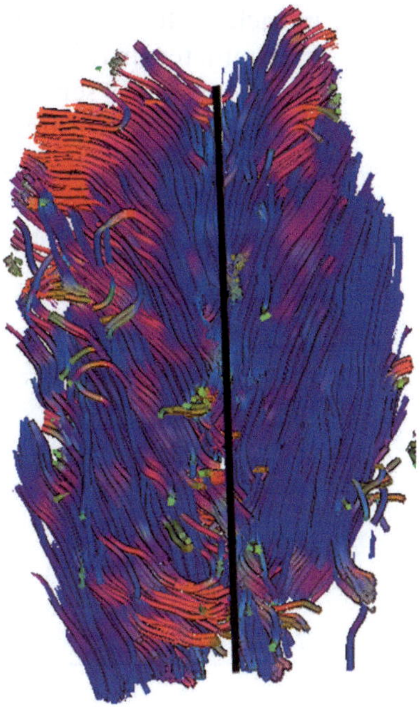

Fig. 8 Diffusion tractogram of m. medial gastrocnemius showing the fiber paths in the muscle as found with tractography. The central black line represents the insertion to the tendon from which the pennation angle can be determined

An example tractogram for *m. medial gastrocnemius* is shown in Fig. 8. The pennation angles can be measured as the angle of the fibers to the central black line representing the tendon insertion.

The time required to collect a full DTI dataset should not be underestimated. The more gradient directions and strengths that are sampled, the longer the participant must remain within the scanner. The example in Fig. 8, collected with a multichannel coil for the calf only, took 16 min to acquire and 10 min to process. It is, however, important to note that, unlike other techniques routinely used for pennation angle measurements, the MRI DTI approach allows for measurements in a truly rested state (*see* **Note 15**). Fiber tractography in muscles requires careful setup and a good understanding of the interplay between different parameters on the MRI system being used. In comparison to diffusion along axons in the brain, for which the majority of tractography sequences are pre-set on clinical systems, fractional anisotropies are in a different range, and b-values and gradient spacings must be carefully chosen to provide data that are suitable for processing if using freely available tools.

1.5.1 Suggested Scan Card

There is significant variation in the protocols used between scanner vendors and centers resulting from hours of invested time optimizing each parameter to local systems. Presented below is a useful starting point for the scientist conducting their first measurements in each of the areas described in this chapter. Some minor changes to parameters may lead to better images on the specific system used, but these should serve as a good starting point.

Scout image—three-slice gradient echo

Sequence	Siemens: FLASH GE: SPGR Phillips: T1-FFE Canon: RF-spoiled FE
Plane	Axial
T_E	3 ms
T_R	8 ms
Flip angle	20°
FOV	500 mm × 500 mm
Slice thickness	6 mm
Matrix size	256 × 192

Localizer image—3 × 3 slice gradient echo

Sequence	Siemens: FLASH GE: SPGR Phillips: T1-FFE Canon: RF-spoiled FE
Plane	Axial, sagittal, coronal
T_E	3 ms
T_R	8 ms
Flip angle	20°
FOV	500 mm × 500 mm
Slice thickness	6 mm
Matrix size	256 × 192

Diffusion tensor EPI

Sequence	Siemens: EPI SE Others: SE EPI
Plane	With muscle fiber direction along the read axis
T_E	42 ms
T_R	7000 ms
FOV	Tight to muscle (~256 mm × width of muscle)
Slice thickness	3 mm

(continued)

Matrix size	128 × 128
b-value	400 s/mm^2
Train length	44
Averages	12
Fat suppression	SPAIR

1.5.2 Software Tools for Fiber Measurements

Unlike volume measurements, there are several steps that must be undertaken to achieve a fiber tractogram or measurements thereof. In brief, these are:

- Registration of the images and distortion correction to ensure that artifacts do not produce false tracts
- Smoothing or other noise reduction methods since tracking is made more accurate by doing so and, given the highly ordered nature of the fiber bundles, can be performed with little loss of accuracy
- Diffusion tensor calculation to work out the dominant diffusion direction
- Fiber tracking, where the paths of the fibers are estimated from the diffusion tensors

These steps may be performed by several pieces of software or by a single tool.

To the authors' knowledge, there is still no software available that is specifically for the purposes of muscle fiber tractography, and thus all the tools that are discussed below are intended for use on brain data sets. While this does not mean that the data cannot be processed, the authors would be remiss not to draw attention to the additional effort which must be put into processing muscle datasets using these tools, since many of them expect parameters to be within certain ranges, which do not lead to good diffusion images of the muscles, as discussed earlier.

Almost all studies undertake manual software creation using *MATLAB*, or something similar, to provide complete control over the process which yields the highest quality results. It is possible, however, to undertake competent tracking using freely available neurological tools. While many of the standard tools (*FSL*, *Freesurfer*, *AFNI*, *SPM*) should be capable of completing the necessary steps, the authors have found two pieces of software to be most useful for those without prior experience in one of the traditional applications. These are *Diffusion Toolkit* and *medInria*. By utilizing distortion correction on the MRI system, the dataset (collected with appropriate parameter and sequence choices) can be loaded directly into these applications, both of which have a grid seeding utility, allowing for seedpoints to be created in a regular grid (1 cm

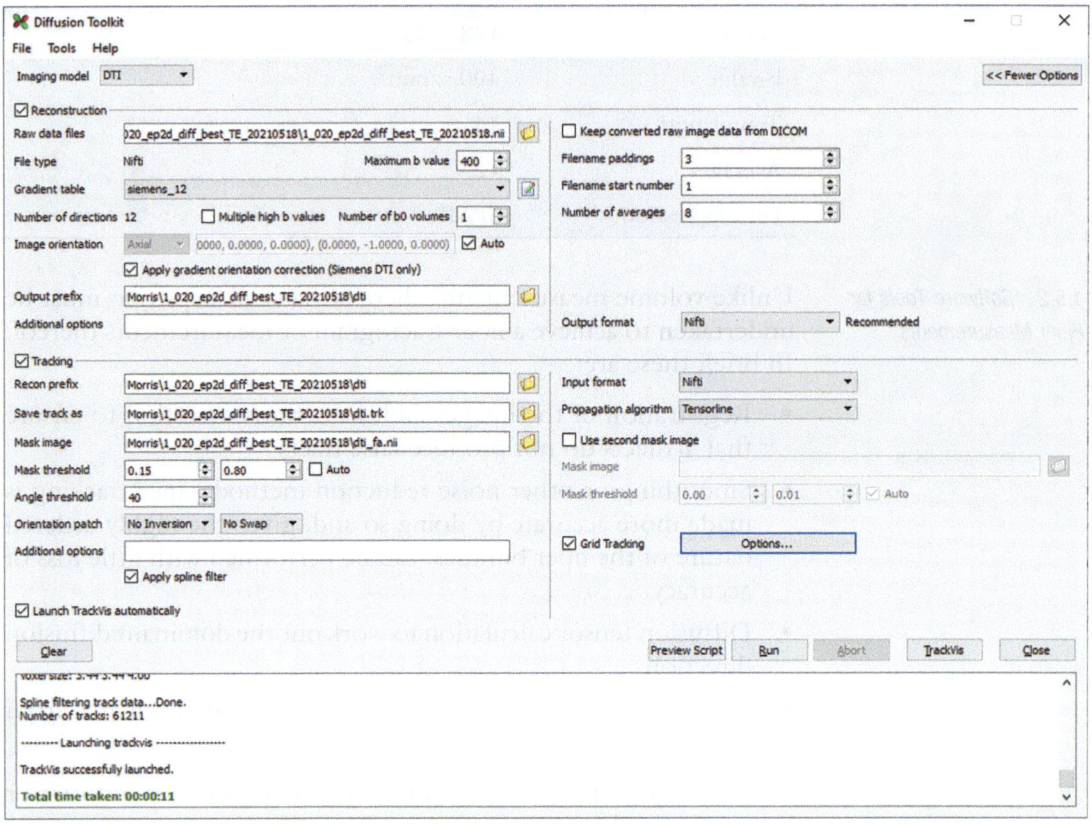

Fig. 9 Screenshot of the Diffusion Toolkit graphical user interface

is appropriate for most muscles) or can undertake seeding from the muscle insertion by using a mask created in *ImageJ* or similar. Tracking is rapid and can be tuned with limited but enough parameters.

A good starting point in *Diffusion Toolkit* is to use the fractional anisotropy map as a mask and to set a threshold of 0.15 for minimum and 0.8 for maximum, to use an angle threshold of around 40 and to turn on grid tracking, with a seeding interval of 1 and a step length of 0.5 (or to use a secondary mask, where a small volume over the insertion has been already created). A screenshot of the graphical user interface is shown in Fig. 9. With the parameters, a script can be made that automatically applies the same processing to many scans, which is good for multiple timepoint studies with many participants. Registration is required to download but the software is free. It can be downloaded here: http://trackvis.org/dtk/.

The steps in medInria begin with starting the application in diffusion mode and working through each of the steps presented in a side panel, a screenshot of which can be seen in Fig. 10. A basic workflow is shown below:

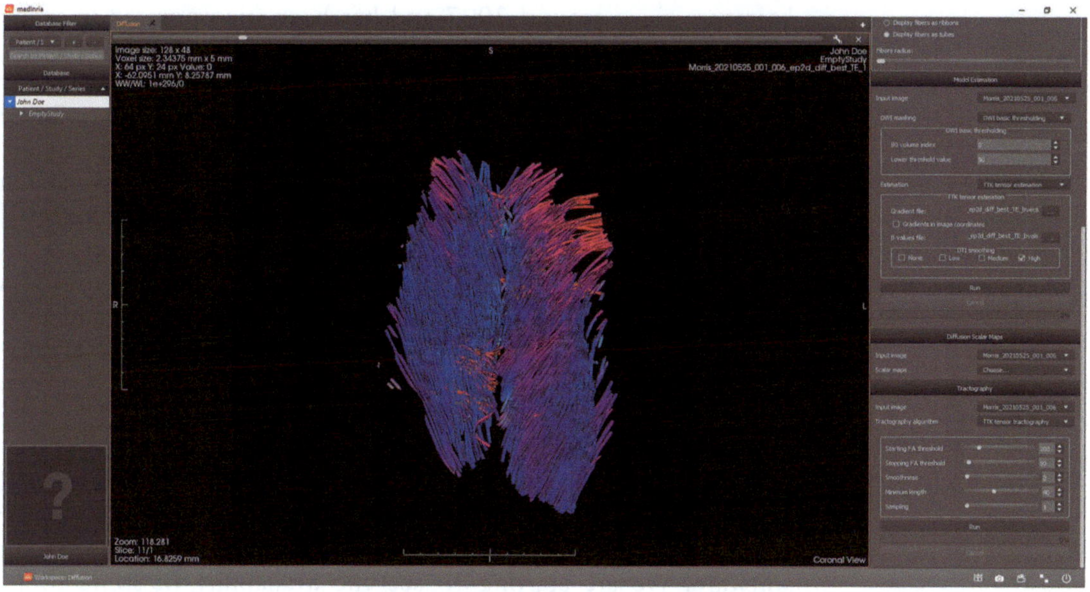

Fig. 10 Screenshot of the medInra interface showing diffusion tractography.

- NIfTI images are loaded into the software.
- DWI basic thresholding— the cursor can be used to show the grayscale value. Pick a value 5–10 points higher than the highest value in the noise.
- Gradient vector and value files are loaded, and TTK tensor estimation is run.
- Tractography is run using TTK tractography algorithm with a stopping FA threshold of 50 and starting FA threshold of 200 as a starting point (the units are nonconventional). Smoothness and sampling can be adjusted from the default after reviewing the results. Exclusion based on a minimum length of 40 mm generally gives good reproduction of muscle fibers.

Various display modes are available in the software for the visualization of the tractograms (Fig. 8 shows tubes, for example). Anatomical images can also be overlaid by dragging them into the work area. The software can be downloaded from here: https://med.inria.fr/

1.6 Looking to the Future of MRI of Muscle Mass

In this chapter we have presented an overview of the generally accepted best practice regarding imaging muscle mass in the human body. The three main measurements presented represent those which are most easily applicable to those whose primary area of expertise is not MRI. There is one main area prevalent in the literature, but it is currently beyond the scope of most studies owing to the likely availability of new pulse sequences on clinical MRI systems: compressed sensing. This has been growing in the literature since around 2007 and has been incorporated in many pulse sequences in the latest generation of MRI systems. It is not, however, universally available and thus has not yet started to show its true potential in the imaging of muscles. The technique can be considered the equivalent of a JPEG image to a bitmap image. Instead of collecting all of the signals required to construct a traditional MRI image, a very small subset of signals are collected in such a way that all of the signals absolutely necessary to create an image are acquired, while the majority are not. This brings about a vast increase in speed, allowing for very rapid scanning and thus high temporal resolution. This opens possibilities for dynamic muscle imaging, whereby muscle loading can be observed in near real time or rapid diffusion tensor imaging allowing for high-quality imaging for muscle fiber tractography in a fraction of the time.

The authors believe that the other main change that is needed for wider scale adoption of MRI, as the paradigm of choice for imaging of muscle mass, is the availability of useful processing software. We are starting to see the availability of some tools, particularly for measurements of adipose tissue volumes and muscle composition, but the same cannot be said for tractography or segmentation. Take, for example, artificial intelligence-based segmentation of muscles, which has exceptional potential to alleviate the burden of this time-consuming process. There are many studies using such methods (e.g., [67–69]), but they remain unavailable to most researchers owing to the challenges involved in implementation. This is beginning to improve with the availability of pretrained artificial neural networks, for example, MuscleSense (http://niftyweb.cs.ucl.ac.uk/program.php?p=MUSCLESENSE), which is available for the segmentation of the calf and thigh using DIXON images. This network is available, as part of a web service, allowing access to the latest algorithms from University College London (UCL) and can be tested by uploading DICOM images through the web interface above. This shows great promise for the future but still requires significant knowledge and experience to incorporate it into a study to improve workflow. This is also true of fiber tracking in muscles, which has great potential for assessing muscle structure and inferring functionality and performance but is largely hampered by the lack of available software for processing the resulting images. Again, numerous algorithms are available in the literature, which will soon begin to filter down into useful tools for the nonexpert.

Even with these advances, the primary barrier to adoption remains the cost of access to MRI. The technique is becoming more widely available, but most clinical systems are unavailable owing to high use, and most research systems are priced outside of the range that is practical for high numbers of participants or a high number of timepoints. More recently systems are becoming available aimed specifically at low-cost MRI of limbs as a viable alternative to limb DXA systems. These include the Magnetica MSK (https://magnetica.com/mri-products/compact-mri-systems/) and the Esaote O-scan (https://www.esaote.com/dedicated-mri/mri-systems/p/o-scan/). These have a significantly lower price point than whole-body MRI systems and will hopefully lead to increased availability of systems for imaging of muscle mass for multiple timepoint studies.

2 Notes

1. One of the biggest frustrations for the newcomer to MRI is the perception that one needs to have a high-level understanding of how MRI works to collect worthwhile data. As with any technique, the more you know, the easier it is to improve the quality of data and propose more advanced experiments, although a high-level understanding is not needed at the outset. The best way to learn MRI is to give it a go. At the outset, time is much better invested in finding a friendly MRI scanner operator to work with to develop your study than it is in gaining a high-level understanding of MRI or trying to predict the T_2 of muscle tissue from first principles.
2. These gradient pulses are responsible for the myriad of different noises that the MRI scanner makes while running a scan. There are no moving parts in an MRI scanner other than the patient table (CT scanners have a very fast-moving gantry, but there is nothing like that in any MRI). Despite this, the sounds the scanner makes come at number 2 (after claustrophobia) in the reasons for aborting a scan session. Explaining what is going to happen to the participant before they even see the MRI scanner goes a long way to ensuring a successful scan session. For these reasons it is also worth having an MRI yourself (if you haven't) to better understand what it is like for your participants.
3. This analogy works really well as a class demonstration if you have a free afternoon and can record some students (or colleagues!) running back and forth around a track. In fact, it also replicates surprisingly well a few of the diffusion artifacts seen in NMR too.
4. This is a very rapidly emerging field, and new tools are released every day. The authors expect that in the not too distant future, the latest wave of new processing tools will be released for the lay operator. Until then, there are academics at most institutions who have the skills necessary to implement the latest approaches and are often very happy to collaborate if you present them with a suitable publication.

5. Shape factors, such as those found in the literature, are an excellent starting point and for similar conditions (participant demographics, MRI vendor, field strength, etc.) will give reasonable results out of the bag. However, it is essential to verify these shape factors are correct for your specific experiment. Time spent running a single participant through your entire protocol including data processing is essential for a successful study and almost always highlights an issue which would have led to the collection of inferior data. Run a gold standard measurement (e.g., manual segmentation of an entire muscle), and check it agrees. If it doesn't, you will need to do a small study to determine the shape factor for your specific study.
6. The choice of gradient echo or spin echo is one of the most complicated that is made in developing MRI protocols. Gradient echo sequences are much faster and impart less RF energy into your participant (which can lead to reduced scan times, particularly for heavier participants since the legal limits of RF energy are dependent upon participant weight), but the faster-switching gradients can cause peripheral nerve stimulation leading to unusual sensations in the extremities of your participant.
7. The parameters that are published in journal articles will be specific to the authors' MRI system. If you have a different MRI system (you probably do!), you should prepare not to be able to set the exact parameters that they have. A lot goes on behind the "normal parameters" that get quoted (T_E, T_R, matrix size, pixel sizes. etc.), which can impact the maximum or minimum values that can be set. The authors of this chapter routinely use a 1.5 T Siemens Avanto MRI. The default TSE sequence, for example, on this system has a minimum possible value for T_E of 12 ms, which is too long for lots of studies in the literature. By going a bit deeper into the less commonly adjusted parameters, for example, the bandwidth, we can get that down to 7.3 ms (we normally increase the bandwidth as high as it will go, bring the T_E down to its minimum value, and then see how low the bandwidth can go without adjusting anything else). If you can't initially set the parameters that the study you are replicating uses, don't give up! Set them as close as you can, and give the scan a go. Most of the time, those parameters are probably okay. From time to time, you will need to tinker with the more advanced parameters to get the exact image you need (it's worth remembering that there is a lot of protection in place on most MRI systems to prevent damage by setting poor parameters so exploring the effect of different parameters is a fairly safe activity). If you use a slightly older scanner, this becomes a more common frustration, since continuing advances lead to shorter gradients and RF pulses.

A final note, if the maximum or minimum settable value is a long way off the value you want, you might have picked a spin echo sequence instead of a gradient echo sequence. This is a good thing to check first (particularly as the vendors all use different names for the same sequences).

8. The key to participant positioning is consistency for the operator and comfort for the participant. A comfortable participant is a still participant who gives you the best image. Try not to run a study with lots of operators where you want comparable data between sessions. Take time to practice different cushions and pads on each other before you enlist participants to agree on how you are going to position them comfortably. Gather together the necessary supports and pads before the participant arrives so that you can get them in and out of the scanner as smoothly as possible.
9. Care must be taken to benchmark the performance of operators who are doing any form of not fully automated segmentation. Even among experienced users of 3D Slicer or similar tools, variations of as much as 20% are commonplace. If multiple operators get different values, outside the tolerance permitted by the constraints of the study, it may be necessary to process all images with multiple operators to take an average providing suitable error ranges or to process the data with a single operator.
10. Sometimes you will genuinely need whole-body compositional analysis. However, sometimes you think you need whole-body compositional analysis so you can change your mind later. Spend time at the beginning of the study planning what you need, including trying data processing. A whole-body data set can take 20 times longer to acquire than a single limb, typically generating data sets around 100 MB with over 100 slices. Even if the automatic processing methods discussed towards the end of the chapter are employed, the person hours needed to verify the method is working on a single data set should not be underestimated. Taking a day to plan out and test the proposed protocol on a single participant can save weeks of effort later on.
11. This raises a useful general point with MRI sequences. Just because there isn't anything in the literature to do specifically what you want to do, doesn't mean it can't be done. The mDIXON sequences (and the equivalents from other vendors) are showing their value for use in other parts of the body even though they were originally intended for the liver.
12. The wraparound artifact is one of the more common artifacts found on MRI scans and is relatively easy to fix in most cases. It is essentially caused by creating areas of phase encoding outside the intended field of view. When the signals are acquired, the

scanner has no way of knowing which phases were intentional and which are accidental, so both appear in the image. This appears as other body parts appearing upside down over the top of the image. The first step in fixing it is to pick the "phase-encoding direction" carefully. This is the direction in which the problem is most likely to occur and thus by setting it to be in a direction where the whole body part can be fully encompassed in the field of view. For example, on a leg setting, the phase-encoding direction along the length of the leg is almost guaranteed to result in a wraparound artifact. Instead, setting it from the anterior–posterior (top to bottom) direction ensures that there is nothing outside the field of view.

13. In addition to gradients caused by RF inhomogeneity discussed in the chapter, images are subject to several other sources of drift. The MRI vendors have implemented lots of very sophisticated routines to ensure that the collected images have as much diagnostic potential as possible. This ensures that the clinicians who read the images stand the greatest chance of diagnosing any problems. However, it can really get in the way of monitoring changes in signal intensities or volumes over more than one scan session. For any scans relying upon consistent numerical data, it is worth including a known reference sample (even something as simple as a cod liver oil capsule, which makes excellent fiducial markers, or a small phantom bottle) in the scans that you can use as a reference to ensure that there is no drift between scan sessions.

14. Please don't be put off by how difficult DTI sounds! In comparison to the other scan protocols, there is a lot that can go wrong and give you an image showing next to nothing, but that doesn't mean it isn't achievable. The results of persisting can be well worth the effort, and once you have achieved it once, it is much easier the next time. Taking some time to optimize the b-value of the system will be time well spent. A value of 400 mm/s^2 or thereabouts is a good starting point, but for some muscles, and some participant groups (particularly if you are looking at injured participants, elite athletes, bariatric participants, or any other nonstandard group), you may need to explore values between 250 mm/s^2 and 600 mm/s^2. If you set the value too high, you will get unduly dark images without sufficient intensity for further processing; if it's too low, you will not encode sufficient information for the tractography routine to work. Educated trial and error is the best approach here: start at 400 mm/s^2 and see how the tractogram looks. If there are lots of very short fibers, the b-value might be too high. If there are lots of fibers which look like they are misplaced, the b-value might need to increase. Try a range of different values, processing the data

all of the way through (if you're not already familiar with a DTI tool, the authors found medInria to be the easiest for this sort of optimization).

15. In the authors' own experience, the force applied when undertaking ultrasound measurements of pennation angle, for example, can change the results by 10 s of degrees. With good positioning of the participants, this can be avoided using MRI. However, it is important to also realize that a misplaced pad can completely change the measured angle if it contacts the location of the muscle; particular care should be taken to avoid this.

Acknowledgments

The authors wish to extend their thanks to the various volunteers whose images appear anonymously in this text and those who kindly gave up their time to proofread the chapter.

Ethical Concerns All images are collected under ethics application 477 approved by the Human Invasive Ethics Committee of Nottingham Trent University.

References

1. Rosenberg IH (1997) Sarcopenia: Origins and clinical relevance. J Nutr 127:990S–991S
2. Aversa Z, Costelli P, Muscaritoli M (2017) Cancer-induced muscle wasting: latest findings in prevention and treatment. Ther Adv Med Oncol 9:369–382
3. Sala D, Zorzano A (2015) Differential control of muscle mass in type 1 and type 2 diabetes mellitus. Cell Mol Life Sci 72:3803–3817. https://doi.org/10.1007/s00018-015-1954-7
4. Challal S, Minichiello E, Boissier MC, Semerano L (2016) Cachexia and adiposity in rheumatoid arthritis. Relevance for disease management and clinical outcomes. Rev du Rhum (Edition Fr) 83:19–25
5. Breen L, Stokes KA, Churchward-Venne TA et al (2013) Two weeks of reduced activity decreases leg lean mass and induces "anabolic resistance" of myofibrillar protein synthesis in healthy elderly. J Clin Endocrinol Metab 98: 2604–2612. https://doi.org/10.1210/JC.2013-1502
6. Cruz-Jentoft AJ, Baeyens JP, Bauer JM et al (2010) Sarcopenia: European consensus on definition and diagnosis. Age Ageing 39:412–423. https://doi.org/10.1093/ageing/afq034
7. Laukkanen P, Heikkinen E, Kauppinen M (1995) Muscle strength and mobility as predictors of survival in 75-84 - year - old people. Age Ageing 24:468–473. https://doi.org/10.1093/ageing/24.6.468
8. Heymsfield SB, Adamek M, Gonzalez MC et al (2014) Assessing skeletal muscle mass: historical overview and state of the art. J Cachexia Sarcopenia Muscle 5:9–18
9. Foster MA, Hutchison JMS, Mallard JR, Fuller M (1984) Nuclear magnetic resonance pulse sequence and discrimination of high- and low-fat tissues. Magn Reson Imaging 2:187–192. https://doi.org/10.1016/0730-725X(84)90004-3
10. Ross R, Rissanen J, Pedwell H et al (1996) Influence of diet and exercise on skeletal muscle and visceral adipose tissue in men. J Appl Physiol 81:2445–2455. https://doi.org/10.1152/jappl.1996.81.6.2445
11. Ross R, Pedwell H, Rissanen J (1995) Effects of energy restriction and exercise on skeletal

muscle and adipose tissue in women as measured by magnetic resonance imaging. Am J Clin Nutr 61:1179–1185. https://doi.org/10.1093/ajcn/61.6.1179
12. Lee SY, Gallagher D (2008) Assessment methods in human body composition. Curr Opin Clin Nutr Metab Care 11:566–572
13. Mitsiopoulos N, Baumgartner RN, Heymsfield SB et al (1998) Cadaver validation of skeletal muscle measurement by magnetic resonance imaging and computerized tomography. J Appl Physiol 85:115–122. https://doi.org/10.1152/jappl.1998.85.1.115
14. McRobbie DW, Moore EA, Graves MJ, Prince MR (2006) MRI from picture to proton. Cambridge University Press
15. Schick F, Machann J, Brechtel K et al (2002) MRI of muscular fat. Magn Reson Med 47:720–727. https://doi.org/10.1002/mrm.10107
16. Dixon WT (1984) Simple proton spectroscopic imaging. Radiology 153:189–194. https://doi.org/10.1148/radiology.153.1.6089263
17. Gotra A, Sivakumaran L, Chartrand G et al (2017) Liver segmentation: indications, techniques and future directions. Insights Imaging 8:377–392
18. Vaidyanathan M, Clarke LP, Velthuizen RP et al (1995) Comparison of supervised MRI segmentation methods for tumor volume determination during therapy. Magn Reson Imaging 13:719–728. https://doi.org/10.1016/0730-725X(95)00012-6
19. Lieber RL, Fridé NJ (2000) Invited review functional and clinical significance of skeletal muscle architecture
20. Morse CI, Degens H, Jones DA (2007) The validity of estimating quadriceps volume from single MRI cross-sections in young men. Eur J Appl Physiol 100:267–274. https://doi.org/10.1007/s00421-007-0429-4
21. Mersmann F, Bohm S, Schroll A et al (2015) Muscle shape consistency and muscle volume prediction of thigh muscles. Scand J Med Sci Sport 25:e208–e213. https://doi.org/10.1111/sms.12285
22. Smeulders MJC, Van Den Berg S, Oudeman J et al (2010) Reliability of in vivo determination of forearm muscle volume using 3.0 T magnetic resonance imaging. J Magn Reson Imaging 31:1252–1255. https://doi.org/10.1002/jmri.22153
23. Bass JJ, Hardy EJO, Inns TB, et al (2021) Atrophy resistant vs. atrophy susceptible skeletal muscles: "aRaS" as a novel experimental paradigm to study the mechanisms of human disuse atrophy. Front Physiol 12. https://doi.org/10.3389/fphys.2021.653060
24. Gacesa JP, Dragnic NR, Prvulovic NM et al (2011) The validity of estimating triceps brachii volume from single MRI cross-sectional area before and after resistance training. J Sports Sci 29:635–641. https://doi.org/10.1080/02640414.2010.549498
25. Amabile C, Moal B, Chtara OA et al (2017) Estimation of spinopelvic muscles' volumes in young asymptomatic subjects: a quantitative analysis. Surg Radiol Anat 39:393–403. https://doi.org/10.1007/s00276-016-1742-6
26. Lehtinen JT, Tingart MJ, Apreleva M et al (2003) Practical assessment of rotator cuff muscle volumes using shoulder MRI. Acta Orthop Scand 74:722–729. https://doi.org/10.1080/00016470310018270
27. Baskin RJ, Paolini PJ (1967) Volume change and pressure development in muscle during contraction. Am J Phys 213:1025–1030. https://doi.org/10.1152/ajplegacy.1967.213.4.1025
28. Irving BA, Weltman JY, Brock DW et al (2007) NIH ImageJ and Slice-O-Matic computed tomography imaging software to quantify soft tissue. Obesity 15:370–376. https://doi.org/10.1038/oby.2007.573
29. Fortin M, Battie MC (2012) Quantitative paraspinal muscle measurements: inter-software reliability and agreement using OsiriX and ImageJ. Phys Ther 92:853–864. https://doi.org/10.2522/ptj.20110380
30. Virzì A, Muller CO, Marret J-B et al Comprehensive review of 3D segmentation software tools for MRI usable for pelvic surgery planning. https://doi.org/10.1007/s10278-019-00239-7
31. Schneider CA, Rasband WS, Eliceiri KW (2012) NIH image to ImageJ: 25 years of image analysis. Nat Methods 9:671–675
32. Ghatas MP, Lester RM, Khan MR, Gorgey AS (2018) Semi-automated segmentation of magnetic resonance images for thigh skeletal muscle and fat using threshold technique after spinal cord injury. Neural Regen Res 13:1787. https://doi.org/10.4103/1673-5374.238623
33. Battaglia PJ, Maeda Y, Welk A et al (2014) Reliability of the Goutallier classification in quantifying muscle fatty degeneration in the lumbar multifidus using magnetic resonance imaging. J Manip Physiol Ther 37:190–197. https://doi.org/10.1016/J.JMPT.2013.12.010

34. Singh RG, Nguyen NN, DeSouza SV et al (2019) Comprehensive analysis of body composition and insulin traits associated with intrapancreatic fat deposition in healthy individuals and people with new-onset prediabetes/diabetes after acute pancreatitis. Diabetes Obes Metab 21:417–423. https://doi.org/10.1111/DOM.13523
35. Kikinis R, Pieper S (2011) 3D Slicer as a tool for interactive brain tumor segmentation. In: Annual International Conference of the IEEE Engineering in Medicine and Biology Society, 6982–6984. https://doi.org/10.1109/IEMBS.2011.6091765
36. Egger J, Kappus C, Carl B, Nimsky C GrowCut-based vertebral body segmentation with 3D slicer
37. Arrieta C, Urrutia J, Besa P et al (2020) Automatic quantification of fat infiltration in paraspinal muscles using T2-weighted images: an OsiriX application. Biomed Signal Process Control 57:101793. https://doi.org/10.1016/J.BSPC.2019.101793
38. Ahmad E, Yap MH, Degens H, McPhee JS (2014) Atlas-registration based image segmentation of MRI human thigh muscles in 3D space:424–435. https://doi.org/10.1117/12.2043606
39. Mansell DS, Frank EG, Kelly NS et al (2020) Comparison of the within-reader and intervendor agreement of left ventricular circumferential strains and volume indices derived from cardiovascular magnetic resonance imaging. PLoS One 15:e0242908. https://doi.org/10.1371/JOURNAL.PONE.0242908
40. Thörmer G, Bertram HH, Garnov N et al (2013) Software for automated MRI-based quantification of abdominal fat and preliminary evaluation in morbidly obese patients. J Magn Reson Imaging 37:1144–1150. https://doi.org/10.1002/JMRI.23890
41. Karampatos S, Papaioannou A, Beattie KA et al (2016) The reliability of a segmentation methodology for assessing intramuscular adipose tissue and other soft-tissue compartments of lower leg MRI images. Magn Reson Mater Phys, Biol Med 29:237–244. https://doi.org/10.1007/s10334-015-0510-7
42. Zoabli G, Mathieu PA, Aubin CE et al (2001) Assessment of manual segmentation of magnetic resonance images of skeletal muscles. Annu Reports Res React Institute, Kyoto Univ 3:2685–2687. https://doi.org/10.1109/IEMBS.2001.1017336
43. Sinelnikov A, Qu C, Fetzer DT et al (2016) Measurement of skeletal muscle area: comparison of CT and MR imaging. Eur J Radiol 85:1716–1721. https://doi.org/10.1016/J.EJRAD.2016.07.006
44. Thomas EL, Parkinson JR, Frost GS et al (2012) The missing risk: MRI and MRS phenotyping of abdominal adiposity and ectopic fat. Obesity 20:76–87. https://doi.org/10.1038/oby.2011.142
45. Marcus RL, Addison O, Dibble LE et al (2012) Intramuscular adipose tissue, sarcopenia, and mobility function in older individuals. J Aging Res 2012. https://doi.org/10.1155/2012/629637
46. Borga M, West J, Bell JD et al (2018) Advanced body composition assessment: from body mass index to body composition profiling. J Investig Med 66:887–895
47. Ogawa M, Lester R, Akima H, Gorgey AS (2017) Quantification of intermuscular and intramuscular adipose tissue using magnetic resonance imaging after neurodegenerative disorders. Neural Regen Res 12:2100–2105
48. Kim HS, Yoon YC, Choi BO et al (2019) Muscle fat quantification using magnetic resonance imaging: case–control study of Charcot–Marie–Tooth disease patients and volunteers. J Cachexia Sarcopenia Muscle 10:574–585. https://doi.org/10.1002/jcsm.12415
49. Pedrosa, Yokoo Publication for the Philips MRI Community mDIXON Quant non-invasively aids in high quality assessment of fatty liver disease UT Southwestern collaborates with Philips on a fast, robust, non-invasive method to measure fat in liver
50. Zhang K, Wang F, Liang W et al (2018) Thermal and mechanical properties of bamboo fiber reinforced epoxy composites. Polymers (Basel) 10:608. https://doi.org/10.3390/polym10060608
51. Leinhard OD, Johansson A, Rydell J et al (2008) Quantitative abdominal fat estimation using MRI. In: Proceedings - International Conference on Pattern Recognition. Institute of Electrical and Electronics Engineers Inc.
52. Maddalo M, Zorza I, Zubani S et al (2017) Validation of a free software for unsupervised assessment of abdominal fat in MRI. Phys Med 37:24–31. https://doi.org/10.1016/j.ejmp.2017.04.002
53. Linge J, Ekstedt M, Dahlqvist Leinhard O (2021) Adverse muscle composition is linked to poor functional performance and metabolic comorbidities in NAFLD. JHEP Reports 3:100197. https://doi.org/10.1016/J.JHEPR.2020.100197
54. Mandić M, Rullman E, Widholm P et al (2020) Automated assessment of regional muscle volume and hypertrophy using MRI. Sci Rep 10.

https://doi.org/10.1038/s41598-020-59267-x

55. Borga M, Ahlgren A, Romu T et al (2020) Reproducibility and repeatability of MRI-based body composition analysis. Magn Reson Med 84:3146–3156. https://doi.org/10.1002/MRM.28360
56. Finch ED, Harmon JF, Muller BH (1971) Pulsed NMR measurements of the diffusion constant of water in muscle. Arch Biochem Biophys 147:299–310. https://doi.org/10.1016/0003-9861(71)90337-7
57. Cleveland GG, Chang DC, Hazlewood CF, Rorschach HE (1976) Nuclear magnetic resonance measurement of skeletal muscle: anisotrophy of the diffusion coefficient of the intracellular water. Biophys J 16:1043–1053. https://doi.org/10.1016/S0006-3495(76)85754-2
58. Damon BM, Ding Z, Anderson AW et al (2002) Validation of diffusion tensor MRI-based muscle fiber tracking. Magn Reson Med 48:97–104. https://doi.org/10.1002/mrm.10198
59. Heemskerk AM, Strijkers GJ, Vilanova A et al (2005) Determination of mouse skeletal muscle architecture using three-dimensional diffusion tensor imaging. Magn Reson Med 53:1333–1340. https://doi.org/10.1002/mrm.20476
60. Damon BM, Buck AKW, Ding Z (2011) Diffusion-tensor MRI-based skeletal muscle fiber tracking. Imaging Med 3:675–687
61. Englund EK, Elder CP, Xu Q et al (2011) Combined diffusion and strain tensor MRI reveals a heterogeneous, planar pattern of strain development during isometric muscle contraction. Am J Physiol 300:R1079. https://doi.org/10.1152/ajpregu.00474.2010
62. Malis V, Sinha U, Sinha S (2020) 3D muscle deformation mapping at submaximal isometric contractions: applications to aging muscle. Front Physiol 11. https://doi.org/10.3389/fphys.2020.600590
63. Scheel M, Prokscha T, Von Roth P et al (2013) Diffusion tensor imaging of skeletal muscle - correlation of fractional anisotropy to muscle power. RoFo Fortschritte auf dem Gebiet der Rontgenstrahlen und der Bildgeb Verfahren 185:857–861. https://doi.org/10.1055/s-0033-1335911
64. Froeling M, Oudeman J, Strijkers GJ et al (2015) Muscle changes detected with diffusion-tensor imaging after long-distance running. Radiology 274:548–562. https://doi.org/10.1148/radiol.14140702
65. Biglands JD, Grainger AJ, Robinson P et al (2020) MRI in acute muscle tears in athletes: can quantitative T2 and DTI predict return to play better than visual assessment? Eur Radiol 30:6603–6613. https://doi.org/10.1007/s00330-020-06999-z
66. Damon BM, Froeling M, Buck AKW et al (2017) Skeletal muscle diffusion tensor-MRI fiber tracking: rationale, data acquisition and analysis methods, applications and future directions. NMR Biomed 30:e3563
67. Lee H, Troschel FM, Tajmir S et al (2017) Pixel-level deep segmentation: artificial intelligence quantifies muscle on computed tomography for body morphometric analysis. J Digit Imaging 304(30):487–498. https://doi.org/10.1007/S10278-017-9988-Z
68. Wang B, Torriani M (2020) Artificial intelligence in the evaluation of body composition. Semin Musculoskelet Radiol 24:030–037. https://doi.org/10.1055/S-0039-3400267
69. Burns JE, Yao J, Summers RM (2020) Artificial intelligence in musculoskeletal imaging: a paradigm shift. J Bone Miner Res 35:28–35. https://doi.org/10.1002/JBMR.3849

Chapter 4

Imaging of Skeletal Muscle Mass: Ultrasound

Martino V. Franchi and Marco V. Narici

Abstract

As the detection and monitoring of skeletal mass are of primary importance in clinical settings and the ones more related to human performance, the need to assess changes in muscle size quickly and reliably is still fundamental. Muscle ultrasound represents a cheap and user-friendly, yet robust and reliable, method to monitor muscle mass in different scenarios. This chapter provides an overview of the different ultrasound-based methodologies employed for the in vivo measurement of muscle mass and architectural properties in human skeletal muscle. The main applications in loading, unloading, aging, and clinical/rehabilitation settings are also discussed.

Key words Muscle size, Muscle CSA, Anatomical cross-sectional area, Muscle volume, Extended field-of-view ultrasound, Muscle architecture, Fascicle length, Pennation angle, Muscle thickness

1 Introduction: Muscle Ultrasound

The skeletal muscle is the largest adipose tissue-free mass in humans and accounts for a substantial portion of the whole-body mass, being thus crucial for locomotion and metabolic health [1]. Thus, it is no surprise that the measurement of muscle mass remains a key requirement in health and clinical settings, from the testing of performance-related factors to the detection of potential injury risks in athletes [2], from the clinical diagnosis of sarcopenia [3] to the monitoring of changes in muscle mass in the ICU environment [4]. The quantification of skeletal muscle mass has been revolutionized by the introduction of various imaging techniques such as computed tomography (CT), magnetic resonance imaging (MRI), and dual-energy X-ray absorptiometry (DXA), which allow the accurate quantification of whole-body and regional muscle masses. Although these techniques have been used in a variety of settings, they can be expensive and often inaccessible, and, in the case of CT and DXA, they involve exposure to ionizing radiation. On the other hand, ultrasound represents a relatively affordable,

Philip J. Atherton and Daniel J. Wilkinson (eds.), *Neuromuscular Assessments of Form and Function*, Neuromethods, vol. 204, https://doi.org/10.1007/978-1-0716-3315-1_4,

noninvasive, accessible in vivo alternative for examining skeletal muscle mass and architecture [5].

Pioneering studies were already conducted in the 1960s by Ikai and Fukunaga [6], where the upper limb muscle cross-sectional area (CSA) was obtained by an ultrasonic scanner placed outside of a circular and transparent water tank in which the arm was placed. However, it was in the 1990s that the use of brightness mode (B-mode) ultrasound imaging was introduced more consistently [7], and since then, it has been extensively implemented for over 25 years to investigate skeletal muscle. In fact, up to the early 1990s, muscle architecture (i.e., the physical arrangement of muscle fibers) and muscle anatomical features were studied mostly in cadaver specimens [8, 9]. This is because such measurements were used to assemble comprehensive databases to provide the foundation of our understanding of muscle structure and its functional significance [10]. Yet, cadaver-based measures are influenced by the premortem age and health status of the subject, as well as the perimortal changes and fixation artifacts. Therefore, to overcome some of these limitations, medical imaging techniques, and especially ultrasound, have been applied to study in vivo muscle morphology at the tissue scale. The B-mode technique represents the most frequently applied ultrasound technique to quantify skeletal muscle mass and evaluate architectural features to relate these parameters to muscle function [11] in healthy and aging populations [12–14] and during static or dynamic tasks [15, 16]. The assessment of these properties results in a fundamental understanding of how the muscle–tendon unit functions as a whole system [17]. In the last 15 years, more refined methodologies, such as extended field of view (EFOV) and 3D ultrasound (as well as diffusion tensor imaging), have been developed and implemented for the study of muscle architectural properties. Clinically, the major advantage is that ultrasound devices are present in most hospital wards and thus are easily accessible; examinations are cheap when compared to DEXA, CT, or MRI; and they do not involve ionizing radiations such as CT or DEXA and as such may be frequently repeated given their safety and low cost [5, 18].

The present chapter aims to provide an overview of the different ultrasound-based methodologies employed for the measurement of muscle mass and architectural properties, as well as to present their main applications in loading, unloading, aging, and clinical scenarios.

1.1 Ultrasound Measurements of Skeletal Muscle

While ultrasound is easily applicable, measurements of muscle morphology and architecture are operator-dependent, which has led several researchers to question the reliability of the technique. However, works by several research groups (including our laboratory) have demonstrated that (as for most technical equipment), when performed by properly trained personnel, measures can be

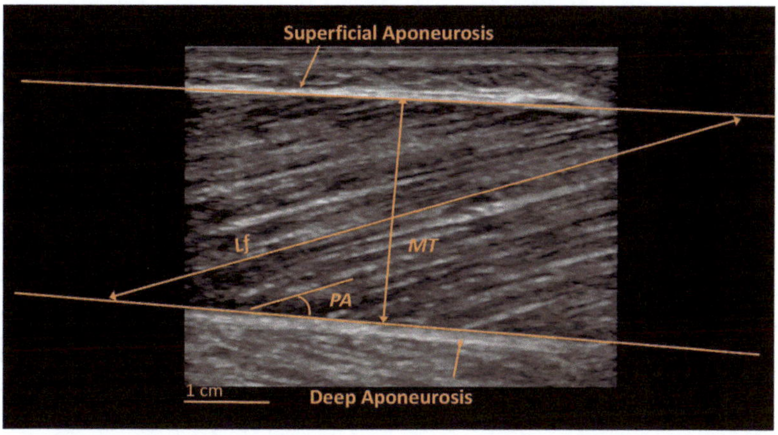

Fig. 1 Longitudinal B-mode ultrasound image of a human vastus lateralis muscle captured at 50% of the total femur length. Highlighted are the deep and superficial tendon aponeuroses and the most common features of muscle architecture: MT muscle thickness, Lf fascicle length, PA pennation angle. Image was obtained using a 4.7 cm linear transducer. (Figure published in *Sarto et al. Sports Med 2021)*

highly reproducible [19], as shown by intraclass correlation coefficients of >0.99 [20] (intraclass correlation coefficient (ICC) of 0.97–0.99 for muscle thickness parameter and 0.97–0.99 for fascicle length) [21–23].

1.1.1 Muscle Dimensions

Muscle dimensions can be evaluated through linear, two-dimensional, and volumetric indexes. Here below we present the three most important measures of muscle dimensions that can be obtained by ultrasound: muscle thickness (MT), muscle anatomical cross-sectional area (ACSA), and muscle volume (VOL).

1.2 Muscle Thickness (MT)

Muscle thickness represents the simplest index of muscle mass, representing a linear measure evaluated as the perpendicular distance between skeletal muscle interfaces (Fig. 1) [18]. The US transducers can be placed transversally [24] or longitudinally [18, 25] in relation to the region of interest chosen for a specific muscle or limb. MT has been shown to be reliable for different muscle groups [18, 26].

Previous studies reported a positive relationship between muscle thickness (MT) and lean mass (measured by DXA) [27, 28], MT and anatomical cross-sectional area (ACSA, measured by MRI) [29], and MT and muscle volume (VOL, measured by MRI) [30–32] at a single time-point. Further, it has been demonstrated that MT is a reliable tool for monitoring local long-term hypertrophic responses (changes in VL skeletal muscle $ACSA_{mid}$) induced by RT [18].

1.3 Anatomical Cross-Sectional Area (ACSA)

US has also been employed to assess larger anatomical structures, such as muscle anatomical cross-sectional area, whereby scans are acquired in the transversal plane (Fig. 2) [5, 33]. The evaluation of muscle ACSA may play a pivotal role in athletic performance or clinical settings. Indeed, ACSA is strongly related to joint torque production and isokinetic strength in different muscle groups [34–36].

1.4 Muscle Volume Reconstruction

Like ACSA, muscle volume is strongly related to muscle function, as it is considered one of the major determinants of joint torque in humans for both upper and lower limbs [37, 38]. In addition, the use of muscle volume has the advantage to account for regional differences in muscle dimension, of particular importance for detecting resistance training-induced adaptations in muscle mass [18]. Muscle volume can be estimated by using the truncated cone formula (shown in Fig. 3), as previously reported [39]. Another way to obtain muscle volumetric reconstructions is through free-hand 3D ultrasound (3DUS), a technique that has been developed and implemented to measure in vivo human muscle volume [40]. Briefly, 3DUS combines B-mode ultrasound imaging with synchronous tracking of the transducer's position and orientation (usually through optical tracking) to form a 3D dataset that describes the muscle volume in the global coordinate system. Muscle slices through this volume can then be reconstructed and viewed [5]. For more information about the 3DUS method, the reader is invited to refer to the work of Raiteri and colleagues [41, 42].

1.4.1 Muscle Architecture

Muscle architecture is defined as the macroscopic arrangement of muscle fibers within a muscle relative to the axis of force generation and has been recognized as one of the primary determinants of muscle function [10]. Indeed, Professor Rick Lieber previously described it as the "structural property of the entire muscle that dominates its function" [10]. Muscle architecture can be assessed by different ultrasound techniques, but the most employed so far is the B-mode modality. When the transducer is correctly aligned with the muscle fascicle plane, we can fully distinguish the orientation of the fascicles relative to the transducer due to the presence of high contrast between connective and muscle tissue [17]. The length of the muscle fascicles (Lf) and the pennation angle (PA) (i.e., the angle formed by the insertion of the fascicle within the deep tendon aponeurosis) represent the architectural parameters that can be detected by B-mode (or even EFOV) US modality [5, 43]. Lf and PA, determined via B-mode US, are generally reliable and valid as shown in a systematic review by Kwah et al [19].

Theoretically, fascicle length (Lf) is dependent on the length and number of sarcomeres in series [44–47], while pennation angle (PA) is mostly seen as a strategy that allows more contractile material to be packed in along the deep aponeurosis of a muscle, seemingly reflecting the number of sarcomeres arranged in parallel

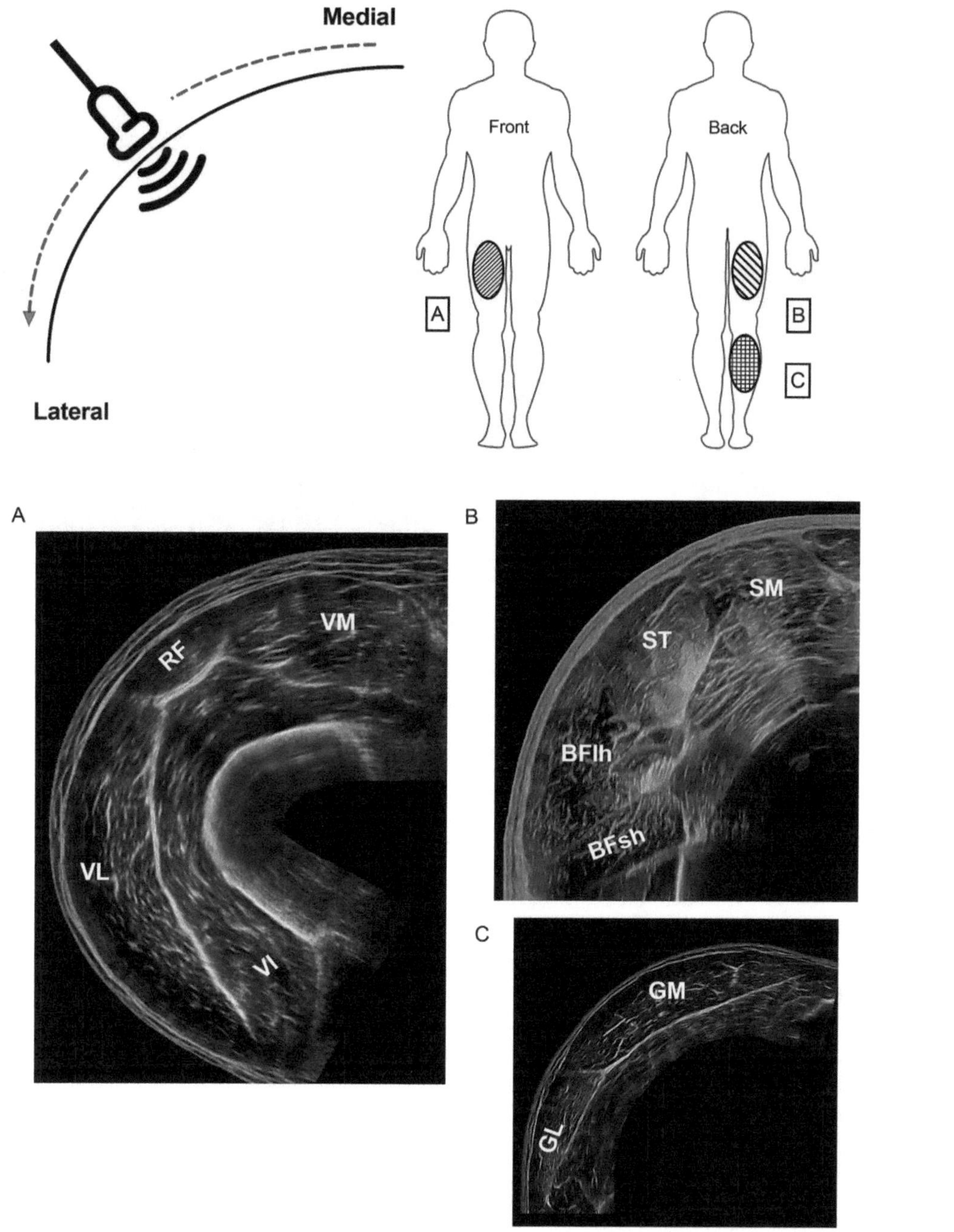

Fig. 2 Transverse EFOV scans for three distinct muscle groups in young healthy volunteers: (**a**) quadriceps femoris, (**b**) hamstrings, (**c**) triceps surae. Images were obtained using a 4.7 cm linear transducer. EFOV extended field-of-view ultrasound technique, VM vastus medialis, RF rectus femoris, VL vastus lateralis, VI vastus intermedius, SM semimembranosus, ST semitendinosus, BFlh biceps femoris long head, BFsh biceps femoris short head, GM gastrocnemius medialis, GL gastrocnemius lateralis. (Figure published in *Sarto et al. Sports Med 2021)*

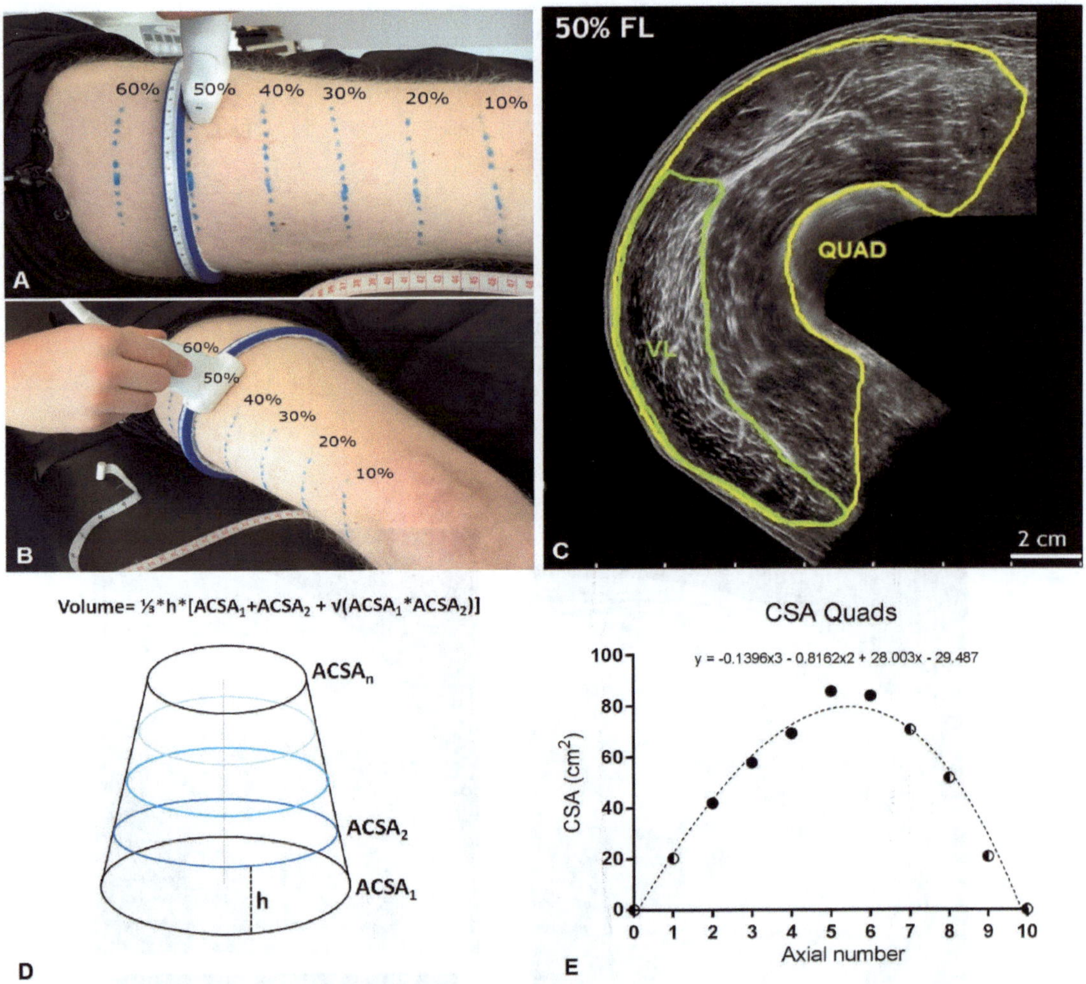

Fig. 3 Ultrasonographic assessment of quadriceps femoris (QF) cross-sectional area (CSA) at different percentages of the femur length (measured from the great trochanter to the mid-patellar point). (**a**, **b**) The regions of interests considered were 20%, 30%, 40%, 50%, and 60% of the femur length. (**c**) Quad and vastus lateralis (VL) CSA obtained at the 50% of the femur length from a representative subject. (**d**) Truncated cone formula. The muscle volume between each couple of contiguous axials was calculated as shown in the formula, where ACSA stands for the anatomical cross-sectional area. (**e**) Mathematical estimation of the lacking quad CSA (10%, 70%, 80%, and 90% of the femur length). A third-order polynomial curve (reported for a representative subject) was fitted through the CSA values that were truly measured. For simplicity, the CSA values corresponding to the 0% and the 100% of the femur length were set as 0. Half and half symbols represent the estimated CSAs. (Figure published in Monti E, Franchi MV et al. Frontiers in Physiol 2020)

[25, 48]. Architectural characteristics can influence muscle force–length and force–velocity relationships [10, 11, 43]. Gans and Bock [48] and Lieber and Fridén [10] explained theoretically that shorter muscle fibers present a lower maximum shortening velocity which likely would influence the velocity of contraction at the entire muscle level. In contrast, the maximum force developed by a muscle is proportional to the number of sarcomeres in parallel and thus to their PA [43].

2 Materials

One of the best features of ultrasound is that one does not need many reagents to work with it—only some ultrasound gel, a good transducer, and a measuring tape. In fact, B-mode ultrasound relies on the transmission of echo pulses from an ultrasound transducer. The capacity of ultrasound waves to penetrate the body is directly proportional to their wavelength while spatial image resolution is inversely proportional to it. For the study of skeletal muscles, frequencies of 5–10 MHz (on rare occasions, up to 17.5 Hz for superficial muscles) are most used. Linear transducers are essential for the methods that will be explained below, and their length and field of view will be further discussed in the methods subsections.

3 Methods

3.1 Identification of Muscle Regions of Interests and Scan Sites

Depending on the purpose of the scientific question, different muscle sites can be scanned: however, there is a need to make these sites consistent over time and/or across study participants. In order to do this, there are usually anatomical landmarks that can be easily identified in order to measure whole muscle length: for example, for the *vastus lateralis* muscle, the femur length is first measured as the distance between the mid-patellar point and the great trochanter [49]. Once the distance has been measured and identified, usually 50% of this length is used to measure muscle architectural parameters [2]. However, not only the length at which the scans are acquired is of importance, but also at what width (or medial to lateral length, transversally)—again, for vastus lateralis, usually scans are acquired at the mid-distance between lateral and medial muscle borders.

3.2 Muscle Thickness (MT): Methodological Aspects

For the images acquired longitudinally, MT is commonly obtained by averaging three measurements across the proximal, central, and distal portions of the acquired image [18] (Fig. 1).

The principal consideration in the assessment of MT is related to the pressure applied by the operator on the skin and the underlying tissue. As muscle tissue can be deformed, an excessive transducer pressure applied by the operator will subsequently result in lower MT values. An appropriate amount of transmission gel should be used to improve the acoustic contact and to reduce the transducer pressure on the skin [2, 5, 18].

3.3 Anatomical Cross-Sectional Area (ACSA): Methodological Aspects

ACSA scans can be obtained by image-stitching technique [22] or by panoramic US technique (i.e., extended field of view imaging) [50, 51]. The extended field-of-view (EFOV) technique was introduced at the end of the 1990s [50] to facilitate the study of larger anatomical structures. This technique relies on texture mapping algorithms that are programmed to merge sequences of images collected during real-time scanning and reconstructing, resulting in larger composite images. As the transducer is moved along the skin and along a plane of interest, successive frames are then recorded and stitched together to obtain a panoramic image. As the software continuously compares the features of the current frame (the real-time portion of the image) with that of the preceding frame (the static portion of the image), each frame serves as a reference to match the subsequent frame and autocorrelation algorithms, and an image processor creates a panoramic image [50]. Due to the absence of respiratory movements and the pulsatility of large vessels, the musculoskeletal system is particularly well-suited for EFOV imaging. Indeed, errors of EFOV distance measurements performed on phantoms of up to 70 cm in length were less than 5% [5, 50]. Indeed, EFOV has been shown to be a valid and reliable method for the assessment of quadriceps [51–53], gastrocnemius [53], and hamstrings [54, 55] ACSA when compared with magnetic resonance imaging (MRI) or computer tomography.

3.4 Muscle Volume Reconstruction: Methodological Aspects

For muscle volume reconstructions involving ACSA acquisition (and thus the employment of the EFOV technique), all the previously mentioned limitations regarding such procedures must be taken into consideration (i.e., the pressure exerted by the operator on the skin through the transducer and the operator's experience). The acquisition of an adequate number of ACSAs (generally 5–12 slices), at different portions of muscle length, enables the estimation of muscle volume via different approaches [56]. In a previous study from our lab in young healthy males (Monti et al., Frontiers Physiol 2020), we obtained partial muscle volume values from the 20% to the 60% of the femur length using the ACSAs data obtained from EFOV ultrasound scans. As no EFOV scans were performed below 20% and above 60% of the femur length marks, the remaining ACSA values for the 10%, 70%, 80%, and 90% axials were estimated by fitting a spline curve (third-order polynomial) through the ACSA values that were truly measured, as described previously from Seynnes and colleagues [23] (Fig. 3).

3.5 Muscle Architecture: Methodological Aspects

The analysis of muscle architecture can be carried out with image analysis software (e.g., ImageJ, NIH; Bethesda, MD, USA) on previously collected scans: briefly, after identifying the superficial and deep aponeuroses that the muscle fascicles originate from and insert into, the lengths of one or several clearly visible muscle fascicles can be traced. Similarly, PA values can be measured by assessing the angles enclosed by fascicles and the aponeurosis they insert into. For a detailed description of such analyses, please refer to the pioneering works of Rutherford and Jones [7], Narici et al. [15], and Fukunaga et al. [16].

Muscle architecture can also be assessed in dynamic conditions, as B-mode ultrasound offers the unique advantage of facilitating examinations of contracting muscle. During contraction, muscle fascicles dynamically change their length and rotate about their origin. In order to facilitate the analysis of image sequences acquired during contraction, computer (mostly optical flow) algorithms may be used for automated fascicle tracking [57].

3DUS, introduced in the previous section related to muscle volume reconstructions, can also be employed to measure muscle belly length [41], fascicle length [58], 3D muscle fascicle architecture [59], and aponeurosis width [42].

4 Notes

4.1 Muscle Thickness (MT): Notes

As previously explained, MT can be evaluated by placing the transducer both transversally and longitudinally with respect to the limb/muscle of interest. In the experience of our laboratory, the longitudinal scan modality probably represents the better of the two. The reason for this is that, for transversal scans, the outlines of muscle borders would be mostly curved and thus the evaluation of MT would be entirely dependent on where MT is measured, thus making it more difficult to standardize. Instead, if the image is correctly acquired in longitudinal scans, the aponeuroses should result parallel (in scans performed in the middle of the muscle belly, the site where MT is usually assessed), and thus the MT will be represented by a simple linear distance between two linear borders [2, 5]. We, therefore, suggest evaluating MT and muscle architecture from the same image to obtain a measure of muscle dimension that would also be related to the fascicle plane in which other architectural parameters can be analyzed. For the images acquired with the latter procedure, MT is commonly obtained by averaging three measurements across the proximal, central, and distal portions of the acquired image [18].

As stated in the introduction, MT is a reliable tool also for monitoring local long-term hypertrophic responses (changes in VL muscle $ACSA_{mid}$) induced by RT [18]. However, for that specific study, the percentage changes of VL MT and muscle volume (calculated by MRI) did not show any significant correlation, thus highlighting the importance of accounting for regional hypertrophy in resistance training. Hence, MT should not be used to estimate changes in muscle volume. These considerations are of primary importance when assessing regional/local vs. whole muscle hypertrophic adaptations, especially in relation to heterogeneous molecular/metabolic responses along the full muscle length, such as when ultrasound scan sites and muscle biopsy sites are not the same [18].

4.2 Anatomical Cross-Sectional Area (ACSA): Notes

Potential limitations for US-based ACSA scans lie in the use of linear transducers for the assessment of curved surfaces and the pressure applied by the operator on the skin and then on the muscle tissue, which may influence measurement results. Nonetheless, although MRI seems to systematically produce larger anatomical cross-sectional area values than ultrasound, as mentioned above, EFOV-based data show good agreement with MRI for different muscle groups [52, 54].

Another potential limitation could be represented by the operator's experience. In order to investigate the degree of the operator's influence on the validity of US-based ACSA measurements, we have previously analyzed the agreement between ACSA values assessed for each muscle of the hamstrings group by US vs. MRI in a population of 45 young athletes [54]. As expected, the agreement was observed to be higher for the trained vs. the novice rater. This methodological aspect acquires great practical impact, as novel operators should refrain from carrying out studies or clinical assessments if not properly trained. In addition, it appears that previous experience with MRI-based analyses could represent an advantage even for trained ultrasonography operators, as, remarkably, the agreement improved for the trained rater when, after carrying out the MRI analyses, a second analysis round was performed for US scans in a randomized fashion [54].

Two further limitations of the technique are that EFOV cannot provide insight into the deformation of muscle during contraction and that it yields (just as standard B-mode ultrasound) only two-dimensional images. To overcome the latter, 3D ultrasound imaging may represent a valuable solution. We will briefly cover this topic in the next section.

Further, although muscle ACSA is usually acquired at the site where the peak of ACSA can be obtained (i.e., usually around the mid-length portion of the muscle group— as done for the quadriceps), multiple scans can be obtained at different muscle lengths, and this is useful to acquire information about the regional distribution of muscle mass and to assess regional changes in ACSA during longitudinal studies in loading or unloading settings [33, 60, 61]. Because of these reasons, our lab and collaborators are currently working on methodological advancements to implement the semiautomatic [62] and automatic [63] assessment of ACSA evaluation for different muscles (vastus lateralis, rectus femoris, gastrocnemius medialis).

4.3 Muscle Volume Reconstruction: Notes

Compared to the EFOV technique, during the 3D ultrasound, images are registered and not acquired on a pixel pattern basis, thus independent of artifactual features [5]. Muscles with highly visible borders and widths smaller than the transducer's field of view are ideal for 3DUS imaging as they can be imaged in their entirety with a single sweep of the ultrasound transducer, improving the quality of the scans and reducing errors associated with the segmentation process. Reducing the frame rate (10–15 Hz) to improve image quality (i.e., through a higher line density) is recommended, as it will in turn improve muscle segmentations, even if this will result in slightly higher scanning times.

During post-processing evaluations, a greater number of segmentations will generally need to be performed at the ends of the muscle because of the more irregular and dissimilar cross sections, whereas segmentations in the mid-belly can be reduced if the cross sections are relatively uniform to decrease post-processing time [41].

4.4 Muscle Architecture: Notes

The two most commonly investigated muscles in humans, the gastrocnemius medialis (GM) and vastus lateralis (VL), present an average *Lf* (as measured at rest and with the ankle and knee held in a neutral and fully extended position, respectively) between ~3.5–5 cm for GM and ~6.5–8 cm for VL, respectively [8, 9, 64]. In other muscles, such as the human sartorius muscle, however, *Lf* may exceed 20 cm [65]. Taking into consideration that cadaver data can be affected by fiber shrinkage, *Lf* values could be expected to be even higher in in vivo examinations. It follows that *Lf* may be easily measured in some muscles using a standard ultrasound transducer, whereas only estimates (based on the extrapolation of visible segments) of *Lf* may be possible in other muscles. Consequently, muscle *Lf* measurements from more proximal leg muscles are inherently subjected to greater measurement inaccuracies. In fact, one key limitation of standard B-mode ultrasound is its relatively small field of view determined by the size of the linear transducer. The field of view commonly ranges between 4 and 6 cm, which can be larger (up to ~9 or 10 cm in some instances). Shorter transducers are ideal for the study of muscles with short Lf (e.g., triceps surae muscle group). However, a limited FOV can affect the evaluation of Lf for muscles that present longer fascicles and more complex architectural arrangements (e.g., quadriceps femoris or hamstrings muscle groups) [26, 66–68]. In these circumstances, longer transducers may be more useful for the evaluation of muscle architecture. Alternatively, panoramic ultrasound can be employed to image a larger portion of muscles and to follow fascicles in their entirety [2, 61, 67, 68].

4.5 Applications of Ultrasound to Assess Muscle Mass

4.5.1 Hypertrophy

The first use of ultrasound in the assessment of muscle hypertrophy was by Rutherford and Jones in 1992 [7]. These investigators measured pennation angle of the vastus lateralis and vastus medialis muscles and found a correlation between fiber pennation and muscle CSA but could not show an increase in pennation angle after 3-month resistance training or a relationship between pennation and force-generating capacity. A year later, Kawakami and colleagues in Japan using B-mode ultrasound showed that the fiber pennation angle of the triceps brachii muscle is greater in hypertrophied than that in normal muscles [25]. In a seminal paper

published a few years later, Aagaard et al. [69] showed a significant increase in the vastus lateralis pennation angle as a result of 14-week heavy resistance strength training in young male adults. Furthermore, they showed a positive, significant relationship between pennation angle and muscle volume, a relationship that was even stronger once fiber CSA data were included. The authors concluded that morphology, architecture, and contractile capacity of the human pennate muscle are interrelated, in vivo. Several authors have now described changes in muscle architecture with strength training both in young and older individuals, most of these reports concern an increase in pennation angle following conventional concentric/eccentric exercise but also an increase in fascicle length following eccentric exercise [21, 45, 49, 70–72].
These adaptations reflect the different contraction types involved in the two exercise modes. While shortening contractions involved in concentric exercise lead to an increase in pennation angle (PA), likely due to the addition of sarcomeres in parallel, eccentric exercise, involving stretching of muscle fibers, leads mostly to an increase in fascicle length, presumably due to the addition of sarcomeres in series [45]. What is particularly noteworthy is that the increases in fascicle length and pennation angle induced by strength training are early biomarkers of muscle hypertrophy as changes are detectable several weeks before the appearance of an increase in muscle cross-sectional area [73]. These observations emphasize the importance of considering hypertrophy as the result of the addition of sarcomeres arranged both in parallel and in series [43]. In this respect, ultrasound offers the unique advantage of enabling the separate assessment of muscle growth through the addition of sarcomere in series, through the measurement of fascicle length, and, in parallel, through the measurement of pennation angle. Furthermore, using panoramic ultrasound (EFOV), it is now possible to accurately assess muscle cross-sectional area, with an accuracy comparable to that of MRI [54] and by acquiring EFOV images along the muscle length, muscle volume may also be estimated [5, 33]. Both techniques have been successfully applied for the study of work-induced muscle hypertrophy [33].

From a functional point of view, a physiological increase in pennation angle by allowing more contractile tissue to be packed along the tendon aponeurosis will lead to an increase in muscle physiological CSA [74] and thus in the maximum force generated by the muscle. Although an increase in pennation angle should predict a reduction in the force component transmitted to the tendon (as tendon force varies with the cosine of pennation angle), this will have in effect a minor impact on the force output since even a 10° increase in pennation angle will lead to a reduction in the tendon force component of less than 5%. Instead, an increase in fascicle length may in theory displace the operating range of muscle fibers away from optimum length (along the ascending limb of the length-tension relation), as the increased number of sarcomeres in series will cause greater shortening fascicles for the same absolute muscle length [75]. This negative effect seems, however, compensated by the increase in tendon stiffness induced by strength training, as a result of which, fascicles undergo lesser shortening to stretch the tendon during maximum force generation [75], since a more compliant tendon requires greater muscle shortening [76].

4.5.2 Atrophy

Ultrasound imaging is an extremely useful and reliable tool for the assessment of muscle atrophy in disuse and disease.

The first application of ultrasound for the assessment of disuse atrophy was by Narici and Cerretelli [77] who showed that muscle atrophy due to limb casting in orthopedic patients is associated with a decrease in pennation angle and fascicle length. Over the last 20 years, this observation has been repeated by many other investigators, and a decrease in pennation angle has now become one of the hallmarks of muscle atrophy (at least in antigravity muscles). Atrophy, being the diametrical opposite condition of hypertrophy, involves a loss of sarcomeres in parallel, reflected by a decrease in pennation angle, but also a loss of sarcomeres in series, reflected by a decrease in fascicle length [78]. For instance, disuse induced by 21-day unilateral lower limb suspension (ULLS) in young males caused a reduction of vastus lateralis fascicle length (8%) and pennation angle (8%), together with a 10% decrease in quadriceps muscle CSA [79]. These early observations have been confirmed by several following studies [80], but what is particularly striking is the speed with which these architectural changes occur. Indeed, in two recent studies, carried out by our laboratory and colleagues, separately involving bed rest and ULLS (in preparation), a significant decrease in pennation angle of 8% and 4%, respectively, has been found after just 10 days of unloading [81]. Instead, no significant changes in fascicle length were observed within 10 days of unloading, which suggests that the time course of a decrease in fascicle length is slower than that of pennation angle, probably because the fiber cannot be completely unloaded due to its insertion into the aponeuroses.

Several studies have also used ultrasound in clinical settings for the evaluation of muscle atrophy caused by pathophysiological conditions (neurodegenerative and neuromuscular disease (ALSD, SMA; DMD, PPS), myopathies, inflammatory disease, malnutrition, diabetes, cancer cachexia, chronic illness). In one of these, rectus femoris muscle CSA measured by ultrasound was found to be reduced by 18% after 10 days of stay in an intensive therapy unit [4]. What is noteworthy is that the decrease in the rectus femoris CSA is greater in patients who experience multiorgan failure compared with single organ failure, emphasizing the usefulness of ultrasound in monitoring the severity and time evolution of pathophysiological conditions in critical illness.

4.6 Aging

The loss of muscle mass associated with aging (sarcopenia) is another condition in which the assessment of muscle mass loss is of critical importance.

In sarcopenia, similar to disuse atrophy, both pennation angle and fascicle length decrease, which implies a loss of sarcomeres both in parallel and in series. Comparing gastrocnemius muscle architecture of young adults (mean age 34.5 years) with those of older individuals (mean age 75.5 years), the older population had lower values of muscle volume (−19%), pennation angle (−13%), and fascicle length (−10%) than those of the younger adults [13]. Several other studies confirmed these early observations [82–85].

Ultrasound measures are now being used for the assessment of sarcopenia in clinical studies using standardized protocols [86–88]. A particularly useful development in muscle ultrasound for the diagnosis of sarcopenia is the recently proposed "ultrasound sarcopenia index" (USI) [3]. This method enables to diagnose sarcopenia based on the ratio between fascicle length (Lf) to muscle thickness (Tm). The Lf/Tm ratio (USI score) was found to increase with the severity of sarcopenia, and based on the USI z-scores, with reference to a young healthy control population, it enables to stratify individuals either as non-sarcopenic, pre-sarcopenic, moderately sarcopenic, sarcopenic, or severely sarcopenic. Apart from being a practical and inexpensive method for diagnosing muscle sarcopenia, a clear advantage of using a ratio rather than absolute values is that it is independent of sex, height, and body mass.

4.7 Spaceflight

Weightlessness presents great challenges for the preservation of muscle mass and the function of crew members during space missions. Despite programs of intensive exercise countermeasures, most astronauts/cosmonauts experience muscle atrophy, particularly of the antigravity muscles of the lower limbs, back, and neck [89]. The severity of atrophy also depends also on genetic variability in the individual response to the lack of gravity and on differences in crew member adherence to exercise protocols.

Although ultrasound has been applied in several studies to assess muscle atrophy in response to simulated spaceflight using bed rest or ULLS, only a few studies so far have investigated using ultrasound the changes in muscle size and architecture in crew members exposed to actual spaceflight.

The first of these studies [90] reports changes in muscle volume (measured by MRI) and architecture (measured by ultrasound) of the calf muscles in two crew members (A and B) exposed to a 6-month spaceflight on board the International Space Station. In crew member A, muscle volume (gastrocnemius medialis) decreased by 18%, pennation angle by 23%, and fascicle length by 7%. In crew member B, muscle volume decreased by 8%, pennation angle by 4%, and fascicle length by 1%. As a result of these changes, physiological muscle CSA (PCSA = volume/fascicle length) decreased by 11% in crew member A and by 6% in crew member B. These findings show that atrophy of key locomotor muscles and alterations in muscle architecture occur in response to spaceflight despite the practice of daily exercise countermeasures. The noteworthy differences in response to spaceflight between the two crew members are likely due to disparities in the adherence to the exercise protocols as well as to interindividual variability in the biological response to weightlessness.

Consistent with the findings above, Koryak [91] reported significant alterations in calf muscle (gastrocnemii medialis and lateralis and soleus) architecture, evaluated by ultrasound, on eight Russian cosmonauts after 213 days of spaceflight. At the neutral ankle angle (0°), the decrease in fascicle length ranged between 12% and 13% between the three muscles while the decrease in pennation angle ranged between 16% and 22%.

Recently, panoramic ultrasound images were collected on 11 astronauts in a NASA telemedicine study using a self-guided image acquisition protocol. A decrease in muscle CSA was found in several muscles of the lower limbs, minimal for the rectus femoris (probably because of its biarticular nature) and greatest for the gastrocnemius medialis and lateralis muscles (15% ca). A noteworthy observation is that no significant differences were found between measurements of muscle CSA obtained by ultrasound and MRI [92].

References

1. Brook MS, Wilkinson DJ, Mitchell WK et al (2016) Synchronous deficits in cumulative muscle protein synthesis and ribosomal biogenesis underlie age-related anabolic resistance to exercise in humans. J Physiol 594(24): 7399–7417. https://doi.org/10.1113/JP272857
2. Sarto F, Spörri J, Fitze DP, Quinlan JI, Narici MV, Franchi MV (2021) Implementing ultrasound imaging for the assessment of muscle and tendon properties in elite sports: practical aspects, methodological considerations and future directions. Sports Med 51:1151. https://doi.org/10.1007/s40279-021-01436-7
3. Narici M, McPhee J, Conte M et al (2021) Age-related alterations in muscle architecture are a signature of sarcopenia: the ultrasound sarcopenia index. J Cachexia Sarcopenia Muscle 12(4):973–982. https://doi.org/10.1002/JCSM.12720
4. Puthucheary ZA, Rawal J, McPhail M et al (2013) Acute skeletal muscle wasting in critical illness. JAMA 310(15):1591–1600. https://doi.org/10.1001/jama.2013.278481
5. Franchi MV, Raiteri BJ, Longo S, Sinha S, Narici MV, Csapo R (2018) Muscle architecture assessment: strengths, shortcomings and new frontiers of in vivo imaging techniques. Ultrasound Med Biol 44:2492. https://doi.org/10.1016/J.ULTRASMEDBIO.2018.07.010
6. Ikai M, Fukunaga T (1968) Calculation of muscle strength per unit cross-sectional area of human muscle by means of ultrasonic measurement. Int Z Angew Physiol 26(1):26–32. https://doi.org/10.1007/BF00696087
7. Rutherford OM, Jones DA (1992) Measurement of fibre pennation using ultrasound in the human quadriceps in vivo. Eur J Appl Physiol Occup Physiol 65(5):433–437. http://www.ncbi.nlm.nih.gov/pubmed/1425649. Accessed March 26, 2018
8. Huijing PA (1985) Architecture of the human gastrocnemius muscle and some functional consequences. Acta Anat (Basel) 123(2): 101–107. https://doi.org/10.1159/000146047
9. Wickiewicz TL, Roy RR, Powell PL, Edgerton VR (1983) Muscle architecture of the human

lower limb. Clin Orthop Relat Res 179:275–283

10. Lieber RL, Fridén J (2000) Functional and clinical significance of skeletal muscle architecture. Muscle Nerve 23(11):1647–1666. https://doi.org/10.1002/1097-4598(200011)23:11<1647::AID-MUS1>3.0.CO;2-M
11. Narici M (1999) Human skeletal muscle architecture studied in vivo by non-invasive imaging techniques: functional significance and applications. J Electromyogr Kinesiol 9:97. https://doi.org/10.1016/S1050-6411(98)00041-8
12. Maganaris CN, Paul JP (1999) In vivo human tendon mechanical properties. J Physiol 521 (Pt 1):307–313. http://www.ncbi.nlm.nih.gov/pubmed/10562354. Accessed March 26, 2018
13. Narici MV, Maganaris CN, Reeves ND, Capodaglio P (2003) Effect of aging on human muscle architecture. J Appl Physiol 95(6): 2229–2234. https://doi.org/10.1152/japplphysiol.00433.2003
14. Narici MV, Maganaris CN (2007) Plasticity of the muscle-tendon complex with disuse and aging. Exerc Sport Sci Rev 35(3):126–134. https://doi.org/10.1097/jes.0b013e3180a030ec
15. Narici MV, Binzoni T, Hiltbrand E, Fasel J, Terrier F, Cerretelli P (1996) In vivo human gastrocnemius architecture with changing joint angle at rest and during graded isometric contraction. J Physiol 496(Pt 1):287
16. Fukunaga T, Ichinose Y, Ito M, Kawakami Y, Fukashiro S (1997) Determination of fascicle length and pennation in a contracting human muscle in vivo. J Appl Physiol 82(1):354–358. https://doi.org/10.1152/jappl.1997.82.1.354
17. Cronin NJ, Lichtwark G (2013) The use of ultrasound to study muscle-tendon function in human posture and locomotion. Gait Posture 37(3):305–312. https://doi.org/10.1016/j.gaitpost.2012.07.024
18. Franchi MV, Longo S, Mallinson J et al (2018) Muscle thickness correlates to muscle cross-sectional area in the assessment of strength training-induced hypertrophy. Scand J Med Sci Sports 28(3):846–853. https://doi.org/10.1111/sms.12961
19. Kwah LK, Pinto RZ, Diong J, Herbert RD (2013) Reliability and validity of ultrasound measurements of muscle fascicle length and pennation in humans: a systematic review. J Appl Physiol 114(6):761–769. https://doi.org/10.1152/japplphysiol.01430.2011
20. Aeles J, Lenchant S, Vanlommel L, Vanwanseele B (2017) Bilateral differences in muscle fascicle architecture are not related to the preferred leg in jumping athletes. Eur J Appl Physiol 117(7):1453–1461. https://doi.org/10.1007/s00421-017-3638-5
21. Reeves ND, Narici MV, Maganaris CN (2004) In vivo human muscle structure and function: adaptations to resistance training in old age. Exp Physiol 89(6):675–689. http://www.ncbi.nlm.nih.gov/pubmed/15328305
22. Reeves ND, Maganaris CN, Narici MV (2004) Ultrasonographic assessment of human skeletal muscle size. Eur J Appl Physiol 91(1): 116–118. Epub 2003 Nov 25
23. Seynnes OR, Maganaris CN, de Boer MD, di Prampero PE, Narici MV (2008) Early structural adaptations to unloading in the human calf muscles. Acta Physiol 193(3):265–274. https://doi.org/10.1111/j.1748-1716.2008.01842.x
24. Weiss LW (1984) The use of B-mode ultrasound for measuring the thickness of skeletal muscle at two upper leg sites. J Orthop Sports Phys Ther 6(3):163–167. https://doi.org/10.2519/jospt.1984.6.3.163
25. Kawakami Y, Abe T, Fukunaga T (1993) Muscle-fiber pennation angles are greater in hypertrophied than in normal muscles. J Appl Physiol 74(6):2740–2744. http://www.ncbi.nlm.nih.gov/pubmed/8365975
26. Pimenta R, Blazevich AJ, Freitas SR (2018) Biceps Femoris long-head architecture assessed using different sonographic techniques. Med Sci Sports Exerc 50(12):2584–2594. https://doi.org/10.1249/MSS.0000000000001731
27. Takai Y, Ohta M, Akagi R et al (2013) Validity of ultrasound muscle thickness measurements for predicting leg skeletal muscle mass in healthy Japanese middle-aged and older individuals. J Physiol Anthropol 32(1):12. https://doi.org/10.1186/1880-6805-32-12
28. Takai Y, Ohta M, Akagi R et al (2014) Applicability of ultrasound muscle thickness measurements for predicting fat-free mass in elderly population. J Nutr Health Aging 18(6): 579–585. https://doi.org/10.1007/s12603-013-0419-7
29. Abe T, Kawakami Y, Suzuki Y, Gunji A, Fukunaga T (1997) Effects of 20 days bed rest on muscle morphology. J Gravit Physiol 4(1):S10–S14. http://www.ncbi.nlm.nih.gov/pubmed/11541170. Accessed February 27, 2017
30. Miyatani M, Kanehisa H, Kuno S, Nishijima T, Fukunaga T (2002) Validity of ultrasonograph

muscle thickness measurements for estimating muscle volume of knee extensors in humans. Eur J Appl Physiol 86(3):203–208. https://doi.org/10.1007/s00421-001-0533-9
31. Kanehisa H, Ito M, Kawakami Y, Fukunaga T, Miyatani M (2004) The accuracy of volume estimates using ultrasound muscle thickness measurements in different muscle groups. Eur J Appl Physiol 91(2–3):264–272. https://doi.org/10.1007/s00421-003-0974-4
32. Sanada K, Kearns CF, Midorikawa T, Abe T (2006) Prediction and validation of total and regional skeletal muscle mass by ultrasound in Japanese adults. Eur J Appl Physiol 96(1): 24–31. https://doi.org/10.1007/s00421-005-0061-0
33. Monti E, Franchi MV, Badiali F, Quinlan JI, Longo S, Narici MV (2020) The time-course of changes in muscle mass, architecture and power during 6 weeks of plyometric training. Front Physiol 11:946. https://doi.org/10.3389/fphys.2020.00946
34. Blazevich AJ, Coleman DR, Horne S, Cannavan D (2009) Anatomical predictors of maximum isometric and concentric knee extensor moment. Eur J Appl Physiol 105(6):869–878. https://doi.org/10.1007/s00421-008-0972-7
35. Evangelidis PE, Massey GJ, Pain MTG, Folland JP (2016) Strength and size relationships of the quadriceps and hamstrings with special reference to reciprocal muscle balance. Eur J Appl Physiol 116(3):593–600. https://doi.org/10.1007/s00421-015-3321-7
36. Masuda K, Kikuhara N, Takahashi H, Yamanaka K (2003) The relationship between muscle cross-sectional area and strength in various isokinetic movements among soccer players. J Sports Sci 21(10):851–858. https://doi.org/10.1080/0264041031000102042
37. Fukunaga T, Miyatani M, Tachi M, Kouzaki M, Kawakami Y, Kanehisa H (2001) Muscle volume is a major determinant of joint torque in humans. Acta Physiol Scand 172(4):249–255. https://doi.org/10.1046/j.1365-201x.2001.00867.x
38. Maden-Wilkinson TM, Balshaw TG, Massey G, Folland JP (2019) What makes long-term resistance-trained individuals so strong? A comparison of skeletal muscle morphology, architecture, and joint mechanics. J Appl Physiol 128:1000. https://doi.org/10.1152/japplphysiol.00224.2019
39. Tracy BL, Ivey FM, Jeffrey Metter E, Fleg JL, Siegel EL, Hurley BF (2003) A more efficient magnetic resonance imaging-based strategy for measuring quadriceps muscle volume. Med Sci Sports Exerc 35(3):425–433. https://doi.org/10.1249/01.MSS.0000053722.53302.D6
40. Barber L, Barrett R, Lichtwark G (2009) Validation of a freehand 3D ultrasound system for morphological measures of the medial gastrocnemius muscle. J Biomech 42(9):1313–1319. https://doi.org/10.1016/j.jbiomech.2009.03.005
41. Raiteri BJ, Cresswell AG, Lichtwark GA (2016) Three-dimensional geometrical changes of the human tibialis anterior muscle and its central aponeurosis measured with three-dimensional ultrasound during isometric contractions. PeerJ 4:e2260. https://doi.org/10.7717/peerj.2260
42. Raiteri BJ, Cresswell AG, Lichtwark GA (2018) Muscle-tendon length and force affect human tibialis anterior central aponeurosis stiffness in vivo. Proc Natl Acad Sci 115(14):E3097–E3105. https://doi.org/10.1073/PNAS.1712697115
43. Narici M, Franchi M, Maganaris C (2016) Muscle structural assembly and functional consequences. J Exp Biol 219(2):276–284. https://doi.org/10.1242/jeb.128017
44. Franchi MV, Reeves ND, Narici MV (2017) Skeletal muscle remodeling in response to eccentric vs. concentric loading: morphological, molecular, and metabolic adaptations. Front Physiol 8:447. https://doi.org/10.3389/fphys.2017.00447
45. Franchi MV, Atherton PJ, Reeves ND et al (2014) Architectural, functional and molecular responses to concentric and eccentric loading in human skeletal muscle. Acta Physiol (Oxf) 210(3):642–654. https://doi.org/10.1111/apha.12225
46. Lichtwark GA, Farris DJ, Chen X, Hodges PW, Delp SL (2018) Microendoscopy reveals positive correlation in multiscale length changes and variable sarcomere lengths across different regions of human muscle. J Appl Physiol 125(6):1812–1820. https://doi.org/10.1152/japplphysiol.00480.2018
47. Lieber RL, Fridén J (2001) Clinical significance of skeletal muscle architecture. Clin Orthop Relat Res 383:140–151. http://www.ncbi.nlm.nih.gov/pubmed/11210948. Accessed January 31, 2017
48. Gans C, Bock W (1965) The functional significance of muscle architecture – a theoretical analysis. Ergeb Anat Entwicklungsgesch 38: 115–142
49. Quinlan JI, Franchi MV, Gharahdaghi N et al (2021) Muscle and tendon adaptations to moderate load eccentric vs. concentric resistance exercise in young and older males.

GeroScience 43:1567. https://doi.org/10.1007/S11357-021-00396-0

50. Weng L, Tirumalai AP, Lowery CM et al (1997) US extended-field-of-view imaging technology. Radiology 203(3):877–880. https://doi.org/10.1148/radiology.203.3.9169720
51. Noorkoiv M, Nosaka K, Blazevich AJ (2010) Assessment of quadriceps muscle cross-sectional area by ultrasound extended-field-of-view imaging. Eur J Appl Physiol 109(4):631–639. https://doi.org/10.1007/s00421-010-1402-1
52. Ahtiainen JP, Hoffren M, Hulmi JJ et al (2010) Panoramic ultrasonography is a valid method to measure changes in skeletal muscle cross-sectional area. Eur J Appl Physiol 108(2):273–279. https://doi.org/10.1007/s00421-009-1211-6
53. Scott JM, Martin DS, Ploutz-Snyder R et al (2012) Reliability and validity of panoramic ultrasound for muscle quantification. Ultrasound Med Biol 38(9):1656–1661. https://doi.org/10.1016/j.ultrasmedbio.2012.04.018
54. Franchi MV, Fitze DP, Hanimann J, Sarto F, Spörri J (2020) Panoramic ultrasound vs. MRI for the assessment of hamstrings cross-sectional area and volume in a large athletic cohort. Sci Rep 10(1):14144. https://doi.org/10.1038/s41598-020-71123-6
55. Kositsky A, Gonçalves BAM, Stenroth L, Barrett RS, Diamond LE, Saxby DJ (2020) Reliability and validity of ultrasonography for measurement of hamstring muscle and tendon cross-sectional area. Ultrasound Med Biol 46(1):55–63. https://doi.org/10.1016/j.ultrasmedbio.2019.09.013
56. Nordez A, Jolivet E, Südhoff I, Bonneau D, De Guise JA, Skalli W (2009) Comparison of methods to assess quadriceps muscle volume using magnetic resonance imaging. J Magn Reson Imaging 30(5):1116–1123. https://doi.org/10.1002/jmri.21867
57. Farris DJ, Lichtwark GA (2016) UltraTrack: software for semi-automated tracking of muscle fascicles in sequences of B-mode ultrasound images. Comput Methods Prog Biomed 128:111–118. https://doi.org/10.1016/J.CMPB.2016.02.016
58. Hug F, Goupille C, Baum D, Raiteri BJ, Hodges PW, Tucker K (1819) Nature of the coupling between neural drive and force-generating capacity in the human quadriceps muscle. Proc R Soc B Biol Sci 2015(282):20151908. https://doi.org/10.1098/rspb.2015.1908
59. Rana M, Hamarneh G, Wakeling JM (2014) 3D curvature of muscle fascicles in triceps surae. J Appl Physiol 117(11):1388–1397. https://doi.org/10.1152/JAPPLPHYSIOL.00109.2013/ASSET/IMAGES/LARGE/ZDG0231412460008.JPEG
60. Monti E, Reggiani C, Franchi MV et al (2021) Neuromuscular junction instability and altered intracellular calcium handling as early determinants of force loss during unloading in humans. J Physiol 599:3037. https://doi.org/10.1113/JP281365
61. Franchi MV, Sarto F, Simunič B, Pišot R, Narici MV (2022) Early changes of hamstrings morphology and contractile properties during 10 days of complete inactivity. Med Sci Sports Exerc 54:1346. https://doi.org/10.1249/MSS.0000000000002922
62. Ritsche P, Wirth P, Franchi MV, Faude O (2021) ACSAuto-semi-automatic assessment of human vastus lateralis and rectus femoris cross-sectional area in ultrasound images. Sci Rep 11(1):1–11. https://doi.org/10.1038/s41598-021-92387-6
63. Ritsche P, Wirth P, Cronin NJ et al (2022) DeepACSA: automatic segmentation of anatomical cross-sectional area in ultrasound images of human lower limb muscles using deep learning. Med Sci Sports Exerc 54(12):2188–2219. https://doi.org/10.1101/2021.12.27.21268258
64. Friederich JA, Brand RA (1990) Muscle fiber architecture in the human lower limb. J Biomech 23:91. https://doi.org/10.1016/0021-9290(90)90373-B
65. Heron MI, Richmond FJR (1993) In-series fiber architecture in long human muscles. J Morphol 216(1):35–45. https://doi.org/10.1002/JMOR.1052160106
66. Ema R, Wakahara T, Miyamoto N, Kanehisa H, Kawakami Y (2013) Inhomogeneous architectural changes of the quadriceps femoris induced by resistance training. Eur J Appl Physiol 113(11):2691–2703. https://doi.org/10.1007/s00421-013-2700-1
67. Franchi MV, Fitze DP, Raiteri BJ, Hahn D, Spörri J (2020) Ultrasound-derived biceps femoris long head fascicle length: extrapolation pitfalls. Med Sci Sports Exerc 52(1):233. https://doi.org/10.1249/MSS.0000000000002123
68. Sarto F, Monti E, Simunič B, Pišot R, Narici MV, Franchi MV (2021) Changes in Biceps Femoris long head fascicle length after 10-day bed rest assessed with different ultrasound methods. Med Sci Sports Exerc 53(7):1529–1536. https://doi.org/10.1249/mss.0000000000002614

69. Aagaard P, Andersen JL, Dyhre-Poulsen P et al (2004) A mechanism for increased contractile strength of human pennate muscle in response to strength training: changes in muscle architecture. J Physiol 534(2):613–623. https://doi.org/10.1111/j.1469-7793.2001.t01-1-00613.x
70. Franchi MV, Wilkinson DJ, Quinlan JI et al (2015) Early structural remodeling and deuterium oxide-derived protein metabolic responses to eccentric and concentric loading in human skeletal muscle. Physiol Rep 3: e12593. https://doi.org/10.14814/phy2.12593
71. Franchi MV, Ruoss S, Valdivieso P et al (2018) Regional regulation of focal adhesion kinase after concentric and eccentric loading is related to remodelling of human skeletal muscle. Acta Physiol 223(3):e13056. https://doi.org/10.1111/apha.13056
72. Franchi MV, Monti E, Carter A et al (2019) Bouncing Back! Counteracting muscle aging with plyometric muscle loading. Front Physiol 10:178. https://doi.org/10.3389/fphys.2019.00178
73. Seynnes OR, de Boer M, Narici MV (2007) Early skeletal muscle hypertrophy and architectural changes in response to high-intensity resistance training. J Appl Physiol 102(1): 368–373. https://doi.org/10.1152/japplphysiol.00789.2006
74. Haxton HA (1944) Absolute muscle force in the ankle flexors of man. J Physiol 103(3): 267–273
75. Reeves ND, Narici MV, Maganaris CN (2006) Myotendinous plasticity to ageing and resistance exercise in humans. Exp Physiol 91(3): 483–498. https://doi.org/10.1113/EXPPHYSIOL.2005.032896
76. Wilkie DR (1949) The relation between force and velocity in human muscle. J Physiol 110(3–4):249–280. https://doi.org/10.1113/JPHYSIOL.1949.SP004437
77. Narici M, Cerretelli P (1998) Changes in human muscle architecture in disuse-atrophy evaluated by ultrasound imaging. J Gravit Physiol 5(1):P73–P74. http://www.ncbi.nlm.nih.gov/pubmed/11542371. Accessed July 6, 2015
78. Narici M, Maganaris C (2007) Muscle architecture and adaptations to functional requirements. Skelet Muscle Plast Heal Dis:265–288. https://doi.org/10.1007/1-4020-5177-8_9
79. de Boer MD, Maganaris CN, Seynnes OR, Rennie MJ, Narici MV (2007) Time course of muscular, neural and tendinous adaptations to 23 day unilateral lower-limb suspension in young men. J Physiol 583(3):1079–1091. https://doi.org/10.1113/jphysiol.2007.135392
80. Campbell EL, Seynnes OR, Bottinelli R et al (2013) Skeletal muscle adaptations to physical inactivity and subsequent retraining in young men. Biogerontology 14(3):247–259. https://doi.org/10.1007/s10522-013-9427-6
81. Narici MV, Monti E, Franchi M et al (2020) Early biomarkers of muscle atrophy and of neuromuscular alterations during 10-day bed rest. FASEB J 34(S1):1–1. https://doi.org/10.1096/fasebj.2020.34.s1.09027
82. Morse CI, Thom JM, Birch KM, Narici MV (2005) Changes in triceps surae muscle architecture with sarcopenia. Acta Physiol Scand 183(3):291–298. https://doi.org/10.1111/j.1365-201X.2004.01404.x
83. Suetta C, Andersen JL, Dalgas U et al (2008) Resistance training induces qualitative changes in muscle morphology, muscle architecture, and muscle function in elderly postoperative patients. J Appl Physiol 105(1):180–186. https://doi.org/10.1152/japplphysiol.01354.2007
84. Strasser EM, Draskovits T, Praschak M, Quittan M, Graf A (2013) Association between ultrasound measurements of muscle thickness, pennation angle, echogenicity and skeletal muscle strength in the elderly. Age (Omaha) 35:2377. https://doi.org/10.1007/s11357-013-9517-z
85. Ticinesi A, Meschi T, Narici MV, Lauretani F, Maggio M (2017) Muscle ultrasound and sarcopenia in older individuals: a clinical perspective. J Am Med Dir Assoc 18(4):290–300. https://doi.org/10.1016/j.jamda.2016.11.013
86. Ticinesi A, Narici MV, Lauretani F et al (2018) Assessing sarcopenia with vastus lateralis muscle ultrasound: an operative protocol. Aging Clin Exp Res 30(12):1437–1443. https://doi.org/10.1007/s40520-018-0958-1
87. Perkisas S, Baudry S, Bauer J et al (2018) Application of ultrasound for muscle assessment in sarcopenia: towards standardized measurements. Eur Geriatr Med 9(6):739–757. https://doi.org/10.1007/s41999-018-0104-9
88. Perkisas S, Bastijns S, Baudry S et al (2021) Application of ultrasound for muscle assessment in sarcopenia: 2020 SARCUS update. Eur Geriatr Med 1:18. https://doi.org/10.1007/s41999-020-00433-9
89. Narici MV, de Boer MD (2011) Disuse of the musculo-skeletal system in space and on earth.

Eur J Appl Physiol 111(3):403–420. https://doi.org/10.1007/s00421-010-1556-x
90. Rittweger J, Albracht K, Flück M et al (2018) Sarcolab pilot study into skeletal muscle's adaptation to long-term spaceflight. NPJ Microgravity 4(1):1–9. https://doi.org/10.1038/s41526-018-0052-1
91. Koryak YA (2019) Architectural and functional specifics of the human triceps surae muscle in vivo and its adaptation to microgravity. J Appl Physiol 126(4):880–893. https://doi.org/10.1152/JAPPLPHYSIOL.00634.2018
92. Scott JM, Downs M, Martin DS et al (2021) Teleguided self-ultrasound scanning for longitudinal monitoring of muscle mass during spaceflight. iScience 24(4):102344. https://doi.org/10.1016/J.ISCI.2021.102344

Chapter 5

Measures of Neuromuscular Function

Michael D. Roberts and Jason M. Defreitas

Abstract

Measures of neuromuscular function are commonly obtained in the research disciplines of aging, disease physiology, and exercise science. The methods used to assess neuromuscular function range in sophistication. Here, we provide comprehensive overviews of strength testing, dynamometer assessments, the use of force plates, and evoked potentials. Our goal is to guide the reader on how neuromuscular function can be assessed according to their research needs.

Key words Strength testing, Evoked contractions, Dynamometer

1 Introduction

Assessments of neuromuscular function have been widely used in clinical and research settings. In the clinical setting, decrements in neuromuscular characteristics (e.g., nerve conduction velocity or electromyographic signals during muscle contraction) may indicate the presence of nerve demyelination or another underlying condition such as accelerated muscle loss with advanced aging. In the research setting, neuromuscular tests are generally performed to characterize how aging, diseases, or various stressors affect attributes such as strength or contractile function. The purpose of this chapter is to discuss techniques that are commonly used to assess various aspects of neuromuscular function. Overviews of strength testing, dynamometer assessments, the use of force plates, and evoked contractions will be described. Additionally, abbreviated methods for each assessment will be provided. Upon concluding the chapter, the reader should possess a better appreciation of how various measures of neuromuscular function can be assessed.

Philip J. Atherton and Daniel J. Wilkinson (eds.), *Neuromuscular Assessments of Form and Function*, Neuromethods, vol. 204, https://doi.org/10.1007/978-1-0716-3315-1_5,

2 Assessments of Neuromuscular Function

2.1 Strength Testing

As simply defined, muscular strength is the maximal amount of force that a muscle group can generate during a particular endeavor. Various factors can affect muscle strength including (but not limited to) the size of the involved muscle, muscle architecture (e.g., fiber orientation relative to the plane of action), and motor unit discharge behavior [1]. While muscular strength can be determined using different methods, the most common assessment method has included the use of free or machine weights. In short, strength tests involve a warm-up protocol followed by a gradual progression towards maximal strength (also termed a one-repetition maximum, or 1RM). Although 1RM testing can be performed using several exercises, it is common for researchers to obtain this metric using exercises that involve the use of multiple joints. Examples of such exercises include bench or chest press for upper body strength, leg press for lower body strength, barbell back squat for lower body strength, and barbell deadlift for lower body strength. Several studies have evaluated the reliability of 1RM testing [2–5], and five general considerations for testing will be summarized below. Additionally, the National Strength and Conditioning Association provides general guidelines for 1RM testing, and the reader is encouraged to use this excellent resource for further guidance [6].

As a first consideration for 1RM testing, the tester should provide a standardized warm-up routine for participants. Warm-up routines enhance performance while minimizing the potential for muscle injury. An excellent study in this area suggests the warm-up prior to 1RM testing should consist of two phases [7], and these will be discussed in greater detail:

(i) A general warm-up routine should be first performed that is more aerobic in nature. If lower body 1RM testing is being performed, an example of a general warm-up for this test would include having the participant cycle for 5 min at a pace or ergometer resistance that would be rated as 5–7 on a 0 (easy)-10 (hard) difficulty scale. If upper body 1RM testing is being performed, an arm ergometer could be used in a similar manner. Beneficial outcomes of a general warm-up routine include increasing muscle temperature and joint mobility. However, longer warm-up routines do increase the risk of prematurely fatiguing the participant and compromising 1RM values.

(ii) A specific warm-up routine should then be performed for 3–5 min following the general warm-up described above, and this routine should mimic the exercise being 1RM tested. In the aforementioned example, participants performed 1RM testing using the leg press exercise. For the specific warm-up, participants first perform one set of eight repetitions at ~50% of their maximum estimated strength. Following a 2-min recovery period, participants perform one set of three repetitions at ~70% of their maximum estimated strength. These two sets are considered the specific warm-up routine, and the 1×8 and 1×3 scheme was shown to optimize 1RM relative to other schemes. Thereafter, participants rest for 3 min, and have up to five attempts to obtain a 1RM value. Between successful attempts, weight is modestly added to the leg press machine, with 3-min rest intervals granted between 1RM attempts.

The aforementioned methodology for 1RM testing is technically sound, given that it contained general and specific warm-up protocols, it progressively challenged participants, and it implemented 2–3-min recovery phases to ensure that residual fatigue did not carry over from prior sets and/or 1RM attempts.

The second consideration for 1RM testing includes the standardization of lifting garments, lifting technique, and range of motion. Devices such as elbow wraps, knee wraps, and lifting belts can significantly enhance performance [8, 9]. Research has also shown 1RMs are greater for partial squats relative to full range of motion squats [10]. Additionally, hand placement deviations on the bar during bench press, alterations in foot placement on the sled platform during leg press, and the location of the bar on the back during a back squat can all affect 1RM outcomes [11]. Lifting tempo can also affect 1RM outcomes [12], where slower lifting tempos typically yield lower values. Thus, testers must ensure that lifting technique and tempo as well as the range of motion executed during 1RM assessments are standardized. In an ideal setting, the following procedures would be implemented to reduce these potential sources of technical error:

(i) The researcher must decide as to whether lifting garments (e.g., weight belts, knee wraps, elbow wraps, straps for deadlifts) will or will not be allowed for participant use. If said devices are allowed, then the tester must ensure that all participants wear the device(s), and participants don the same garments during subsequent 1RM tests.

(ii) The tester should instruct participants about proper hand or foot placement. In this regard, standardized markings can be used such as tape markings on bars for bench press, or tape markings on a sled platform or floor for leg press or back squat, respectively.

(iii) The tester should have participants execute a pre-determined range of motion for the exercise of interest. Although this may pose challenges, there are methods that can be used to standardize the range of motion. For instance, researchers can opt to have participants don wearable knee goniometers for a real-time range of motion monitoring during the leg press or back squat exercises. For the bench press exercise, participants should be instructed to touch (not bounce) the bar on a certain portion of the torso between the eccentric and concentric phases of the lift.

(iv) Finally, a metronome can be used to guide the lifting tempo during the concentric and eccentric phases of the lift. While the lifting tempo is ultimately up to the discretion of the researcher, it is notable that a 2 s – 0 s – 2 s (eccentric-transition-concentric) lifting protocol has been shown to produce significantly greater 1RM values relative to a 4 s – 0 s – 2 s protocol [12].

The third consideration for 1RM testing is that multiple RM testing (e.g., the assessment of a 3RM, for instance) can be used to estimate 1RM values. However, the scientific literature suggests that the tester should strive to assess 2RM–5RM values when extrapolating 1RM values, given that 10RM–20RM tests have poorer predictability [3].

The fourth consideration for 1RM testing includes implementing a familiarization session prior to a "true" 1RM testing session, especially if the population being tested is naïve to resistance exercise training. Although an in-depth discussion of this recommendation is beyond the scope of this chapter, it should be noted that significant neuromuscular adaptations (and, potentially, strength adaptations) have been shown to occur in response to a single testing session in untrained individuals [13]. Researchers have commonly termed this phenomenon as a "learning effect" to the testing protocol. Thus, in an ideal situation, a familiarization protocol would occur ~3–7 days prior to the 1RM testing day.

The fifth and final consideration for 1RM testing involves blinding participants to the load being used during the assessment. While this is perhaps one of the more overlooked aspects of 1RM testing, it can profoundly influence measures according to a well-controlled research study [14]. Briefly, the researchers divided previously trained male participants into two groups, and each group performed four 1RM bench press tests where each test was separated by 24–48 h. One group was blinded to loads during each test, and the other was not. The researchers reported that the blinded group obtained similar 1RM values across all four tests, whereas the non-blinded group significantly improved their performance by ~7% when comparing the fourth to the first test. While blinding exercise tests can be challenging, there are certain precautions that can be taken. First, weight stacks on machines can be visually obscured, so participants cannot observe mass progression during testing. When using barbells, larger weights can be first placed on bars to act as a visual blockade for the subsequent addition of smaller weights. The tester can also instruct the participant to step away from the testing apparatus during rest periods, and, during this time, the addition of weight can occur in a fashion to prevent participant observation. While more uncommon, the participant can don wearable devices during testing (e.g., goggles with side-binding mechanisms) to minimize observation.

Below we briefly describe methods to perform maximal strength testing. Here, three-repetition maximum (3RM) back squat testing was performed by Vann et al. [15]. Participants were instructed to report to the laboratory under well-hydrated, 4-h fasted conditions. Testing first began with two rounds of calisthenics (jumping jacks, single-leg lunges). Participants then performed one warm-up set of ten repetitions using the bar only (20 kg) followed by one set of ten repetitions at 60 kg. Participants then performed three repetitions at a self-estimated 70% 1RM and began maximal 3RM attempts thereafter. Approximately 3 min of rest was allowed between 3RM attempts, and repetitions were considered successful when a squat depth was accomplished where the hip crease was below the knee joint. 1RM values were estimated from 3RM values using the NSCA's most recent text [6]. Notably, a 20-kg barbell (York Barbell; York, PA, USA) and STS Power Rack (York Barbell) were used for testing. Additionally, strict technical parameters were employed during back squat assessments under the direct supervision of the study coordinator who holds a Certified Strength and Conditioning Specialist Certification from the National Strength and Conditioning Association. As a final note related to 1RM testing, Grgic et al. [16] authored an excellent meta-analysis suggesting that 1RM testing possesses excellent test–retest reliability regardless of the gender, resistance training experience, and/or age of participants. However, the authors did mention that various aspects discussed in this section should be considered to optimize testing procedures.

2.2 Dynamometer Testing

Another common method of strength testing involves the use of dynamometers ("dye-nuh-maa-muh-ter"), which are devices designed to quantify rotational kinematics and kinetics (i.e., they are able to measure rotational motion and forces). In terms of human testing, these rotational measures are in fact the functional movements of joints. There are two types of dynamometers that are commonly employed in both clinical and research settings, and these include handgrip dynamometers and isokinetic dynamometers.

Maximal handgrip strength is a reliable and surprisingly robust outcome predictor [17]. It is considered an indicator or proxy of total body strength and has become part of standardized measures of functional ability in older adults. It has also been shown to be associated with clinical morbidities and all-cause mortality [18]. In fact, the widely used NIH Toolbox for Motor Function has five domains (gait speed, balance, dexterity, strength, and endurance), and maximal handgrip is the test used to assess strength in all patients aged 3 years and older [19].

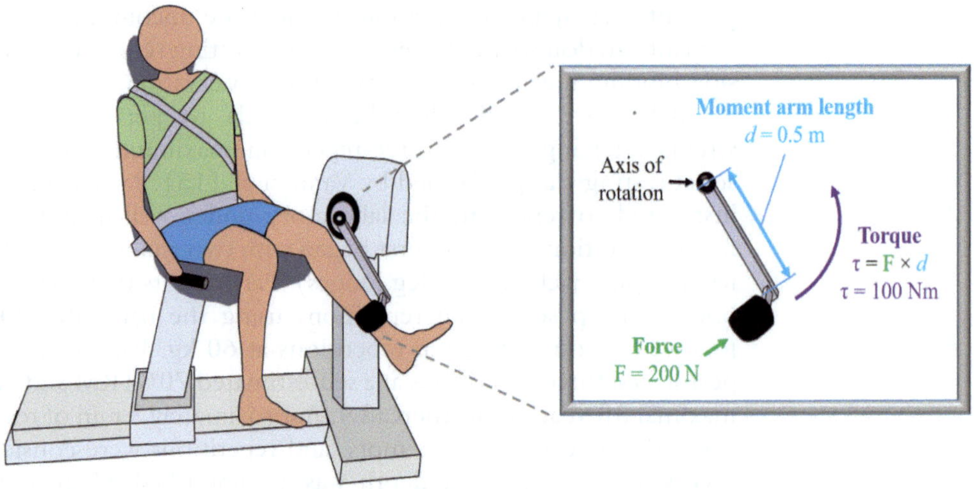

Fig. 1 Visual representation of an isokinetic dynamometer. The illustrated subject is performing a concentric knee extension with the left leg. Note that the axis of rotation of the dynamometer's lever arm is in line with the axis of rotation of the left knee. Once the knee is fully extended (180°), the leg can be returned to 90° (the presumed starting point) by either performing a concentric knee flexion, or an eccentric knee extension

It is worth noting that the many associations with functional outcomes are not the sole reasons for handgrip strength's wide usage, as it has many other advantages. Handgrip dynamometers are relatively inexpensive, highly mobile, and extremely safe and require very little training to use correctly. After a brief familiarization, each trial consists of 3–5 s bouts of maximal isometric effort, and the total assessment, including instructions, may last ~3 min. Only the proper posture or positioning of the patient and how to adjust the dynamometer for varying hand sizes need to be understood to perform this test. A convenient by-product of grip strength's vast usage is that it is relatively easy to find normative values for a wide range of populations.

The second type of commonly used dynamometers is the isokinetic dynamometer. Isokinetic dynamometers are extremely agile devices that can simultaneously measure and control the torque (rotational force), range of motion, and/or velocity of the movements. It is worth noting that while there are multi-joint dynamometers (e.g., isokinetic squat machines), the vast majority used in clinical and research settings are designed for single-joint assessments. Although only a single joint can be measured at a time, the dynamometer head, lever arm, and chair can be repositioned in numerous ways to allow for measurement of any of the major joints (i.e., the shoulder, elbow, wrist, hip, knee, and ankle of both the left and right limbs). Figure 1 shows a visual depiction of a knee extension in an isokinetic dynamometer.

A unique feature of isokinetic dynamometers is the ability to control the velocity of the movement. If the tester sets the velocity at a slow setting, 30°/s, for example, then the dynamometer will produce a counter-torque to prevent the limb from going any faster than the setting no matter how much torque the patient produces. The ability to strength train while limiting the range of motion and velocity can be quite useful in at-risk clinical populations (e.g., recent anterior cruciate ligament (ACL) reconstruction patients). Another benefit is that a brake/stop option can be engaged to prevent movement, thereby allowing the lever arm to be fixed at any joint angle for isometric strength assessments. One of the major criticisms of isokinetic assessments is that natural, real-world movements never occur at constant, fixed velocities. Therefore, the results are not always considered transferable or applicable to, for instance, athletic performance. There are exceptions; for example, the hamstring-to-quadriceps ratio ("H:Q," i.e., the ratio of knee flexion strength to knee extension strength) has been shown to have predictive value [20]. A summary of the advantages and disadvantages of isokinetic dynamometers is shown below.

Advantages of Isokinetic Dynamometers

- Can measure and control rotational limb velocity from 0 to 500°/s.
- Can measure torque at all joint angles and for practically any major joint.
- Can perform concentric, isometric, and eccentric muscle actions.
- Can measure limb position or angle, which in combination with torque and velocity, and can be used to calculate additional, functional variables (rate of torque development, limb acceleration, power, etc.).
- Can measure joint movement in both directions, including antagonist pairs (e.g., quadriceps and hamstrings) or concentric/eccentric actions of the same muscle group, within the same trial.
- Can set movement to a fixed torque, which allows for assessment of the passive range of motion (°) and musculotendinous stiffness (Nm/°).

Disadvantages of Isokinetic Dynamometers

- Expensive.
- Extremely heavy and immobile (i.e., it must be used where it is installed).
- Requires personnel training.
- Of the various strength training assessments, it is potentially the least applicable to, or predictive of, real-world performance.

As a final note, it is worth mentioning that the terminology of the strength assessments differs for dynamometers. As mentioned previously, the maximal value for a strength test involving free or machine weights is the one-repetition maximum, or 1RM. However, when performing maximal isometric muscle actions against a load cell, strain gauge, or any maximal muscle action (concentric, isometric, or eccentric) with a dynamometer, the trial is termed a maximal voluntary contraction (MVC). The most reported variable from an MVC trial in an isokinetic dynamometer is peak torque (PT; N × m).

2.3 Force Plates

Force plates are used to measure ground reaction forces during activities ranging from body-weighted tasks (e.g., walking or jumping) to weightlifting exercises. Regarding the latter, assessments such as an isometric squat or isometric mid-thigh pull can be performed with force plates and stationary racks. Force plates can also be used during more dynamic exercises such as single- or dual-leg jumps. One advantage of force plates is the ability to move them in and out of laboratory space. Additionally, an advantage of using force plates relative to isometric dynamometry is the ability to monitor attributes during multi-joint tasks. To this end, dynamometers can be reliably used to capture peak torque or rate of torque development of the arm or knee extensor and flexor muscles. However, the results from such tests are seldom relevant to athletic endeavors given that sports-specific tasks involve the action of several muscles across multiple joints. Researchers interested in a more holistic approach may therefore opt to implement force plate testing during activities such as the jump squat or complex weightlifting movements (e.g., power clean, snatch, clean and jerk, push presses). Moreover, it has been suggested that peak force captured via force plates during isometric mid-thigh pulls may be the most generalizable indicator of whole-body strength [21]. A meta-analysis conducted in this area also suggests force plate variables captured during multi-joint isometric exercises show a moderate to strong correlation with dynamic movements that involve large amounts of force and explosive power [22]. Thus, there are clear advantages in implementing force plate testing in the laboratory setting relative to some of the other tests mentioned herein. Figure 2 provides visual examples of the isometric mid-thigh pull using force plates.

Materials needed to perform force plate testing are discussed here. These methods were executed by Mobley et al. [23], and the appropriate ethical committee approved all procedures described prior to data collection. Participants were instructed to report to the laboratory under well-hydrated, 4-h fasted conditions. During this testing session, participants completed an isometric mid-thigh pull test. The testing apparatus was constructed as seen in Fig. 2.

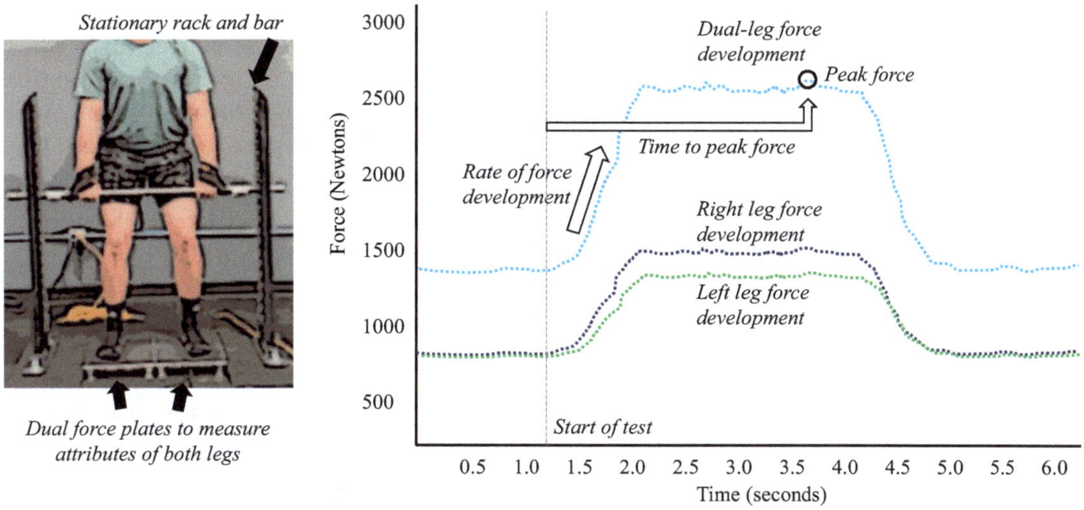

Fig. 2 Examples of force plate testing during the isometric mid-thigh pull. The left inset shows the experimental set-up of the isometric mid-thigh pull. The right inset shows typical data yielded from this test. In this instance, dual force plates can be used to monitor the rate of force development as well as peak force developed in the left and right leg. Such data may yield information regarding leg dominance and/or bilateral deficits

Briefly, the tester demonstrated how to perform the test to each participant. Thereafter, the participant mounted the apparatus containing dual force plates (one under each leg), a 20-kg barbell (York Barbell), and STS Power Rack (York Barbell). Force plates were OR6-7 force plates (AMTI; Watertown, MA, USA) with dual Gen 5 amplifiers (AMTI) sampling at 500 Hz which were used to measure vertical force production in Newtons (N). Upon mounting the device, the participant was instructed to grab the bar using a shoulder width grip. The participant's hands were then taped onto the bar using athletic tape. Once hands were secured, knee and hip angles ($125 \pm 5°$ and $175 \pm 5°$, respectively) were measured using a standard goniometer (Fabrication Enterprises; White Plains, NY, USA). Thereafter, participants were verbally encouraged to perform the repletion during a 5-s period. Each participant was allowed at least two attempts and up to four attempts if the difference in vertical peak force between trials was >250 N. Three minutes of recovery were allowed between attempts. After completion of the test, software was used to calculate vertical peak force during the testing sessions, and a custom-written MATLAB script (Natick, MA, USA) was used to identify the greatest vertical force produced in N for each trial. Two trials within 250 N were used to calculate an average peak force across trials and functioned as a metric for whole-body maximal voluntary force production.

2.4 Evoked Contractions

The techniques for assessing neuromuscular function that we have discussed up to this point have all been of voluntary contractions. Voluntary contractions are certainly the preferred method if trying to assess or predict how someone will perform or function in natural scenarios. However, when a clinician or researcher is trying to understand the underlying physiology, or what the neuromuscular system can do, voluntary contractions have many implicit limitations. The primary issue is that the function of the peripheral nerves and muscles is inherently dependent on supraspinal factors that are upstream of what is being measured (i.e., the performance is also dependent on the brain). Therefore, it is difficult to distinguish whether poor performance is truly indicative of neuromuscular deficits or could instead be the result of psychological factors, such as motivation, mental fatigue, alertness, anxiety, and more. Evoked contractions involve introducing a stimulus (often electrical, magnetic, or mechanical) which causes the muscles to contract relatively independent of psychological factors. It is worth noting that there are evoked techniques for assessing the central nervous system, such as transcranial magnetic stimulation (TMS) and spinal reflexes (e.g., patellar tendon taps and H-reflexes). However, these assessments are submaximal, and this section will focus on the evoked contraction techniques employed to assess the capabilities of the peripheral nervous and muscular systems.

One of the most common techniques involves electrically stimulating a peripheral nerve and assessing the electromyographic (EMG) responses (by placing surface electrodes on the skin over the muscles innervated by that nerve). When the muscle fibers under the electrodes are activated via electrical stimulation, their action potentials summate creating a predictable waveform deflection in the EMG signal. This waveform is often called an M-wave in the muscular research literature and a CMAP (compound muscle action potential) in clinical neurology settings. The maximal M-wave represents the summation of all the motor units activated simultaneously. The amplitude (mV), waveform duration (ms), and latency (ms) all provide useful information and can be seen in Fig. 3. In research labs, the force or torque output of the muscle is often simultaneously recorded. The force output from a peripheral nerve stimulation is referred to as twitch force (or peak twitch force for the maximal value). It is worth noting that peak twitch force, even under constant stimulation (referred to as high-frequency trains), is still substantially lower than the force output during a maximal voluntary contraction trial.

When performing peripheral nerve stimulation, a very low-current stimulus is introduced at first. Then, gradually, the current is slowly increased in a stepwise fashion. This process is continued until the M-wave no longer increases in amplitude because of an increase in current (i.e., the maximal M-wave, or CMAP, is found). The lowest perceptible M-wave is also of interest,

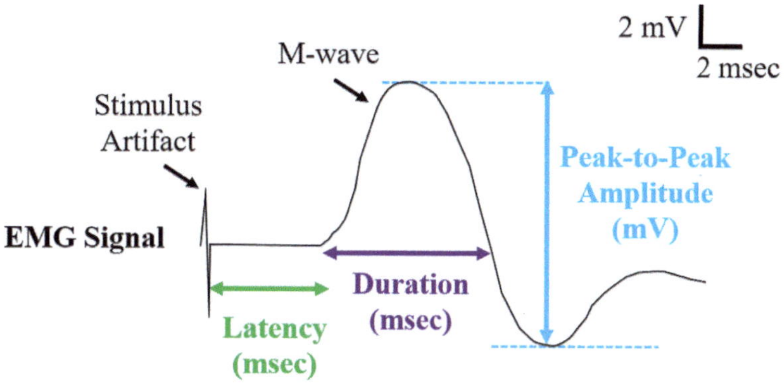

Fig. 3 An illustration of an M-wave from a peripheral nerve stimulation. The latency is measured as the start of the stimulation to the start of the M-wave. The latency is dependent on the distance between the stimulation site and the recording electrodes, as well as the nerve conduction velocity. The amplitude of the signal is often measured as peak-to-peak, which is the difference between the maximal and minimum values. This is important because the shape of the M-wave can vary greatly across muscles. In some instances, the positive deflection may be small, and the M-wave would be characterized by an overwhelmingly large negative deflection. The M-wave is dependent on many factors, including muscle mass, interelectrode distance (the distance between the two recording electrodes), stimulation intensity, and more

and the smallest increment in amplitude is referred to as a SMUP (single motor unit potential). The amplitude of the SMUP can be made relative to the CMAP to estimate the number of functioning motor units. This is referred to as MUNE, or motor unit number estimation. MUNE and similar assessments, such as MUNIX (motor unit number index), have been shown to be sensitive to muscle deterioration due to aging and various clinical disorders (e.g., ALS) [24, 25].

As a final note, evoked and voluntary contractions are not mutually exclusive. In fact, quite a few tests require their implementation simultaneously. One of the most useful is called the interpolated twitch technique (ITT). This involves introducing a supramaximal electrical stimulation during a MVC trial (the current used is slightly higher than the current that led to the maximal M-wave, such as 120%). The stimulation occurs once the torque or force signal has plateaued (i.e., they reached their max). If the twitch causes the MVC force to increase, then it is believed that the participant was unable to activate all their motor units voluntarily. In other words, maximal voluntary effort still left a reserve capacity of strength that the electrical stimulation was able to activate. A few seconds after relaxation, a second supramaximal stimulus is introduced to evoke a resting twitch. These stimulations during an MVC trial are depicted in Fig. 4.

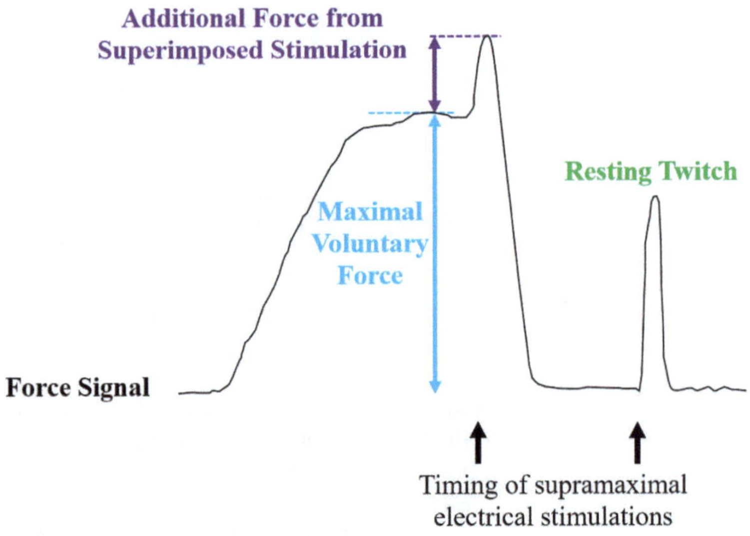

Fig. 4 An illustration of a force signal during the interpolated twitch technique. The interpolated twitch technique involves introducing a supramaximal electrical stimulation to the peripheral nerve while the muscle is maximally voluntarily contracting, and again immediately after while the muscle is relaxed. The amount of additional force or torque produced from the stimulation during the MVC can provide useful information regarding the degree to which the muscle was maximally activated in the first place. The increase is often expressed relative to the resting twitch for normalization. Highly motivated and trained individuals are capable of performing true maximal contractions in which the additional electrical stimulation provides no increase in force

The superimposed twitch and the resting twitch can be used to calculate variables such as voluntary activation and central activation ratio to estimate the percent of muscle that could be activated volitionally [26]. The interpolated twitch has even been used simply to verify or validate that maximal effort was given during MVC trials [27].

3 Conclusions

Various neuromuscular testing techniques can be applied in research settings. 1RM testing is more commonly performed given that the equipment needed is widely available. However, relative to other methods, this test does not provide detailed information about neuromuscular attributes (e.g., activation of different muscles, estimated number of motor units active during contraction). Dynamometry can provide more detailed information on the maximal isometric or isokinetic torques produced by an individual muscle group. However, again, this test is limited in that it does not provide information related to certain neuromuscular attributes (e.g., estimated number of motor units active during contraction

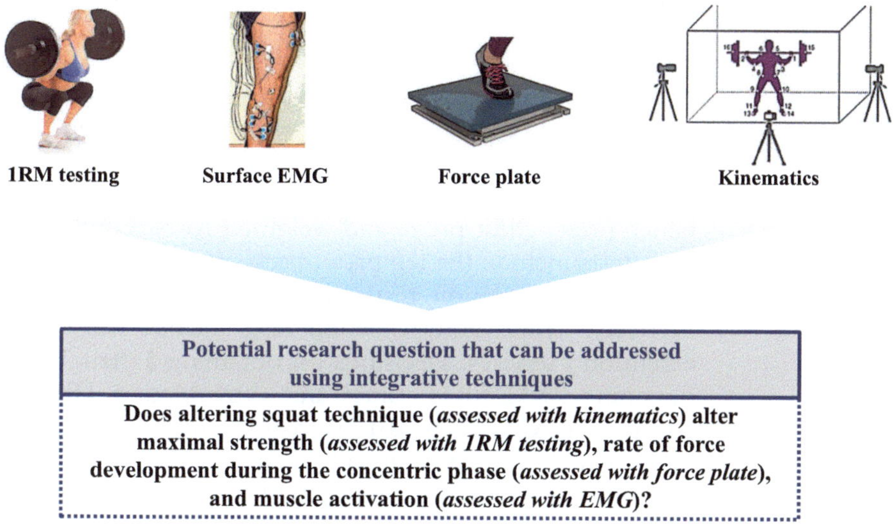

Fig. 5 The integration of multiple techniques. This schematic illustrates how individual techniques discussed in this chapter can be used to develop a more advanced research question relative to the utilization of one individual technique

or motor unit firing rates). Given the limitations of some of the individual techniques, the integration of techniques, when possible, may provide novel and invaluable insights. Moritani and deVries (1979) published a landmark research study illustrating this concept [28]. Briefly, the researchers subjected younger adults to 8 weeks of elbow flexor training (3× weekly). Isometric arm flexion values using a dynamometer were obtained prior to training, every 2 weeks during training, and following training. The researchers also acquired surface EMG signals from the biceps muscles as well as arm circumference values during these testing sessions to estimate the neural versus hypertrophic contributions to strength gains as training progressed. Other studies have since adopted similar integrative approaches. For instance, James et al. [29] had participants perform jump squats on a force plate while wearing wireless surface electrodes to acquire EMG signals from various leg muscles. This approach allowed the researchers to obtain data on jump height as well as the rate of force development and muscle activation during the jump squat tests. Figure 5 provides a contextual example of how an integrated approach using multiple devices described herein can yield insightful information to enrich a research study design.

In conclusion, we foresee that measures of neuromuscular function will continue to be utilized in a widespread fashion in the research disciplines of aging, disease physiology, and exercise science. For additional perspectives on the discussed techniques, readers should refer to these excellent scientific reviews: (i) strength testing with free weights [16], (ii) dynamometer testing [30], (iii) evoked potentials [31], and (iv) force plate testing [21].

4 Notes

There are various factors to consider when performing each of the tests described in this chapter. This section lists these considerations.

For RM testing, upper-body, multi-joint tests include the bench press, chest press, and weighted rows. Lower-body, multi-joint tests include the leg press, back squat, and deadlift exercises. Upper-body isolation tests include the arm curl and extension exercises. Lower-body isolation tests include the knee curl and extension exercises. Decisions on obtaining a "true" 1RM versus obtaining 2RM–5RM values and extrapolating 1RM should be considered. RM values past five repetitions lose validity in predicting a "true" 1RM. Additionally, a familiarization session should be scheduled 3–7 days prior to the RM test day where testing mimics what would be performed. During the familiarization and RM test days, a general warm-up consisting of 5-min upper- or lower-body ergometry followed by exercise-specific warm-up using two sets of weight at 50–70% of estimated strength should be executed. During the familiarization and RM test days, a 3-min recovery period between the warm-up protocol and test initiation, as well as 3 min between RM attempts, is desirable. The researcher should also aim to obtain the final 1RM or multiple RM values within five attempts to minimize residual fatigue from more attempts. Mass should be added in a progressive manner, and (when possible) the participant should be blinded to the amount of mass being used to test. A metronome can be used to promote a 2 s–0 s–2 s eccentric–transition–concentric lifting tempo during testing. Additionally, testers should ensure hand or foot placement is consistent during testing, and the use of wearable technologies (e.g., goniometers) can ensure a consistent range of motion. Finally, the use of weightlifting garments (e.g., knee straps, weight belts) should be consistent given that they can enhance 1RM values.

For maximal handgrip strength (HGS) testing, measurements can be obtained with the subject standing or seated. The supine position is only used in individuals unable to maintain an upright posture. Evidence has shown that HGS is higher from the standing position [32]. However, it should be noted that the seated posture may be preferred, or even necessary, in some clinical or at-risk populations. There do not seem to be any consistent standards for elbow position across HGS studies. However, most of the studies perform HGS trials with the elbow at either a 90° or 180° joint angle (e.g., see [33]). Furthermore, for the studies using 90°, some may support the subject's arm with an elbow rest. It is worth noting, however, that multiple studies suggest that the position of the elbow may affect the maximal strength value [32]. Therefore, it is imperative that the elbow joint angle, as well as whether it was supported or not, is reported in the methodology. Most of the handgrip dynamometers are adjustable for hand size. It is important to ensure that the dynamometer is adjusted for the subject's hand size and personal comfort prior to starting the testing. Having the finger placement too far or close may affect the outcome of the trials. HGS testing with the dominant hand is the most common and provides the highest maximal strength values (see [34] for a summary). However, in some clinical populations, such as stroke or injury rehab patients, the asymmetry between the two sides might be of particular importance. A neutral grip seems to be the most used hand position. However, one study suggests that a slight wrist extension may lead to higher HGS values [35]. Longitudinal studies often report the absolute values of HGS while studies with between-group comparisons often normalize the values relative to body mass. The type of analysis should dictate whether normalization is prudent, but it might be safer to simply report both absolute and relative values. Some criteria for classifying patients or subjects (e.g., HGS has been used to classify sarcopenic patients) may have cutoffs of absolute values for different levels based on body mass index (BMI) [33]. In other words, the absolute cutoff for someone with a low BMI may differ from the cutoff for someone with a higher BMI.

For testing involving isokinetic dynamometry, there are three primary ways to utilize the velocity features of an isokinetic dynamometer. The first is to fix the velocity of the movement. Isokinetic dynamometers can provide a counter-resistance to the torque being produced and can adjust it quickly enough that no matter how fast the subject moves the leg, the counter-resistance will keep it from moving faster than the desired velocity. It is common to assess torque at various movement velocities, which can differ due to the force/velocity relationship (force production decreases as velocity increases) and training specificity (training at a particular velocity leads to better gains at that same velocity than other velocities). Commonly used velocities, from slow to fast, include 30°/s, 60°/s, 180°/s, and 300°/s. The second technique is to place the dynamometer in "isotonic mode," which fixes the resistance. In this case, the velocity component is the variable that has degrees of freedom (i.e., velocity can be assessed at different resistances). The third and final technique is to assess maximal or peak movement velocity (V_{MAX}; °/s). This assessment requires no counter-torque other than the natural weight of the lever arm. To accomplish this, the dynamometer must be fixed at the maximal velocity setting of 500°/s. Multiple studies from our laboratory show that most subjects are unable to reach that velocity, and so the counter-torque never engages [36, 37]. The peak of the velocity signal can then be taken as the maximal movement velocity. Another feature of several isokinetic dynamometers is antagonistic muscle pairs can be simultaneously assessed by putting the dynamometer in the concentric or eccentric mode for both directions. A concentric flexion can be immediately followed by a concentric extension to return the lever arm to its original position. Another option is to alternate concentric and eccentric modes for the same muscle group (i.e., concentric mode in one direction followed by eccentric to return). Most dynamometers are equipped with a stop button that engages a brake on the lever arm. When this is engaged at the desired position, the lever arm cannot move, but torque can still be recorded from isometric contractions. The axis of rotation for the joint that is being isolated should be in line with the axis of rotation of the dynamometer head. This allows for optimal transfer of torque production from the muscles to transfer to the lever arm and be registered by the device. Additionally, the joint angles of the other joints could be of importance. For example, the peak torque of a knee extension may differ with the hip angle set at 90° compared to 180°, or elbow torques may be dependent on shoulder joint angles. As such, it is important to report the position of other joints and to be consistent in the positioning across subjects and trials. Hand position may also be important to control during lower body assessments. Results may differ between having the hands placed on the opposite shoulder (crossing the arms is a common technique) compared to allowing the subjects to grip the handles on the sides of the seat. The primary variables provided by most isokinetic dynamometers are torque, position, and velocity. For many protocols, two of the three variables are often controlled and, therefore, only the third (i.e., the only variable with degrees of freedom) is recorded. For example, in an isometric contraction, position and velocity are fixed (with velocity being zero), and so only the torque signal can vary over time. Another example is to fix the torque and velocity and record the position signal for a range of motion assessments. Since these primary variables can be recorded across time, additional variables can also be calculated. For example, while peak torque ($N \times m$) is commonly assessed, how fast someone produced the torque may also be of interest. This is known as the rate of torque development (RTD; Nm/s) and is related to the better-known variable, the rate of force development (RFD; N/s). Other variables frequently calculated are acceleration (i.e., the rate of change in velocity) and power. It is worth noting that isokinetic dynamometry is often paired with electromyography (EMG), and so additional variables may include measures of muscle activation and rate of activation. Most isokinetic dynamometers have an analog output on the back, and those signals are sent to an external piece of equipment for recording. This equipment has an analog-to-digital converter, which provides a digital representation of the original, analog signal. The equipment used, as well as the frequency at which it records a sample of the signal (i.e., the sampling rate), should always be reported. Once the signals have been digitized (see signal acquisition above), there are many ways that the signals can then be analyzed. For instance, torque signals are commonly smoothed to remove noise while maintaining the overall pattern over time. The two most common methods of smoothing are low pass filters or moving averages. It is important, however, not to over-smooth the signal as it can reduce the peak values and change the outcomes. The length of the analysis windows is important as well as it needs to be reported which parts of the signal are being examined. For example, with zero representing the onset of the contraction, common windows for assessing the rate of torque development (RTD) include 0–50 ms, 0–100 ms, 100–200 ms, and 0–peak. It is worth noting that smoothing the signal is not always recommended, and different analysis windows may represent distinct physiological mechanisms [30].

When performing electrically evoked contractions, most studies use a constant current stimulator, but the type and manufacturer are important to report. The properties of the pulses can greatly affect the outcomes and must be rigorously controlled. The number of pulses for each stimulation is typically either singlets (single pulse), doublets, or octets (eight pulses). If performing a train of pulses, then the frequency is another factor to consider. From our lab experience, it is worth noting that longer-duration pulses and higher-frequency trains both seem to increase the level of discomfort for the subject. Another factor to consider is where the stimulation is occurring. Most analyses use peripheral nerve stimulation, but that depends on the muscle group of interest. Not all muscles are innervated by a nerve that is superficial and easy to stimulate. Therefore, muscle stimulation is also performed. A disadvantage of muscle stimulation is that it is likely to create a large stimulation artifact in the EMG signal. A disadvantage of nerve stimulation, however, is that it is likely to evoke contractions in multiple muscles at a time (i.e., stimulate all the muscles that the nerve innervates). Lastly, the distance between the cathode and anode (positive and negative stimulating electrodes) can affect the efficacy of the stimulation. The analog-to-digital converter and the sampling rate need to be reported. Due to the Nyquist theorem, the sampling rate for EMG is typically at least 1000–2000 samples/s. Furthermore, the electrode position and interelectrode distance can also greatly affect the amplitude of the EMG signal. For more details, please refer to the Electromyography chapter. Filtering evoked contractions usually consists of the standard EMG band-pass filtering (e.g., 10–500 Hz) combined with filters to remove stimulus artifacts. A technique that can help with the separation of the stimulation artifact from the physiological waveforms is to trigger automatic windows based on stimulation. If an additional signal is recorded showing the triggers (i.e., when stimulations occurred), then separate windows of the EMG signals (e.g., 200 ms) can be extracted and viewed immediately following the triggers. Any deflections in EMG occurring within 10 ms of a stimulation trigger are likely an artifact and not physiological in origin.

Acknowledgments

The research described herein at Auburn University in the laboratory of MDR was supported by BioNutritional Research Group, Hilmar Ingredients, Lockwood LLC, JW Nutritional, The Peanut Institute, Florida A&M University, and Renaissance Periodization. The research described herein at Oklahoma State University in the laboratory of JMD was supported by the Oklahoma Center for the Advancement of Science and Technology (OCAST), the Stillwater Medical Center, and the NSCA Foundation.

References

1. Fitts RH, McDonald KS, Schluter JM (1991) The determinants of skeletal muscle force and power: their adaptability with changes in activity pattern. J Biomech 24(Suppl 1):111–122
2. Verdijk LB et al (2009) One-repetition maximum strength test represents a valid means to assess leg strength in vivo in humans. J Sports Sci 27(1):59–68
3. Reynolds JM, Gordon TJ, Robergs RA (2006) Prediction of one repetition maximum strength from multiple repetition maximum testing and anthropometry. J Strength Cond Res 20(3):584–592
4. Seo DI et al (2012) Reliability of the one-repetition maximum test based on muscle group and gender. J Sports Sci Med 11(2): 221–225
5. Benton MJ, Raab S, Waggener GT (2013) Effect of training status on reliability of one repetition maximum testing in women. J Strength Cond Res 27(7):1885–1890
6. Haff GG, Triplett NT (2016) Essentials of stregnth training and conditioning, 4th edn. Human Kinetics, Champaign
7. Barroso R et al (2013) The effects of different intensities and durations of the general warm-up on leg press 1RM. J Strength Cond Res 27(4):1009–1013
8. Machek SB et al (2021) Neoprene knee sleeves of varying tightness augment barbell squat one repetition maximum performance without improving other indices of muscular strength, power, or endurance. J Strength Cond Res 35 (Suppl 1):S6–S15
9. Smith EB et al (1996) The effects of lumbosacral support belts and abdominal muscle strength on functional lifting ability in healthy women. Spine (Phila Pa 1976) 21(3):356–366
10. Bazyler CD et al (2014) The efficacy of incorporating partial squats in maximal strength training. J Strength Cond Res 28(11): 3024–3032
11. Murray N et al (2013) Effects of foot position during squatting on the quadriceps femoris: an electromyographic study. Int J Exerc Sci 6(2): 114–125
12. Headley SA et al (2011) Effects of lifting tempo on one repetition maximum and hormonal responses to a bench press protocol. J Strength Cond Res 25(2):406–413
13. Kamen G, Knight CA (2004) Training-related adaptations in motor unit discharge rate in young and older adults. J Gerontol A Biol Sci Med Sci 59(12):1334–1338
14. Grandolfi K, Sosciarelli V, Polito M (2018) Resistance training and the effect of load blinding in multiple repetition maximum tests. Int J Sports Physiol Perform 15(3):441–443
15. Vann CG et al (2020) Skeletal muscle protein composition adaptations to 10 weeks of high-load resistance training in previously-trained males. Front Physiol 11:259
16. Grgic J et al (2020) Test-retest reliability of the One-Repetition Maximum (1RM) strength assessment: a systematic review. Sports Med Open 6(1):31
17. Rantanen T et al (1999) Midlife hand grip strength as a predictor of old age disability. JAMA 281(6):558–560
18. Kim Y et al (2017) Independent and joint associations of grip strength and adiposity with all-cause and cardiovascular disease mortality in 403,199 adults: the UK Biobank study. Am J Clin Nutr 106(3):773–782
19. Reuben DB et al (2013) Motor assessment using the NIH Toolbox. Neurology 80(11 Suppl 3):S65–S75
20. Orchard J et al (1997) Preseason hamstring muscle weakness associated with hamstring muscle injury in Australian footballers. Am J Sports Med 25(1):81–85
21. Beckham G, Suchomel T, Mizuguchi S (2014) Force plate use in performance monitoring and

sport science testing. New Stud Athl 29(3): 25–37
22. Juneja H, Verma SK, Khanna GL (2010) Isometric strength and its relationship to dynamic performance: a systematic review. J Exerc Sci Physiother 6(5):60–69
23. Mobley CB et al (2017) Effects of whey, soy or leucine supplementation with 12 weeks of resistance training on strength, body composition, and skeletal muscle and adipose tissue histological attributes in college-aged males. Nutrients 9:972
24. Drey M et al (2013) The Motor Unit Number Index (MUNIX) in sarcopenic patients. Exp Gerontol 48(4):381–384
25. Nandedkar SD, Barkhaus PE, Stalberg EV (2010) Motor unit number index (MUNIX): principle, method, and findings in healthy subjects and in patients with motor neuron disease. Muscle Nerve 42(5):798–807
26. Krishnan C, Williams GN (2010) Quantification method affects estimates of voluntary quadriceps activation. Muscle Nerve 41(6): 868–874
27. Psek JA, Cafarelli E (1993) Behavior of coactive muscles during fatigue. J Appl Physiol (1985) 74(1):170–175
28. Moritani T, deVries HA (1979) Neural factors versus hypertrophy in the time course of muscle strength gain. Am J Phys Med 58(3): 115–130
29. James LP et al (2018) The impact of strength level on adaptations to combined weightlifting, plyometric, and ballistic training. Scand J Med Sci Sports 28(5):1494–1505
30. Maffiuletti NA et al (2016) Rate of force development: physiological and methodological considerations. Eur J Appl Physiol 116(6): 1091–1116
31. Siddique U et al (2020) Determining the sites of neural adaptations to resistance training: a systematic review and meta-analysis. Sports Med 50(6):1107–1128
32. Xu ZY et al (2021) The effect of posture on maximum grip strength measurements. J Clin Densitom 24:638
33. Sousa-Santos AR, Amaral TF (2017) Differences in handgrip strength protocols to identify sarcopenia and frailty – a systematic review. BMC Geriatr 17(1):238
34. Bohannon RW (2003) Grip strength: a summary of studies comparing dominant and nondominant limb measurements. Percept Mot Skills 96(3 Pt 1):728–730
35. O'Driscoll SW et al (1992) The relationship between wrist position, grasp size, and grip strength. J Hand Surg Am 17(1):169–177
36. Mackey CS et al (2018) Comparison of fatigue responses and rapid force characteristics between explosive- and traditional-resistance-trained males. Eur J Appl Physiol 118(8): 1539–1546
37. Hester GM et al (2019) Age does not attenuate maximal velocity adaptations in the ipsilateral and contralateral limbs during unilateral resistance training. J Aging Phys Act 27(1):1–8

Chapter 6

Neuromuscular Function: High-Density Surface Electromyography

Eduardo Martinez-Valdes and Francesco Negro

Abstract

Surface electromyography (EMG) is a noninvasive technique that has the potential to provide information on how the nervous system controls movement. Nevertheless, conventional EMG methods with low spatial sampling resolution do not provide the necessary level of detail to study motor behavior. During the last decade, the emergence of high-density surface EMG (HDsEMG) systems has enabled a more detailed analysis of the firing patterns of alpha motoneurons to the muscle. Access to this information has improved our understanding of the neural mechanisms responsible for the control and generation of muscle force. This chapter will focus on the recording of HDsEMG signals and the processing techniques enabling the identification of motor unit firings from surface EMG recordings by employing HDsEMG.

Key words Electromyography, Motor unit, Motor neuron, High-density surface electromyography, Decomposition, Blind source separation

1 Introduction

Motor units are responsible for transforming synaptic activity into forces that generate movement. Therefore, the number of active motor units (motor unit recruitment) and the frequency at which motoneurons discharge action potentials (discharge rate or rate coding) will determine the amount of force generated by a muscle. For many decades, surface electromyography (EMG) has been used to infer the neural determinants of the generation of muscle force [1]. This is due to the relationship between EMG signal amplitude and the amount of torque (or force) measured about a joint [1]. Indeed, surface EMG amplitude can indirectly estimate the net motor unit activity required to perform a task since EMG amplitude is related to discharge rates and the recruitment of active motor units. Nevertheless, a number of biophysical, anatomical, and neurophysiological factors (see [2] for a review) can affect the EMG–force relation. Therefore, EMG amplitude is usually considered a poor determinant of the neural mechanisms responsible for

Philip J. Atherton and Daniel J. Wilkinson (eds.), *Neuromuscular Assessments of Form and Function*, Neuromethods, vol. 204, https://doi.org/10.1007/978-1-0716-3315-1_6,

the neural control of movement [2, 3]. Recordings enabling direct estimation of motor unit activity are required to assess the mechanisms responsible for the control of force. Previously, invasive intramuscular (or subcutaneous) EMG methods were required to record motor unit activity [4]. Due to their selectivity, intramuscular methods permit the isolation of motor unit action potentials, which enables the identification of single-motor unit firing times required to calculate discharge properties such as rate coding and discharge rate variability. Although this methodology has been employed for many years, it has important disadvantages: first, a very small number of motor units can be reliably identified during voluntary contractions; second, only low to moderate forces can be analyzed; and third, electrodes cannot be repositioned across sessions, which does not allow to track the same populations of motor units following, i.e., a training intervention. Recently, high-density surface electromyography (HDsEMG) emerged to solve these issues. HDsEMG is comprised of electrode arrangements involving tens or hundreds of electrodes allowing the recording of EMG activity from multiple muscle compartments simultaneously. These electrodes can be used for two purposes: first, to assess regional changes in muscle activity [5] and, second, to identify single-motor unit activity from the surface EMG, in a process named "HDsEMG motor unit decomposition" [6]. Regarding the latter, when HDsEMG recording systems are used in combination with blind source separation signal processing techniques, the action potentials of single motor units can be separated from the interference EMG, allowing automatic identification of motor unit firing times [7, 8]. Given their low selectivity and higher spatial sampling, HDsEMG recordings can identify a greater number of motor units (usually tens of motor units depending on the muscle), while higher forces can be assessed (up to the maximum voluntary contraction force [9]), and the same populations of motor units can be followed across testing sessions, which allows evaluating changes in motor unit plasticity during longitudinal studies [10, 11]. The larger samples of motor units identified with HDsEMG methods permit the determination of the common synaptic inputs responsible for force generation [12, 13] since the summation of activity of multiple motor units allows accurate prediction of the resultant force/torque output exerted at the joint [3, 14, 15]. Similarly, the discharge information of several motor units concurrently identified during isometric contractions allows the reliable deconvolution of the force signal and the estimation of contractile motor unit properties [16–18]. Moreover, the assessment of higher forces allows the comparison of firing activity between motor units of different recruitment thresholds (low vs. high force recruitment threshold), providing new insights into the mechanisms of motor unit control in various conditions such as pain [19], training [10], and fatigue [20]. Finally, the tracking of the same motor units across testing sessions has improved the understanding of the neural mechanisms responsible for the increase in force production, following resistance [21] and high-intensity interval training interventions [10, 22], and has the potential to assess the progression of neuromuscular/musculoskeletal disorders [23]. Taken together, these advantages show the importance of HDsEMG motor unit decomposition techniques for the study of the neurophysiology of movement. This chapter will, therefore, focus on the methods required to record HDsEMG signals and the signal processing tools necessary to identify and analyze motor unit properties from the HDsEMG.

2 Materials

Following are the materials required to record HDsEMG signals:

- Laptop or desktop computer.
- HDsEMG amplifier.
- HDsEMG electrode connectors, reference electrodes, and pre-amplifiers (if available).
- HDsEMG electrodes.
- Double-sided adhesive foam (if wet electrodes are used).
- Conductive paste for wet electrodes and water for dry electrodes.
- Skin-cleaning materials (abrasive paste, water) and razors (if needed).
- Dry electrode array (optional).

2.1 HDsEMG Amplifiers

The recording of HDsEMG signals for motor unit decomposition requires EMG amplifiers that are able to record tens of surface EMG channels simultaneously. OT Bioelettronica (Turin, Italy, https://otbioelettronica.it/), Biosemi (Amsterdam, Netherlands, https://www.biosemi.com/), and TMSi (Twente, Netherlands, https://www.tmsi.com/) are among the most popular brands which focus on the manufacture of HDsEMG amplifiers. Depending on the model and year of manufacture, multichannel surface EMG amplifiers may have an analog-to-digital (AD) converter of 12, 16, or more bits. When the number of bits is low, some systems include an adjustable gain to account for the full range of EMG amplitudes. In these cases, caution should be used in the selection of the gain to avoid saturation of the amplifier. The sampling frequency required to record motor unit data is similar to that used by conventional surface EMG recordings (minimum of 1000 Hz). However, sampling frequencies of 2000 Hz or higher are recommended. Analog filters should be set at 10–20 Hz for high-pass and 500 Hz for low-pass. Where possible, the HDsEMG amplifier should be set to record signals in monopolar/unipolar mode as this allows more flexibility during offline analysis (e.g., different spatial filters can be employed). However, single-differential configurations can also be used when the power-line noise interference is relatively high.

2.2 HDsEMG Electrodes

There is a wide variety of electrode grids with a different number of electrodes (32 or 64 channels), spatial arrangements (8 × 8, 13 × 5, 8 × 4 rows and columns, respectively), and interelectrode (IED) distances (2.5, 5, 8, 8.5, and 10 mm) (*see examples in* Fig. 1). The selection of the electrode depends upon the investigated muscle. For example, grids with an IED distance of 2.5 mm are recommended for hand muscles, while grids with an IED distance of 8 mm or 10 mm would be recommended for lower limb muscles. It is important to note that the larger the interelectrode distance is, the less spatial information will be sampled, and this can compromise the identification of motor unit activity from surface EMG. Although a configuration of a few tens of electrodes is the minimum necessary for motor unit decomposition using standard blind source separation techniques, a larger number of electrodes may allow the identification of a greater number of motor units in some muscles [8, 24].

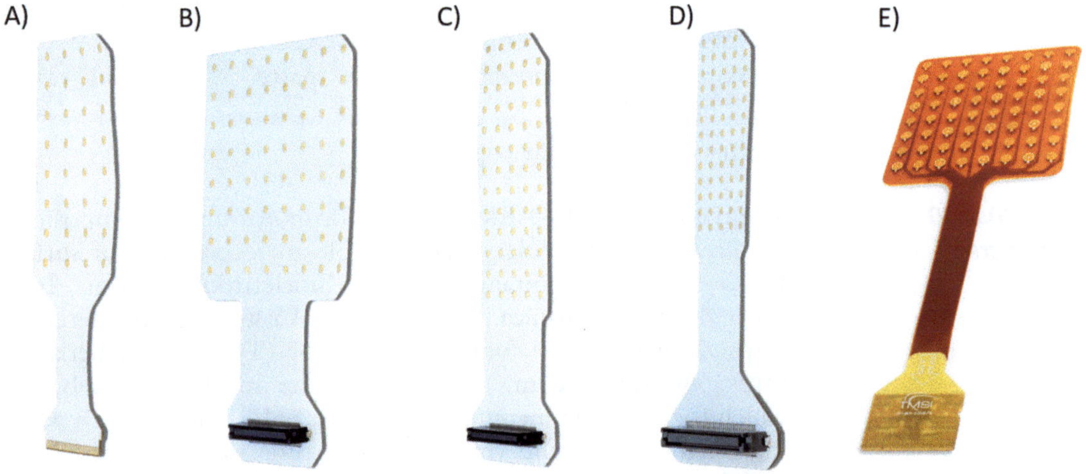

Fig. 1 High-density surface electromyography (HDsEMG) electrodes. HDsEMG electrodes of different sizes and arrangements can be seen in the figure. (**a**) OT Bioelettronica (Torino, Italy) 32-channel, 4 × 8 electrode grid (10 mm interelectrode distance (IED)), (**b**) OT Bioelettronica (Torino, Italy) 64-channel, 5 × 13 electrode grid with a missing electrode on the upper right corner (8 mm IED), (**c**) OT Bioelettronica (Torino, Italy) 64-channel, 8 × 8 electrode grid (10 mm IED), (**d**) OT Bioelettronica (Torino, Italy) 64-channel, 5 × 13 electrode grid with missing electrode on the upper right corner (2.5 mm IED), (**e**) TMSI (Enschede, Netherlands) 64-channel 8 × 8 electrode grid (8.5 mm IED)

2.3 Skin Preparation and Electrode Placement

Skin impedance should be minimized before placing the electrodes. Therefore, hair needs to be shaved, and the skin should be cleaned with abrasive paste in order to remove dirt and dead skin cells. Residuals from the abrasive paste need to be removed with water to allow improved electrode conduction. If available, it is recommended to check action potential propagation during a brief voluntary contraction using a linear nonadhesive dry electrode array [11]. This involves observing the region where the innervation zone is located and better positioning the columns of the HDsEMG grid with the muscle fibers. The innervation zone has been defined as the origin of the propagation of action potentials during both voluntary and electrically induced contractions [25]. When HDsEMG electrodes are aligned in parallel to the muscle fibers and a differential derivation is used, the innervation zone is represented as an EMG channel with low amplitude and the adjacent channels (i.e., above and below the innervation zone) typically show action potentials with greater amplitude propagating in opposite directions [26] (Fig. 2). Although aligning of the

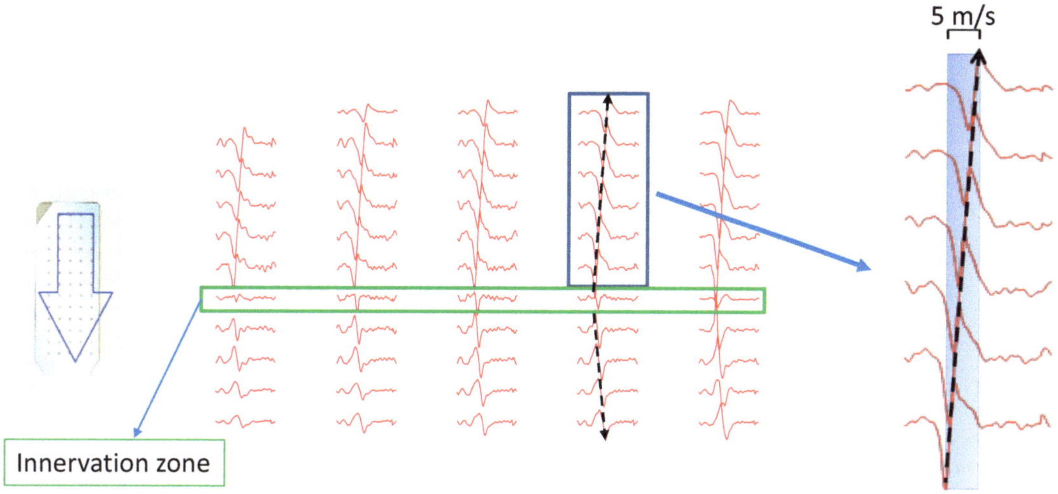

Fig. 2 Innervation zone localization and muscle fiber conduction velocity calculation. The firing times of each motor unit can be used to trigger surface EMG signals and observe the 2D action potential representation on the surface EMG (middle of the figure). In this case, action potentials are shown in double differential mode (5 × 11 matrix of electrodes). If the HDsEMG electrode grid (left side of the figure) is properly aligned with the muscle fibers, the action potential propagation across the electrodes can be observed clearly. Action potential propagation originates from the innervation zone (green box) and projects cranially and caudally (opposite directions, dashed arrows in the middle of the figure). Conduction velocity can be calculated from any column of the electrode grid; however, it is important to select the column with the greatest number of channels with similar shape (quantified by cross-correlation) and clearest propagation of action potentials (blue box). The selection of a greater number of channels ensures a more reliable assessment of conduction velocity estimates (three single-differential channels is the minimum number to calculate conduction velocity). Once these channels have been identified, the delay in action potential transmission from one channel to the other can be calculated (right side of the figure)

electrode grid is not strictly necessary to obtain good decomposition results, it is useful in muscles with fibers that run parallel to the skin (e.g., vastus lateralis, biceps brachii) as this affords calculation of muscle fiber conduction velocity (see methods and Fig. 2) [22]. In addition, grid positioning over the innervation zone is recommended because the variation in the direction of motor unit action potential propagation above and below the innervation zone induces changes in motor unit action potential shapes which help in the separability of different motor units (see methods). When the HDsEMG electrode grid is positioned outside the innervation zone, all motor unit action potentials will likely propagate in the same direction, increasing the redundancy of the signals and making it more difficult to discriminate motor units. The preparation of the electrode grid depends upon the type of electrode used. For instance, Ag/Cl pin-type grids only require water to record EMG signals, while Cu or Au electrodes need to be prepared with conductive paste or gel. For the latter, a double-sided adhesive foam with cavities for each electrode is used to enable filling the cavities with conductive paste and attaching the electrode to the skin of the participant. For longitudinal studies, it is important that the electrodes are positioned in a similar location across sessions, as this enables the tracking of the same motor units across the intervention (see below). For this reason, consider marking the region where the electrode was positioned on the skin of the participants with an indelible pen. If this is not possible, the use of anatomical landmarks is recommended.

2.3.1 HDsEMG Electrode Cleaning

Au or Cu HDsEMG electrodes can be used multiple times if they are adequately cleaned (approximately up to 5–6 times). Therefore, any residual conductive cream or gel should be removed from the electrode immediately after use. If electrodes are exposed to any liquids or cream for an extended period, this can oxidize the surface of the electrodes, which will increase contact impedance and decrease signal quality. A dry cloth can be used to remove conductive cream and dry the electrodes. Alcohol can also be used to clean and disinfect the matrices after use.

3 Methods

3.1 Tasks

HDsEMG motor unit decomposition algorithms have been mainly validated to identify motor units during isometric contractions. This type of contraction is ideal for decomposition since motor unit action potentials remain relatively stable, which improves the ability of blind source separation algorithms to identify single motor units. Nevertheless, a number of studies have shown that motor units can still be decomposed in more challenging conditions such as isometric "sinusoidal" torque-varying contractions

[27], ballistic contractions [28], sustained fatiguing contractions [20], and even shortening and lengthening dynamic contractions [29, 30], showing that HDsEMG methods provide a wide range of research possibilities to accommodate different contraction types. However, it is important to note that the number of identified units decreases in more challenging contraction types since the stability of action potentials is compromised due to the relative movement of the muscle fibers with respect to the recording electrodes. Therefore, caution is required when attempting to record/analyze motor unit data in dynamic conditions. If such recordings are attempted, unless specific decomposition algorithms able to compensate for changes in motor unit action potential shapes [30] are used, it is recommended to assess a relatively small range of motion (i.e., 20° to 30°) and employ an isokinetic dynamometer to control the angular velocity, as these contractions should be performed at slow angular velocities (i.e., up to 20°/s) [31]. Besides contraction type, contraction intensity can also affect the ability to identify motor units. Due to a greater number of recruited motor units at higher contraction intensities, the separation of these units is more challenging than at lower intensities. Therefore, the number of identified units is commonly reduced at higher forces [32]. Additionally, blind source separation algorithms are biased toward motor unit action potential trains with clear spatial representation, high signal-to-noise ratios, and significant sparsity. In most conditions, this results in favoring the identification of higher-threshold motor units. Therefore, it is recommended to perform contractions at different intensities (i.e., 10%, 40%, and 70% of the maximum voluntary force) to assess motor unit properties of different populations of motor units.

3.2 Target Muscle

Another important factor to consider when designing a study on motor unit behavior is the investigated muscle. Multiple studies have shown that muscles such as tibialis anterior or medial gastrocnemius, which have fibers with large pennation angles and complex architecture, show better decomposition results than muscles with smaller pennation angles (i.e., muscles whose fascicles run parallel to the skin) and less complex architecture (i.e., vastus lateralis and biceps brachii) [33]. This could be partially explained by the fact that muscles with complex architecture exhibit motor units with a more unique spatial representation on surface EMG, in contrast to muscles with more parallel fibers. Nevertheless, it is important to note that there are a number of muscles with complex architecture in which it is still challenging to identify motor units (i.e., rectus femoris, gastrocnemius lateralis, paravertebral muscles). Therefore, the different motor unit yield observed in muscles with diverse architecture still requires further investigation. A final point to consider is muscle depth. HDsEMG systems are able to record activity from the most superficial motor units only since the activity of the deeper motor units has a low signal-to-noise ratio [7]. Therefore, intramuscular EMG methods are required when deeper musculature needs to be analyzed (see chapter "Neuromuscular Function: iEMG").

3.3 EMG Signal Preprocessing Steps

Following the recording session, signals are filtered with a bandpass digital filter with corner frequencies at 10 and 500 Hz. Then, each EMG channel is inspected to assess quality. High-quality signals are required prior to running decomposition programs since any artifacts and noise can affect the ability of the algorithm to separate individual motor units from the interference EMG. Therefore, it is important to remove any channels with a poor signal-to-noise ratio, i.e., where noise is more than one-half of the power of the signal [33]. A threshold can be applied to remove channels automatically; however, visual inspection of all EMG channels is recommended. It is important to mention that the ability to detect motor units decreases with the number of removed channels. Therefore, it is important to preserve as many EMG channels as possible during the recording session. Since monopolar recordings are prone to power-line interference contamination, it is recommended to apply a notch filter appropriate to the region where the recordings were made, e.g., 50 Hz (Europe, Asia, Pacific) or 60 Hz (USA). When power line noise is significant, higher harmonics (i.e., 100, 150, and 200 Hz) can also be removed. Once signals have been filtered and channels with poor signal quality have been removed, the next step is to select the signal segments that are going to be decomposed. It is recommended to limit the duration of the signal to approximately ≤50 s for light/low fatiguing contractions or shorter for high fatiguing contractions since motor unit action potentials change over long intervals, and this affects the accuracy of the decomposition algorithm to follow motor units throughout the duration of the contraction. If motor unit activity needs to be analyzed during long-duration contractions, it is recommended to compensate for the changes in the motor unit action potentials by, for example, decomposing signals in multiple overlapped segments and then tracking the units by matching discharge times by cross-correlation [20].

3.4 HDsEMG Motor Unit Decomposition

Once HDsEMG signals have been recorded and preprocessed, the next step is to run the motor unit decomposition algorithm. A summary of the full decomposition procedure (from the recording of HDsEMG signals to the extraction of motor unit spike trains) can be seen in Fig. 3. The high spatial information of multichannel EMG provides the possibility to use convolutive blind source separation to decompose the recorded signals into individual motor unit spike trains. This method and similar techniques have been recently proposed by a number of authors [7, 8, 34, 35] and extensively validated in several experimental conditions [19–21,

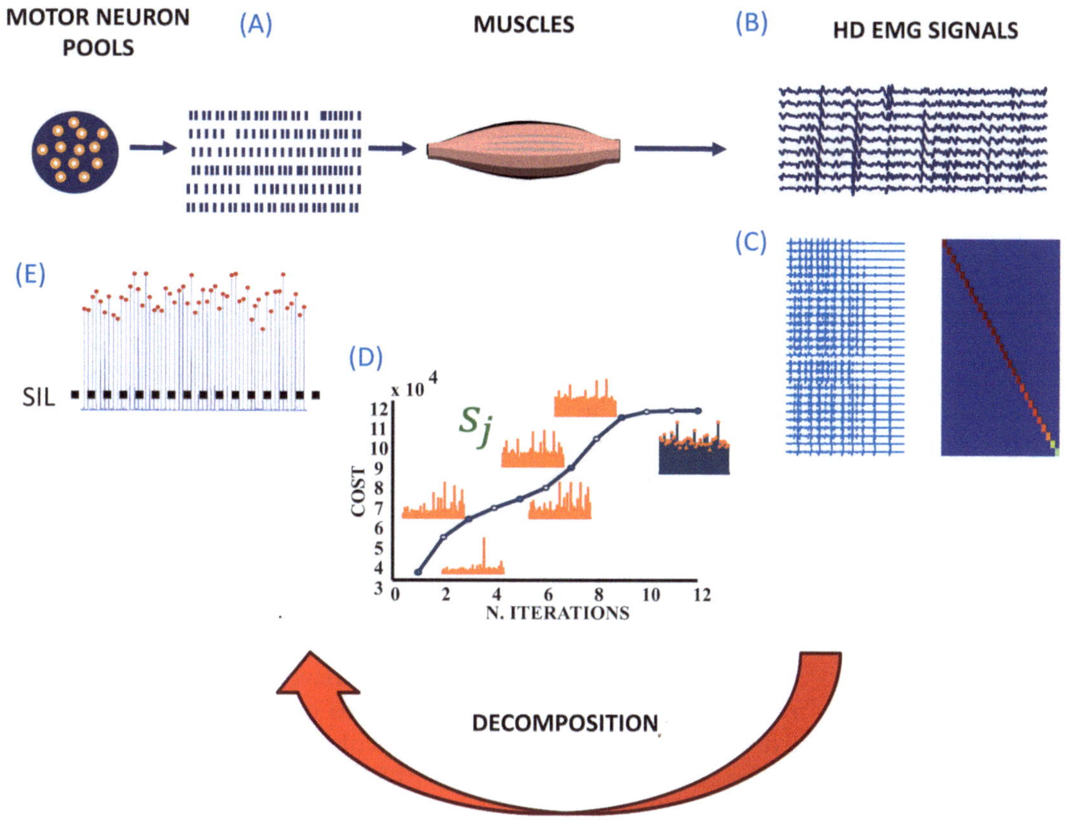

Fig. 3 Schematic representation of the convolutive blind source separation approach. Motoneuron pools generate multiple action potentials (**a**), which can be recorded from the muscle. Motor unit firings of single motor units need to be separated from the interference EMG (**b**) with blind source separation techniques. The HDsEMG signals are first extended to convert the convolutive model into an instantaneous one. A whitening transformation is then applied to the extended measurements to compensate for the spatial and temporal characteristics of the motor unit potential shape (**c**). Afterward, the solutions of the inverse problem are extracted by maximizing iteratively a specific cost function selected to identify sparse sources (**d**). Clustering methods and accuracy measures (i.e., silhouette measure, SIL) are then used to identify the discharge times of the decomposed motor unit (**e**)

27–29]. In general, since motor unit spike trains are inherently sparse, these methods iteratively optimize contrast functions to find solutions of the inverse model (recovering motor unit spike trains from the EMG signals) that maximize this statistical measure and accurately detect individual motor unit discharge patterns. For these reasons, this methodology is not entirely "blind," but it exploits specific statistical properties of the sources (motor unit spike trains). Before performing this procedure, the HDsEMG signals are first extended and then whitened. The extension is performed, including delayed copies of the original signals into the model and effectively transforming the original convolutive model into an instantaneous one. The whitening procedure is applied to the extended EMG measurements to decorrelate them and remove redundant information. In other words, the decorrelation of the extended measurements aims at compensating for the volume conductor filtering effects and providing a more selective spatial–temporal representation of the original signals. The motor unit action potentials in this transformed domain have a more localized representation over the muscle and narrower temporal support. This procedure simplifies the inversion of the mixing model and the separation of the individual motor unit spike trains. At each iteration, the optimization procedure is applied to the whitened extended measurements, starting from a suitable initial estimate and converging to a local maximum of the selected contrast function. The output of this decomposition step is a pulse train that represents the probability of motor unit discharges at specific time instants. In subsequent iterations, different methods can be used to avoid the convergence to the same sources (see, e.g., [8, 20]). Since the identification of pulse trains is based on very general statistical properties of the sources, there is no guarantee that they will belong to individual motor units or combinations of multiple motor units that were not separated correctly. For this reason, different measures of accuracy have been introduced on the result of the decomposition process to quantify the quality of the extraction and correctness of the identified motor unit spike trains; the pulse-to-noise ratio (PNR) [32] and the silhouette measure (SIL) [8]. These measures rely on the relative amplitude of the source spikes compared to the baseline noise. They have been demonstrated to be correlated to the correctness of the extracted motor unit spike trains by two-source validation methods (intramuscular motor unit decomposition vs. HDsEMG motor unit decomposition [8]). Specifically, a PNR > 30 dB and SIL > 0.90 suggest 90% accuracy of the decomposed spike trains. However, these procedures only provide a general quantification of the accuracy of the motor unit spike trains, and localized errors in the identification of the discharge times can still be present. For these reasons, manual editing by an experienced operator with knowledge of motor unit physiology should always be performed. Briefly, the editing process includes a visual assessment of the decomposed motor unit pulse trains and local optimizations of the decomposition extraction parameters based on excluding erroneous spikes or the inclusion of missed ones. The typical errors found after automatic decomposition procedures are missed discharges producing instantaneous discharge rates significantly lower than the average discharge rate or doublets resulting in instantaneous discharges significantly above the mean discharge rate [33]. It is important to note that decisions regarding the inclusion or exclusion of discharge times should be mainly based on the signal-to-noise ratio of the decomposed spikes before or after local optimization and not merely on the instantaneous discharge rate of the inspected

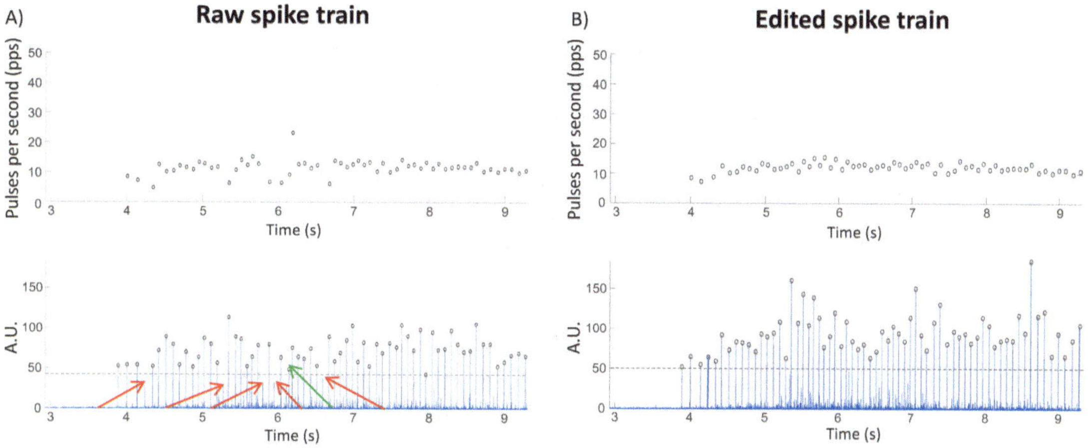

Fig. 4 Motor unit spike train editing. Motor unit spike train editing needs to be performed following the automatic decomposition results as missing firings and additional firings can provide unreliable spike train results. (**a**) Instantaneous firing rate (top of the figure) calculated from the spike train obtained from the automatic decomposition results (bottom of the figure). The red arrows represent missing firings while the green arrow represents an additional firing. These missing and additional discharges increase the variability of motor unit firing (top of the figure). (**b**) Following the addition of missing firings, removal of the additional firing, and recalculation of spike train results, the variability in instantaneous firing rate is reduced (top of the figure)

motor unit pulse train. The procedure for motor unit spike train editing and its effects on motor unit firing properties can be seen in Fig. 4.

3.5 Analysis of Motor Unit Firing Data

After the removal of inaccurate motor units and the editing of the firings of selected motor units, the next step is to assess the main statistical properties of the spike trains. Motor unit variables extracted from HDsEMG systems can be mainly divided into motor unit discharge/firing characteristics and muscle fiber properties. Mean discharge rate, discharge rate at recruitment/de-recruitment, peak discharge rate, discharge rate variability, and recruitment threshold can be regarded as discharge or firing characteristics, while conduction velocity of action potentials is usually regarded as a peripheral or muscle fiber property.

3.6 Motor Unit Firing Characteristics

Recruitment/de-recruitment thresholds, mean discharge rate, discharge rate at recruitment/de-recruitment, peak discharge rate, and discharge rate variability are the most common motor unit firing characteristics reported. All these motor unit firing variables can be quantified directly from the discharge times, inter-spike intervals (ISIs), instantaneous firings (inverse of the ISI: 1/ISI), or smoothed signal. Recruitment/de-recruitment thresholds are defined as the force (%MVC, Newtons (N) or Nm) or joint angle at which a motor unit begins and ends discharging action potentials, repetitively, and this parameter provides information about the type of motor unit under analysis (i.e., lower threshold or higher threshold motor unit). According to the onion skin phenomenon [36], low-threshold motor units are recruited early during a contraction and present higher firing rates than high-threshold motor units. Therefore, motor units with different thresholds can have different discharge rates. It is important to report recruitment thresholds since the assessment of different populations of motor units (low vs. high threshold), i.e., during longitudinal studies, can provide an equivocal interpretation of firing rate results. Moreover, recent studies assessing different conditions such as pain [15, 19] and fatigue [37] have shown that motor units of different thresholds can manifest diverse effects in discharge rate. Therefore, it is recommended to analyze effects across motor units of multiple thresholds at diverse forces. As a note of caution, it is also important to consider that the recruitment threshold can be altered by changing contraction velocity (i.e., rate of force development speed). Therefore, contraction speed should be maintained across testing sessions or contractions when we want to assess the effects of a particular intervention on the recruitment threshold (i.e., by using increasing force ramps with a fixed rate of force development of 10%MVC/s). The mean discharge rate is another parameter of interest in motor unit studies. This variable provides an estimation of the neural drive received by the muscle (since the discharge rate increases together with force). If this parameter is reported, it is important to mention the window in which the mean firing rate was calculated (i.e., ascending force ramp or plateau in force). It is recommended that this parameter is calculated during sustained contractions (or the sustained phase of a contraction), where force or muscle activity is constant, in order to avoid sharp increases/decreases in discharge rate that can affect the estimation. In terms of discharge rate at recruitment and de-recruitment, studies classically average the first and last few firings (i.e., six [37, 38]) to assess the firing frequency when a unit was recruited and de-recruited. Peak discharge rate is usually calculated during ramped contractions without a sustained phase (i.e., contractions that increase and decrease in force), i.e., [37]. Peak discharge rate may be quantified as (1) the average rate over ≤6 discharges, (2) the average of the five shortest inter-spike intervals, or (3) estimated from a function fitted to the inter-spike intervals. Finally, discharge rate variability estimates fluctuations in firing rate over time, and, like any firing characteristic, it is important to indicate when variability was calculated (i.e., ascending ramp or plateau in force). Variability can be calculated directly from discharge times, ISIs, mean of inverse ISI (1/ISI), or a smoothed signal. If the latter, the filter or windowing used on the time series of ISIs should be reported. The most popular way to report discharge rate variability is by reporting the coefficient of variation for the ISI interval (SD ISI/mean ISI × 100); however, this could also be calculated from variations in instantaneous discharge rate (i.e.,

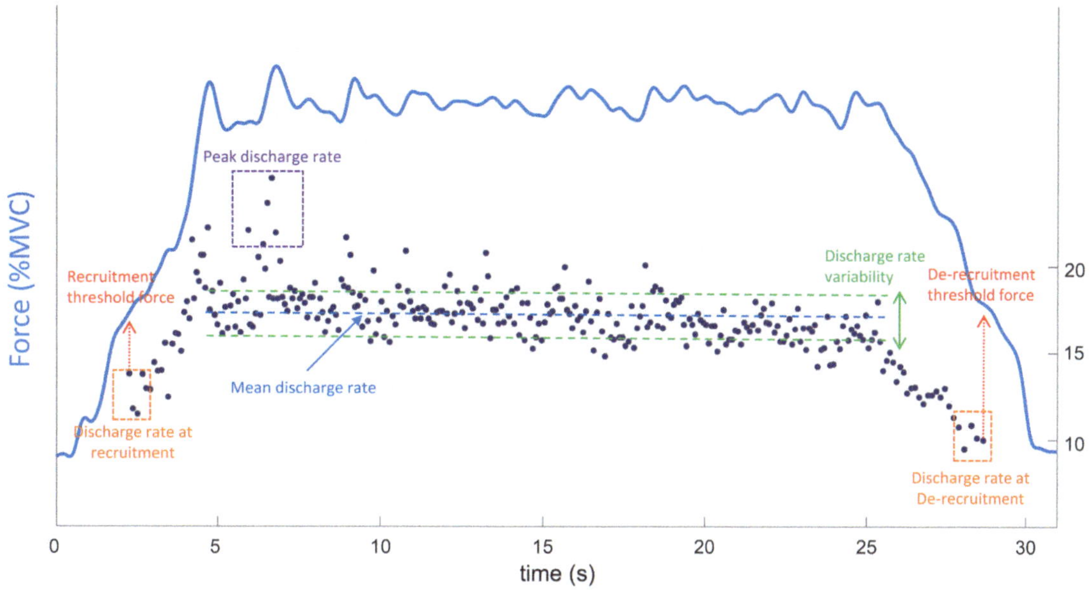

Fig. 5 Calculation of motor unit firing parameters. The most common motor unit firing variables can be seen in the figure. Recruitment and de-recruitment threshold force (red) represents the force/torque (in % of the maximum voluntary contraction or N/Nm) at which a motor unit begins and ends firing action potentials. Discharge rate at recruitment and de-recruitment (orange) represents the average discharge rate calculated from the first and last few firings (i.e., 6). Mean discharge rate (blue dashed line) represents the average firing rate on the sustained part of the contraction. Discharge rate variability (green dashed lines) represents the fluctuations in firing frequency and can be calculated as the standard deviation or coefficient of variation in instantaneous firing rate. Finally, the peak discharge rate is calculated from a few firings (i.e., 6) around the highest discharge rate. Note that the peak discharge rate was presented here for graphical purposes only as this parameter should not be quantified during sustained contractions

standard deviation of 1/ISI). Discharge rate variability should also be calculated during a sustained steady contraction when force or muscle activity (EMG) is constant. An additional parameter that should be reported is the number of identified motor units since this provides information about the number of units included in all calculations. Nevertheless, it is important to note that the number of identified motor units does not represent the number of recruited units during a task. Since HDsEMG systems are only able to assess a subset of the total number of active motor units and the total number of sampled units can vary across different contractions, it is not possible to infer changes in recruitment with HDsEMG motor unit decomposition. Figure 5 depicts most of the motor unit firing parameters that can be calculated from motor unit recordings.

3.7 Muscle Fiber Properties

The spatial representation of motor unit action potentials from the HDsEMG grid electrode can be used to quantify action potential propagation speed along the muscle fibers (muscle fiber conduction velocity). When researchers want to assess muscle fiber conduction velocity, the most important step prior to processing HDsEMG signals is aligning the electrode grid with the muscle fibers since any rotation of the electrode with respect to the muscle fibers will impair the ability to detect the correct propagation of action potentials. Techniques such as ultrasound imaging can help to visualize the orientation of the fascicles; however, the most used technique to align HDsEMG electrodes is by using a dry linear electrode array prior to the placement of the wet-adhesive HDsEMG electrode [22]. Once the action potential propagation is visualized clearly, the skin is then marked to place the adhesive electrode. It is important to consider that it is only possible to record reliable estimates of muscle fiber conduction velocity from muscles that have fibers parallel to the skin (i.e., biceps brachii, vastus medialis, vastus lateralis, pectoralis major). Following the recording of HDsEMG signals and motor unit decomposition, motor unit discharge times can be used to trigger surface EMG signals via spike-triggered averaging [39]. Once the 2D action potential waveforms from single motor units have been identified, it is important to differentiate (single or double differential) the signal (if it was recorded in monopolar mode) to observe clear action potential propagation. After this differentiation, conduction velocity can be quantified with both time [40] and frequency–domain [41] methods (as reviewed by [42]). Simply put, conduction velocity is estimated by calculating the time that it takes for an action potential to propagate from one electrode to the other (conduction velocity = interelectrode distance/time); for this reason, it is important to select channels with the most similar action potential waveforms. This is usually performed by assessing differences between channels with cross-correlation methods [42]. Channels with similar action potential waveforms and high cross-correlation coefficients (i.e., >0.7) are included in the calculation of conduction velocity. Three single-differential channels is the minimum number of channels that can be used to calculate conduction velocity; however, a larger number of selected channels will increase the reliability of conduction velocity estimations. Once the channels with the most similar waveforms have been identified and conduction velocity has been calculated, it is important to check if conduction velocity values are within the physiological range. Normally, conduction velocity values should be between 2 and 8 m/s [43]. Any value outside this range should be discarded from the conduction velocity calculations. An example of a 2D motor unit action potential representation from the vastus medialis muscle (13 × 5, 8 mm IED, 64-channel electrode grid) and subsequent conduction velocity calculation can be seen in Fig. 2. In this figure, the innervation zone (electrode area where action potential propagation direction is inverted) can be clearly appreciated.

3.8 Motor Unit Longitudinal Tracking

One of the main advantages of HDsEMG methods is their ability to track multiple motor units across different sessions. This feature is possible since HDsEMG provides a 2D spatial sampling of motor unit action potentials over the skin plane and a large number of channels allow precise discrimination between different motor units [24]. This motor unit spatial "signature" can be used for longitudinal tracking since it can be detected in an almost identical manner once the electrode grid is placed in a similar location over the skin and the joint position has not changed between recording sessions. The ability to accurately track the same motor unit increases with the number of channels used to record HDsEMG signals since the probability that two motor units detected in different sessions present exactly the same spatial action potentials over tens of channels is negligible. The normalized cross-correlation between motor unit action potential waveforms can be used as a measure of similarity to identify the matched motor units [44]. When employing this technique, only motor units with a normalized cross-correlation higher than 0.8 are identified as the same. If multiple matches with a cross-correlation >0.8 are found, the pair of motor units with the highest cross-correlation should be used. Waveforms with time support of 50 ms are commonly used for computing similarity measures between multichannel action potentials. Besides using cross-correlation analysis to confirm the matching, it is also very important to inspect the matched action potentials visually to ensure that the motor units were tracked correctly (i.e., overlapping action potentials recorded from two different sessions on top of each other; see Fig. 6). Studies have shown that this technique could be used to reliably track motor units following a half marathon [17], 2 weeks of high-intensity interval and moderate-intensity continuous training [10, 22], and 4 weeks of resistance training [21]. Although the application of this technique for training interventions showing significant changes in muscle morphology is yet to be determined, it is likely that this tracking procedure would still be possible if changes in muscle volume/architecture were not too substantial. An example of the tracking procedure following an 8-week resistance training intervention in a 65-year-old individual with type II diabetes can be seen in Fig. 6; these recordings were made 12 weeks apart.

4 Conclusion

The emergence of HDsEMG techniques has improved our understanding of motor unit function in both health and disease. This method has gained popularity during the last 5 years given that it is noninvasive and relatively easier to use in comparison to intramuscular methods. Despite these advantages, there are a number of factors that need to be considered when performing experiments with HDsEMG. This guideline aimed to provide recommendations

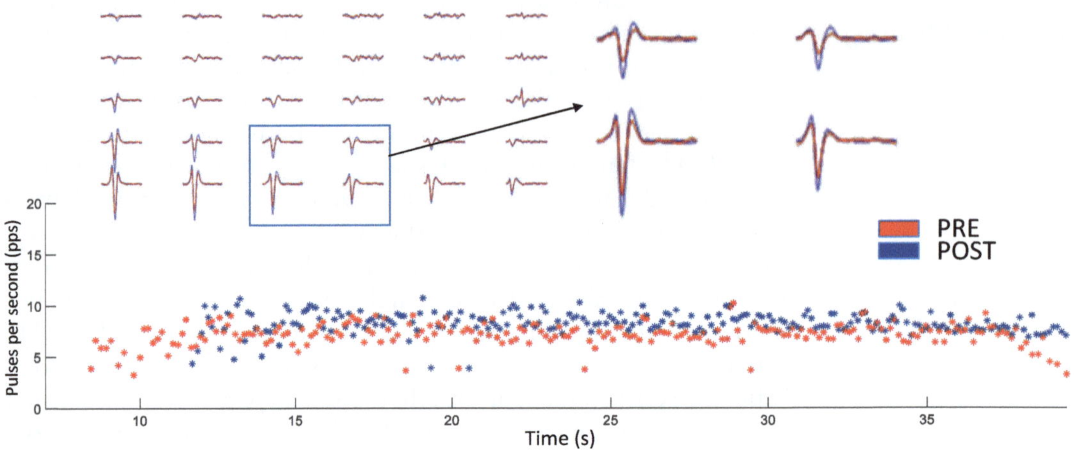

Fig. 6 Motor unit tracking procedure. Vastus lateralis motor unit action potentials, obtained from HDsEMG recordings, are tracked before (red) and after (blue) an 8-week resistance training intervention (65-year-old individual with type II diabetes). Instantaneous firing rate results are presented below the figure. Four channels are zoomed (square) to show the similarity in action potential waveforms. The cross-correlation coefficient for this unit was 0.92, which confirms the reliability of the tracking

on how to record, process, and analyze HDsEMG signals. In addition, we also mentioned and provided potential solutions to the most common problems observed with HDsEMG recordings during experimental sessions. This guideline does not aim to replace formal training with HDsEMG recordings; however, we hope that the information provided can help researchers interested in implementing this method to obtain accurate and physiologically meaningful results.

5 Notes

5.1 Signal Saturation

In certain cases, amplifiers with limited resolution (i.e., A/D converters of 12 bits or lower) could show signal saturation during high-intensity contractions. Moreover, the level of saturation can vary from participant to participant. In these cases, it is recommended to perform brief high-intensity contractions before the experiment and adjust to a gain that provides the best signal-to-noise ratio without saturating (clipping) the signals.

5.2 Electrode Repositioning

HDsEMG shows excellent reliability when used across different testing sessions [45]; however, to achieve this, it is important that the electrodes are positioned in a similar location across sessions. If possible, it is recommended to mark the electrode area with an indelible pen. Otherwise, researchers could use a transparent sheet to record the position of the electrode and/or use precise anatomical landmarks and keep these consistent across testing sessions.

5.3 Short Circuits Between Channels

One common problem with HDsEMG signals is the generation of short circuits between channels due to the excessive application of conductive paste, which can create undesirable bridges between channels. This issue is more common in HDsEMG electrodes with short interelectrode distances (i.e., 2.5 mm). If large noise is observed between adjacent channels, it is recommended to remove the electrode grid and apply a smaller amount of conductive paste.

5.4 Disconnected Channels

Some EMG channels could be disconnected when HDsEMG electrode grids are placed over muscles which show large changes in geometry (i.e., biceps brachii), during contractions that generate displacement of the electrode over the skin (i.e., dynamic contractions) and when an insufficient amount of conductive paste (or water in the case of dry arrays) is applied. In the first two cases, it is recommended to put adhesive tape over the electrode (i.e., surgical adhesive tape) in order to prevent detachment and maximize electrode–skin contact. For the latter case, it is recommended to dismount the electrode and insert more conductive paste into the electrode cavities.

5.5 Long Signals: Nonstationarity/Stationarity

The decomposition of HDsEMG signals for the identification of motor units requires that EMG signals maintain a certain level of stationarity. During long contractions, EMG signal characteristics vary during the contraction, which makes the tracking of the same motor unit difficult. It is recommended to decompose signals with an approximate duration of <50 s. If it is required to follow motor units during contractions of a longer duration, it is recommended to decompose the contraction in multiple overlapped segments and match firings by cross-correlation [20] or update the filters (motor unit action potential profiles) used to identify the motor units during the contraction [46].

5.6 Population Under Study

The population under study is an important factor to consider. Most of the studies using HDsEMG motor unit decomposition specifically assess young, male, healthy, and physically active participants. One reason for this is related to the low-pass filter effect of subcutaneous tissue and skin on EMG signals [47]. Individuals with greater adipose tissue thickness show "blurred" surface EMG motor unit spikes, which can significantly affect decomposition data. Young, physically active, and healthy males have less adipose tissue thickness and greater muscle cross-sectional area compared to females and older individuals. Therefore, the motor unit yield obtained in these individuals is significantly greater than in other populations. Despite this limitation, there are a number of studies showing physiologically relevant results in women, older cohorts, and patients with neuromuscular/musculoskeletal disorders [48, 49].

References

1. Luca CJD (1997) The use of surface electromyography in biomechanics. J Appl Biomech 13(2):135
2. Farina D, Merletti R, Enoka RM (2004) The extraction of neural strategies from the surface EMG. J Appl Physiol (1985) 96(4): 1486–1495
3. Negro F, Holobar A, Farina D (2009) Fluctuations in isometric muscle force can be described by one linear projection of low-frequency components of motor unit discharge rates. J Physiol 587(Pt 24):5925–5938
4. Duchateau J, Enoka RM (2011) Human motor unit recordings: origins and insight into the integrated motor system. Brain Res 1409:42–61
5. Vieira TM, Botter A (2021) The accurate assessment of muscle excitation requires the detection of multiple surface electromyograms. Exerc Sport Sci Rev 49(1):23–34
6. Holobar A et al (2009) Estimating motor unit discharge patterns from high-density surface electromyogram. Clin Neurophysiol 120(3): 551–562
7. Holobar A, Zazula D (2007) Multichannel blind source separation using convolution kernel compensation. IEEE Trans Signal Process 55(9):4487–4496
8. Negro F et al (2016) Multi-channel intramuscular and surface EMG decomposition by convolutive blind source separation. J Neural Eng 13(2):026027
9. Czesnik D et al (2015) Increased HCN channel driven inward rectification in benign cramp fasciculation syndrome. Brain 138(Pt 11): 3168–3179
10. Martinez-Valdes E et al (2017) Differential motor unit changes after endurance or high-intensity interval training. Med Sci Sports Exerc 49(6):1126–1136
11. Martinez-Valdes E et al (2017) Tracking motor units longitudinally across experimental sessions with high-density surface electromyography. J Physiol 595(5):1479–1496
12. Farina D, Negro F (2015) Common synaptic input to motor neurons, motor unit synchronization, and force control. Exerc Sport Sci Rev 43(1):23–33
13. Negro F, Yavuz US, Farina D (2016) The human motor neuron pools receive a dominant slow-varying common synaptic input. J Physiol 594(19):5491–5505
14. Farina D et al (2016) Principles of motor unit physiology evolve with advances in technology. Physiology (Bethesda) 31(2):83–94
15. Martinez-Valdes E et al (2021) Pain-induced changes in motor unit discharge depend on recruitment threshold and contraction speed. J Appl Physiol (1985) 131(4):1260–1271
16. Cudicio A et al (2021) The force-generation capacity of the tibialis anterior muscle at different muscle-tendon lengths depends on its motor unit contractile properties. Eur J Appl Physiol 122:317
17. Cogliati M et al (2020) Half marathon induces changes in central control and peripheral properties of individual motor units in master athletes. J Electromyogr Kinesiol 55:102472
18. Negro F, Orizio C (2017) Robust estimation of average twitch contraction forces of populations of motor units in humans. J Electromyogr Kinesiol 37:132–140
19. Martinez-Valdes E et al (2020) Divergent response of low- versus high-threshold motor units to experimental muscle pain. J Physiol 598(11):2093–2108
20. Martinez-Valdes E et al (2020) Inability to increase the neural drive to muscle is associated with task failure during submaximal contractions. J Neurophysiol 124(4):1110–1121
21. Del Vecchio A et al (2019) The increase in muscle force after 4 weeks of strength training is mediated by adaptations in motor unit recruitment and rate coding. J Physiol 597(7):1873–1887
22. Martinez-Valdes E et al (2018) Early motor unit conduction velocity changes to high-intensity interval training versus continuous training. Med Sci Sports Exerc 50(11): 2339–2350
23. Negro F et al (2020) Impaired firing behavior of individually tracked paretic motor units during fatiguing contractions of the dorsiflexors and functional implications post stroke. Front Neurol 11:540893
24. Farina D et al (2008) Detecting the unique representation of motor-unit action potentials in the surface electromyogram. J Neurophysiol 100(3):1223–1233
25. Nishihara K et al (2013) Investigation of innervation zone shift with continuous dynamic muscle contraction. Comput Math Methods Med 2013:174342
26. Merletti R, Holobar A, Farina D (2008) Analysis of motor units with high-density surface electromyography. J Electromyogr Kinesiol 18(6):879–890
27. Del Vecchio A et al (2018) Central nervous system modulates the neuromechanical delay in a broad range for the control of muscle

force. J Appl Physiol (1985) 125(5): 1404–1410
28. Del Vecchio A et al (2019) You are as fast as your motor neurons: speed of recruitment and maximal discharge of motor neurons determine the maximal rate of force development in humans. J Physiol 597(9):2445–2456
29. Oliveira AS, Negro F (2021) Neural control of matched motor units during muscle shortening and lengthening at increasing velocities. J Appl Physiol (1985) 130:1798
30. Glaser V, Holobar A (2019) Motor unit identification from high-density surface electromyograms in repeated dynamic muscle contractions. IEEE Trans Neural Syst Rehabil Eng 27(1):66–75
31. Yokoyama H et al (2021) Robust identification of motor unit discharges from high-density surface EMG in dynamic muscle contractions of the tibialis anterior. IEEE Access 9:123901–123911
32. Holobar A, Minetto MA, Farina D (2014) Accurate identification of motor unit discharge patterns from high-density surface EMG and validation with a novel signal-based performance metric. J Neural Eng 11(1):016008
33. Del Vecchio A et al (2020) Tutorial: analysis of motor unit discharge characteristics from high-density surface EMG signals. J Electromyogr Kinesiol 53:102426
34. Chen M, Zhou P (2016) A novel framework based on FastICA for high density surface EMG decomposition. IEEE Trans Neural Syst Rehabil Eng 24(1):117–127
35. Dai C, Hu X (2019) Independent component analysis based algorithms for high-density electromyogram decomposition: experimental evaluation of upper extremity muscles. Comput Biol Med 108:42–48
36. De Luca CJ et al (1982) Control scheme governing concurrently active human motor units during voluntary contractions. J Physiol 329: 129–142
37. Farina D et al (2009) Adjustments differ among low-threshold motor units during intermittent, isometric contractions. J Neurophysiol 101(1):350–359
38. Martinez-Valdes E et al (2018) Surface electromyographic amplitude does not identify differences in neural drive to synergistic muscles. J Appl Physiol (1985) 124(4):1071–1079
39. Farina D et al (2002) Assessment of single motor unit conduction velocity during sustained contractions of the tibialis anterior muscle with advanced spike triggered averaging. J Neurosci Methods 115(1):1–12
40. Farina D, Fortunato E, Merletti R (2000) Noninvasive estimation of motor unit conduction velocity distribution using linear electrode arrays. IEEE Trans Biomed Eng 47(3): 380–388
41. Farina D, Negro F (2007) Estimation of muscle fiber conduction velocity with a spectral multidip approach. IEEE Trans Biomed Eng 54(9):1583–1589
42. Farina D, Merletti R (2004) Methods for estimating muscle fibre conduction velocity from surface electromyographic signals. Med Biol Eng Comput 42(4):432–445
43. Beretta-Piccoli M et al (2019) Reliability of surface electromyography in estimating muscle fiber conduction velocity: a systematic review. J Electromyogr Kinesiol 48:53–68
44. Vecchio AD, Farina D (2019) Interfacing the neural output of the spinal cord: robust and reliable longitudinal identification of motor neurons in humans. J Neural Eng 17(1): 016003
45. Martinez-Valdes E et al (2016) High-density surface electromyography provides reliable estimates of motor unit behavior. Clin Neurophysiol 127(6):2534–2541
46. Rossato J et al (2022) Less common synaptic input between muscles from the same group allows for more flexible coordination strategies during a fatiguing task. J Neurophysiol 127(2): 421–433
47. Holobar A, Farina D (2014) Blind source identification from the multichannel surface electromyogram. Physiol Meas 35(7):R143–R165
48. Watanabe K et al (2016) Age-related changes in motor unit firing pattern of vastus lateralis muscle during low-moderate contraction. Age (Dordr) 38(3):48
49. Miller LC et al (2014) High-density surface EMG decomposition allows for recording of motor unit discharge from proximal and distal flexion synergy muscles simultaneously in individuals with stroke. Annu Int Conf IEEE Eng Med Biol Soc 2014:5340–5344

Chapter 7

Neuromuscular Function: Intramuscular Electromyography

Mathew Piasecki and Daniel W. Stashuk

Abstract

Intramuscular electromyography (iEMG) enables the detection of propagating action potentials in contracting skeletal muscles and has provided a wealth of information on the neural input to muscle and its response to intervention for research purposes. As an indwelling electrode, it is able to record individual motor unit potentials (MUPs) in a range of muscles and muscle depths, recording close to the signal origin and unconstrained by the filtering effects of intramuscular and subcutaneous tissue. This proximity of electrode to muscle fiber provides additional information with regard to the de- and repolarization of individual muscle fibers. This chapter will focus on the recording, decomposition, and interpretation of iEMG signals from human skeletal muscle using concentric needle electrodes.

Key words Electromyography, Motor unit, Neuromuscular junction, Neuromuscular

1 Introduction

A motor unit (MU) comprises a motor neuron cell body, its axon and axonal branches, and all muscle fibers innervated via neuromuscular junctions (NMJ). Motor units are responsible for transforming incoming synaptic activity into muscle contractions which generate movement. As such, the investigation of the MU via various methods of electromyography (EMG) in clinical research has proven essential in the diagnosis and treatment of a number of neuromuscular disorders [1] and has also proven useful in nonclinical human research studies with a focus on aging [2–6], exercise [7–13], and sex [14].

Motor neuron activation results in a concurrent activation of all muscle fibers within the MU (MU fibers). The exchange of ions across muscle fiber membranes, as a result of propagating action potentials (AP), results in fluctuations in localized extracellular voltages, and these can be detected in vivo during muscle contractions with appropriately placed EMG electrodes. Surface EMG (sEMG) has been widely used and typically involves the placing of a single EMG channel over the belly of a muscle of interest and

Philip J. Atherton and Daniel J. Wilkinson (eds.), *Neuromuscular Assessments of Form and Function*, Neuromethods, vol. 204, https://doi.org/10.1007/978-1-0716-3315-1_7, © Springer Science+Business Media, LLC, part of Springer Nature 2023

measuring features of the resultant signal, such as EMG amplitude (root mean square) and frequency [15]. However, these features are limited by the influence of the distance between the recording electrode and the muscle fiber origins of the signal due to volume conduction through the skin and subcutaneous tissue, and these limitations should be noted with regard to individual study requirements [16]. Although conventional sEMG has been surpassed by the use of high-density sEMG (HD-sEMG, see Chapter 6), the inability to accurately sample MUs from deeper muscles or deeper regions of muscles may limit its use in certain cohorts where subcutaneous tissue is greatest (e.g., thigh muscles of the elderly, diabetic, or obese). These limitations can be overcome with the use of intramuscular EMG (iEMG).

The recording of individual MU potentials (MUP) with iEMG is a minimally invasive technique that is commonly applied using concentric needle electrodes. Fine wire electrodes operate in a similar fashion and in some cases offer the advantage of reducing participant discomfort, yet the inability to accurately position or reposition the wire electrodes in-between contractions may render it less useful in routine clinical practice and certain research study designs. This chapter will focus on the practical application of iEMG using concentric needles, the materials required, and the parameters attainable and signal processing methods required to obtain them.

2 Materials

2.1 Concentric Needle Electrode

A concentric needle electrode is based on a disposable hollow needle with a beveled tip measuring approximately 0.3–0.45 mm in diameter (24–28 gauge), smaller than those used for drawing blood which measure 0.6–0.8 mm (21–23 gauge). The hollow needle contains an insulated wire which is exposed at the bevel and serves as the active recording surface, and this is referenced to the shaft of the needle. This close proximity between the active and reference sites minimizes noise and interference common to both sites and reduces contributions from fibers distant from the recording surfaces. It also enables the recording of MUPs during extremely low contraction intensities (<2% MVC) in some muscles, when appropriately positioned. It is estimated the concentric needle records from fibers within 2.5 mm of the needle tip [17], and as a rigid structure, the needle can be repositioned in between or during contractions at a range of muscle depths. This may be viewed as important given features of recorded MUPs, and fiber-type composition differs from deep to superficial regions of some muscle groups [10, 18]. To accommodate this depth range, or to sample from muscles covered by substantial subcutaneous tissue (e.g., from elderly quadriceps, obese participants), a range of needle lengths (up to 100 mm) are commercially available and should be selected to suit a particular cohort or study. A reference electrode should be placed close to the site of needle insertion, but not over the muscle (e.g., over the patella with vasti muscle needle insertion).

2.2 Acquisition Equipment

EMG outputs are almost exclusively viewed, analyzed, and interpreted as numerical sequences, more commonly referred to as digital signals. The voltage detected by a needle electrode is an analog signal, being *analogous* to the flow of ions throughout the extracellular volume around fiber membranes which determine the signal recorded. Therefore, all applications of iEMG acquisition require an analog-to-digital convertor, and many are commercially available. Crucial in determining the selection of acquisition equipment is the sampling frequency capability, which should be a minimum of 25 KHz and ideally 50 KHz for iEMG. The extracellular voltage signals generated by contracting muscle fibers are also very small (microvolts, μV) and must be amplified to be viewed online and analyzed offline. Commercial signal amplifiers are widely available and enable direct amplification at a desired setting, in addition to offering variable signal filtering capabilities.

The signals detected by a needle electrode are composed of complex superpositions of sinusoidal wave components of mixed frequencies, including those that are nonphysiological in origin such as those from power supplies (50 or 60 Hz). Filter settings enable the select amplification of signal frequency components most likely to be physiological in origin; high-pass filters reduce amplification of low-frequency signal components, and low-pass filters reduce amplification of high-frequency signal components, with band-pass filtering allowing a combination of the two. Needle EMG band-pass filter settings typically range between 10–30 Hz and 10–20 KHz. However, in certain research settings with additional electrical equipment acting to increase noise, the bandwidth may need to be narrowed (e.g., increasing low-frequency and/or decreasing high-frequency settings). Audio speakers may also be used during acquisition to convert the signal to an audio output, providing a useful aid for positioning the needle tip close to the muscle fiber signal origins and, therefore, detecting suitable MUPs.

3 Methods

3.1 Needle Insertion

Among the many benefits of iEMG is the ability to sample from multiple muscle regions of varying depths without incurring reductions in signal quality. To date, MUP characteristics using concentric needle EMG have been reported in healthy volunteers in the vastus lateralis [3, 11, 14, 19], tibialis anterior [4, 9, 13], biceps brachii [8, 11], soleus [5], trapezius [20], and multiple hand

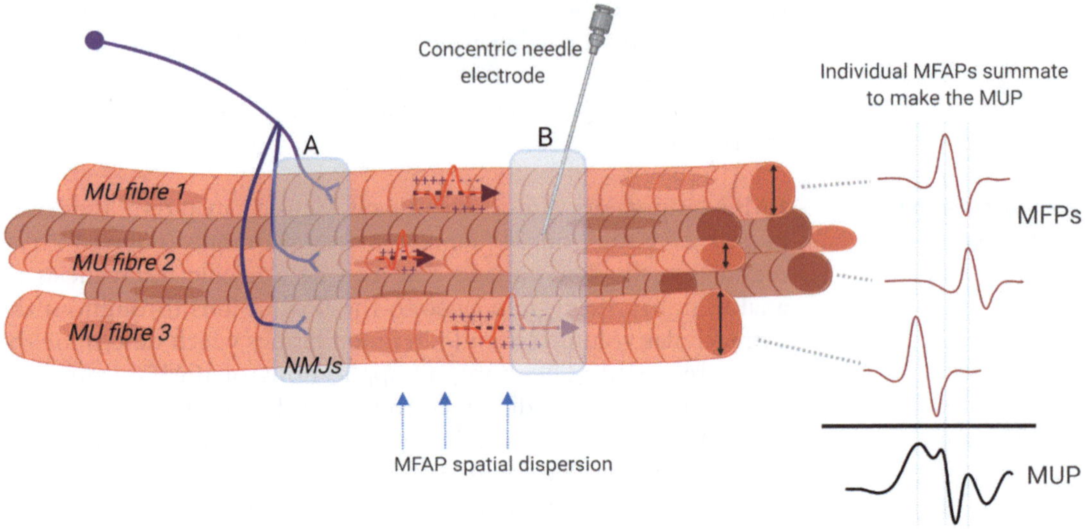

Fig. 1 Schematic representation of the recording of a MUP with contributions from three MU fibers. Muscle fiber action potentials (MFAP) originate at the NMJ (**a**) and propagate along muscle fibres. Corresponding muscle fiber potentials (MFPs) may be distinctly detected by the electrode (**b**) due to the spatial dispersion and variability in detection time of MFPs from MU fibers 1, 2, and 3. This variability can occur as a result of variable NMJ locations and transmission times and variable MFAP conduction times, which will influence the final MUP shape. Abbreviations: MUP motor unit potential, MU motor unit, MFAP muscle fiber action potential, NMJ neuromuscular junction, MFP muscle fiber potential

muscles [21, 22]. Although muscle selection will be study specific, the location of the needle insertion site relative to the estimated location of the NMJ is an important factor, particularly in larger muscles. If the needle tip is positioned distal or proximal to the NMJ, contributions from action potentials propagating along individual muscle fibers are more likely to be distinctly detected by the recording electrode, due to differences in NMJ locations and transmission times and in conduction times along individual fibers (*see* Fig. 1). Therefore, it can be beneficial to record MUPs as close to their point of origin as possible (e.g., around the NMJ). This location can be crudely estimated via searching for the muscle motor point using low-intensity percutaneous electrical stimulation [23] or observing the direction of propagating action potentials with surface linear array electrodes [24].

As the needle insertion is invasive, the skin around the insertion site should be cleaned with appropriate skin-cleansing materials such as alcohol, and the person inserting and manipulating the needle should wear protective gloves. The direction of the needle relevant to the skin surface is entirely subject-specific; however, it is wise to aim for a relatively large volume of muscle (e.g., muscle belly), particularly in small muscles such as those of the hand. If simultaneous surface EMG recordings and spike-triggered averaging are desired, such as that used for motor unit number estimates

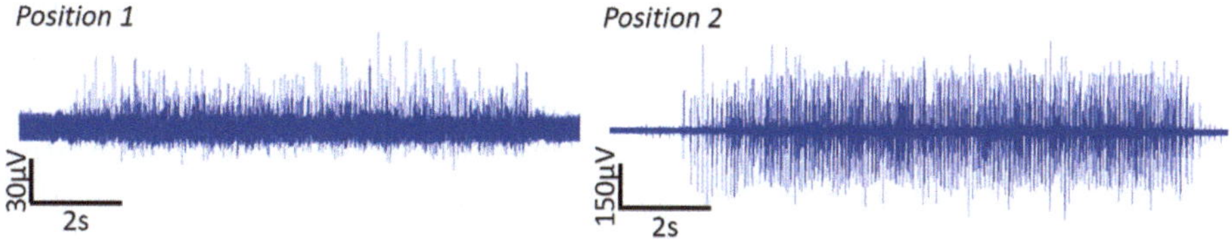

Fig. 2 Example iEMG data from the vastus lateralis of a 23-year-old female, obtained during two knee-extensor contractions at a force level equivalent to 10% of maximum voluntary contraction. In needle position 1, there are visible spikes with a poor signal-to-noise ratio. Needle adjustment to position 2, including altering depth and direction closer to the signal origin, improved the signal-to-noise ratio resulting in more visible spikes. Data visualized via Spike2, Cambridge Electronic Designs. Abbreviations: µV microvolts, s seconds

(MUNE, see [25] for review), the needle should be inserted at a slight angle to place the needle tip beneath the surface electrode, which should be placed over the motor point. Once positioned, the participant should produce a slight contraction and the resulting iEMG signal observed to estimate its signal to noise ratio (SNR). Where this is poor, the needle should be repositioned to improve the SNR (*see* Fig. 2). Audio feedback of the signal can be incredibly useful in quickly selecting a suitable needle position, as the amplitude and pitch frequency of the sound will increase with decreases in distance between the recording electrode and signal origin.

The following details an example protocol for iEMG-acquired data from the vastus lateralis during isometric knee extensor contractions. The maximum voluntary isometric contraction of the participant should be determined prior to needled insertion, under identical experimental conditions as the intended iEMG will be acquired (i.e., participant position, leg angle, force recording). Following this, visual force target lines should be displayed, and the participant is instructed to reach the target line and hold as close to it as possible. The level of contraction is dependent upon the study, and it should be acknowledged that higher contractions (>50%) may become uncomfortable and generate signals with a higher pulse per second (PPS) and are more difficult to successfully decompose (*see* **Note 4**). Signals should be inspected during contractions and needle adjusted where necessary (*see* Fig. 2, **Note 2**). Following data acquisition, the needle should be removed, and as soon as possible, all data should be saved for offline analysis. In many cases, iEMG signals may need to be exported, and it is advised only the "contraction segments" are exported to minimize file size. Following decompositions and MUP/MUPT extraction, data may be interpreted based on a selection of parameters detailed below.

3.1.1 Signal Decomposition

A MUPT is a set of MUPs detected across repeated activations of the same MU. An iEMG signal is composed of several MUPTs, representing the individual activities of concurrently active MUs, recorded during a single contraction. Decomposition of an iEMG signal extracts individual MUP trains (MUPTs) from the composite signal. Computationally, the decomposition of an iEMG signal is a difficult task which relies heavily on signal quality, which is dependent on suitable initial needle positioning (ensures a high signal to noise ratio, SNR; *see* **Note 1**), minimal needle movement (maintains signal to noise ratio and MUP shapes, *see* **Note 2**), and contraction intensity (determines the complexity of the iEMG, i.e., the number of MUPTs comprising the iEMG signal) and length (ensures sufficient numbers of MUP occurrences per train). Several systems are available with some providing online outputs during data acquisition, which are more favored in clinical/diagnostic settings (e.g., Nicolet Viking Quest, Cadwell Sierra II Wedge EMG). Typically, for reliable MUP identification, a contraction lasting a minimum of 5 s should be used. Based on a firing rate of 5–10 Hz, this would produce 25–50 MUPs in a MUPT. Although shorter durations are possible, a MUPT that is too short (too few MUPs) will preclude several measures which rely on the variability of consecutive observations (see below and [26]). Signal decomposition is an efficient time-saving tool, yet it is only the first step in data interpretation, and all individual MUPs and their corresponding MUP trains should be visually inspected and edited where necessary.

3.1.2 iEMG Parameters/Features

Individual motor unit potentials and their corresponding MUP trains can provide a wealth of information on the structure and function of MUs. These can be separated into features of the MUP template (i.e., a MUP that reflects the set of MUPs comprising a MUPT) and features separately derived from the MUP train, i.e., from consecutive observations of the MUPs generated by the same MU. As many parameters are sensitive to the contraction level at which they were recorded, all recordings and outputs from iEMG should be reported alongside an estimate of contraction level or signal intensity.

MUP amplitude is the difference between the negative and positive peaks of the MUP and is reported in μV. MUP duration is measured between the onset and end of the MUP and is reported in ms (*see* Fig. 3). MUP area is the integral of the absolute value of the MUP values between the onset and end of the MUP, multiplied by the sampling time interval and reported in units of μVms. MUP area, amplitude, and duration are all related to the number and size of MU fibers contributing to the potential, and the area and duration also reflect the temporal dispersion of fiber APs. MUP complexity may be reported via the number of phases or turns. Phases are defined as the number of baseline crossings plus one within the MUP duration, with a minimal amplitude change (20 μV) to avoid excessively high counts from fluctuations around the baseline. Similarly, turns are quantified as the number of significant slope reversals (+20 μV) within the MUP duration (*see* Fig. 3). Increased MUP complexity reflects greater temporal dispersion between MU fiber potentials.

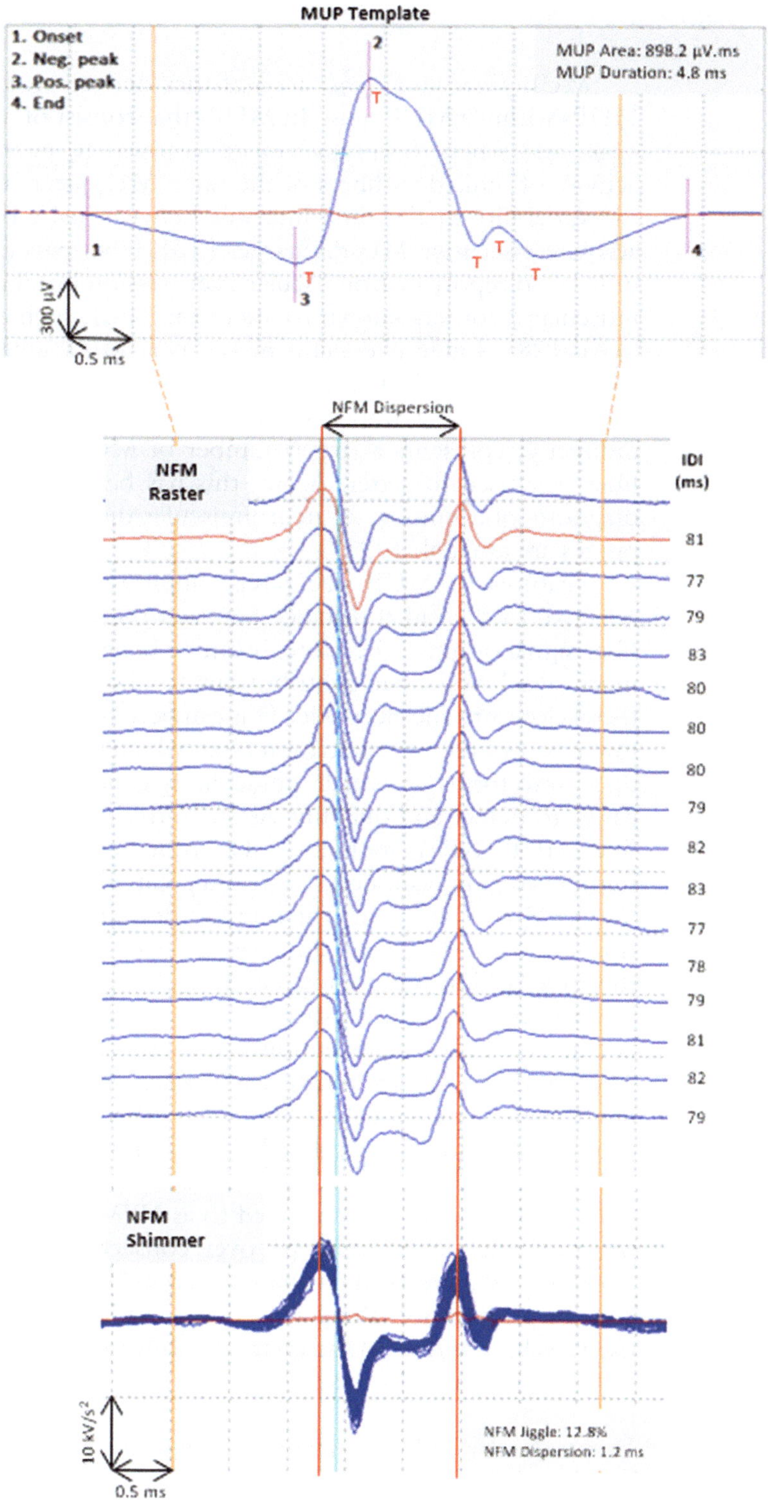

Fig. 3 An individual MUP template and corresponding NFM raster and shimmer plot. The MUP template shows the onset (1), negative peak (2), positive peak (3), and end (4) of the MUP. The corresponding NFM has two distinct peaks, shown as a raster plot of consecutive observations and overlaid as a shimmer plot. The inter-discharge interval (IDI) of each observation is shown on the right in ms, corresponding to a firing rate of approximately 12.6 Hz. Abbreviations: MUP motor unit potential, µV microvolts, ms milliseconds, NFM near-fiber MUP

MUP *jitter* and *jiggle* are statistics derived from consecutive MUPs within a MUP train. In MUPs that consist of more than one peak, and when the peaks can be assumed to be related to the activity of individual fibers of the same MU, jitter is calculated as the variability in the time intervals between pairs of these peaks across consecutive MU discharges [27]. Fiber-pair jitter has been utilized to report neuromuscular transmission instability in neuromuscular disorders known to affect the NMJ, such as myasthenia gravis [28]. Jiggle is dependent on local MUP amplitude differences between MUPs recorded across consecutive MU discharges caused by varying dispersion between MU fiber potentials and indirectly represents a larger number of MU fibers than a single fiber pair used for jitter. Again, this has been used to aid in the diagnosis of a number of neuromuscular disorders, and reference ranges are available [29].

Features of MUP size (area, amplitude, duration) are also sensitive to the radial distance between the electrode and muscle fiber signal origins. Therefore, it can be beneficial to exclude those more distal MU fiber contributions to allow greater detail from those closest to the electrode. This can be achieved with the use of a low-pass-double-differentiation filter, which when applied can produce waveforms in which single muscle fiber contributions are easier to detect and quantify. As such, this is referred to as a near fiber MUP (NFM, *see* Fig. 3). The near fiber count refers to the number of significant peaks (>5-fold greater than baseline noise) in the NFM and represents the number of individual fiber contributions. The NFM area and duration are calculated from the NFM in the same fashion as those for the MUP and reported in V/s and ms, respectively. The NFM dispersion is the time between the first and last detected fiber contribution and is reported in ms (*see* Fig. 3). The NFM can be particularly useful when calculating features of a MUP train, as individual NFMs do not suffer from the same degree of contamination from other MUs and noise and are, therefore, easier to isolate and assess across consecutive MU discharges. Jiggle calculated from NFMs is referred to as NFM jiggle and represents the variability in NFM shape over consecutive MU discharges, which would increase in response to variability in NMJ transmission and/or propagation along MU fibers. NFM measures have been used to characterize MU adaptations/differences in response to age [9], sex [14], and disease [30].

3.1.3 Concluding Remarks

Intramuscular EMG is a relatively inexpensive and minimally invasive procedure which provides a wealth of neuromuscular electrophysiological information. It is capable of recording MUPs from a range of muscles of varying sizes and at varying depths to provide data related to MU structure and function. Features of MUP and NFM templates can reflect the size and structure of MUs, whereas features of consecutive MUPs/NFMs can be used to assess the variability of NMJ transmission times and/or MU fiber action potential propagation times. All of which can prove useful in human research studies.

4 Notes

1. *Excessive noise and/or poor signal to noise ratio (SNR)*: Noise interference is common in all aspects of EMG and, frustratingly, may differ on a day-to-day basis. Nonphysiological 50 Hz or 60 Hz interference most commonly originates from the immediate area, and additional grounding may be required. This may include grounding of the force transducer and amplifier, EMG acquisition equipment, and structural testing equipment in contact with the participant. In many cases this may involve brief "trial and error," with the elimination of different potential sources of interference in a stepwise manner.
2. *Needle movement*: Needle movement can alter the shape of successive MUPs rendering them difficult to identify/assign to a MUPT. A tight grip of the needle shaft often makes movement more severe, and it is advised to lightly support the needle once it has been positioned and allow it to initially move with the muscle at the onset of contraction.
3. *Poor signal quality*: When baseline noise is acceptable, yet spikes appear visibly small, this likely reflects recording distal to the fibers from which the signal originates. This is commonly muscle-dependent and will occur more frequently in larger muscles and during lower contraction levels. This may be remedied by repositioning the needle and/or increasing the contraction intensity.
4. *Unable to reliably identify MUPs with visibly good SNR*: This may reflect an excessively high PPS in the signal, a measure of all MUPTs recorded within 1 s and reported as pulses per second (PPS). This is muscle dependent, and in our experience MUPs cannot be reliably identified from iEMG signals with a PPS >90. This may be remedied by reducing the contraction level, likely reducing the number of, and discharge rate of, MUs contributing to the signal.

References

1. De Carvalho M, Turkman A, Swash M (2014) Sensitivity of MUP parameters in detecting change in early ALS. Clin Neurophysiol 125: 166–169. https://doi.org/10.1016/j.clinph.2013.06.014
2. Piasecki M, Ireland A, Piasecki J et al (2018) Failure to expand the motor unit size to compensate for declining motor unit numbers distinguishes sarcopenic from non-sarcopenic older men. J Physiol. https://doi.org/10.1113/JP275520
3. Piasecki M, Ireland A, Stashuk D et al (2016) Age-related neuromuscular changes affecting

human vastus lateralis. J Physiol 594:4525–4536. https://doi.org/10.1113/JP271087
4. McNeil CJ, Doherty TJ, Stashuk DW, Rice CL (2005) Motor unit number estimates in the tibialis anterior muscle of young, old, and very old men. Muscle Nerve 31:461–467. https://doi.org/10.1002/mus.20276
5. Dalton BH, McNeil CJ, Doherty TJ, Rice CL (2008) Age-related reductions in the estimated numbers of motor units are minimal in the human soleus. Muscle Nerve 38:1108–1115. https://doi.org/10.1002/mus.20984
6. Swiecicka A, Piasecki M, Stashuk D et al (2020) Relationship of anabolic hormones with motor unit characteristics in quadriceps muscle in healthy and frail aging men. J Clin Endocrinol Metab 105:e2358–e2368. https://doi.org/10.1210/clinem/dgaa100
7. Power GA, Dalton BH, Behm DG et al (2010) Motor unit number estimates in masters runners. Med Sci Sports Exerc 42:1644–1650. https://doi.org/10.1249/MSS.0b013e3181d6f9e9
8. Power GA, Dalton BH, Behm DG et al (2012) Motor unit survival in lifelong runners is muscle dependent. Med Sci Sports Exerc 44:1235–1242. https://doi.org/10.1249/MSS.0b013e318249953c
9. Piasecki J, Inns TB, Bass JJ et al (2021) Influence of sex on the age-related adaptations of neuromuscular function and motor unit properties in elite masters athletes. J Physiol 599:193–205. https://doi.org/10.1113/JP280679
10. Jones EJ, Piasecki J, Ireland A et al (2021) Lifelong exercise is associated with more homogeneous motor unit potential features across deep and superficial areas of vastus lateralis. Geroscience. https://doi.org/10.1007/s11357-021-00356-8
11. Piasecki M, Ireland A, Piasecki J et al (2019) Long-term endurance and Power training may facilitate motor unit size expansion to compensate for declining motor unit numbers in older age. Front Physiol 10:449. https://doi.org/10.3389/fphys.2019.00449
12. Guo Y, Piasecki J, Swiecicka A et al (2021) Circulating testosterone and dehydroepiandrosterone are associated with individual motor unit features in untrained and highly active older men. Geroscience. https://doi.org/10.1007/s11357-021-00482-3
13. Piasecki M, Ireland A, Coulson J et al (2016) Motor unit number estimates and neuromuscular transmission in the tibialis anterior of master athletes: evidence that athletic older people are not spared from age-related motor unit remodeling. Physiol Rep 4. https://doi.org/10.14814/phy2.12987
14. Guo Y, Jones EJ, Inns TB et al (2022) Neuromuscular recruitment strategies of the vastus lateralis according to sex. Acta Physiol. https://doi.org/10.1111/apha.13803
15. Cavalcanti Garcia MA, Vieira TMM (2011) Surface electromyography: why, when and how to use it. Revista Andaluza de Medicina del Deporte 4:17–28
16. Vigotsky AD, Halperin I, Lehman GJ et al (2018) Interpreting signal amplitudes in surface electromyography studies in sport and rehabilitation sciences. Front Physiol 8:985
17. Stålberg E, Andreassen S, Falck B et al (1986) Quantitative analysis of individual motor unit potentials: a proposition for standardized terminology and criteria for measurement. J Clin Neurophysiol 3. https://doi.org/10.1097/00004691-198610000-00003
18. Henriksson-Larsen KB, Lexell J, Sjostrom M (1983) Distribution of different fibre types in human skeletal muscles. I. Method for the preparation and analysis of cross-sections of whole tibialis anterior. Histochem J 15:167–178
19. Swiecicka A, Piasecki M, Stashuk DW et al (2019) Frailty phenotype and frailty index are associated with distinct neuromuscular electrophysiological characteristics in men. Exp Physiol 104:1154–1161. https://doi.org/10.1113/EP087579
20. Ives CT, Doherty TJ (2014) Influence of needle electrode depth on DE-STA motor unit number estimation. Muscle Nerve 50:587–592. https://doi.org/10.1002/mus.24208
21. Boe SG, Stashuk DW, Brown WF, Doherty TJ (2005) Decomposition-based quantitative electromyography: effect of force on motor unit potentials and motor unit number estimates. Muscle Nerve 31:365–373. https://doi.org/10.1002/mus.20266
22. Tsang P, MacDermid JC, Miller TA et al (2020) Assessment of age-related differences in decomposition-based quantitative EMG in the intrinsic hand muscles: a multivariate approach. Clin Neurophysiol 131. https://doi.org/10.1016/j.clinph.2020.06.017
23. Botter A, Oprandi G, Lanfranco F et al (2011) Atlas of the muscle motor points for the lower limb: implications for electrical stimulation procedures and electrode positioning. Eur J Appl Physiol 111:2461–2471. https://doi.org/10.1007/s00421-011-2093-y
24. Martinez-Valdes E, Laine CM, Falla D et al (2016) High-density surface electromyography provides reliable estimates of motor unit

behavior. Clin Neurophysiol 127:2534–2541. https://doi.org/10.1016/j.clinph.2015.10.065

25. Gooch CL, Doherty TJ, Chan KM et al (2014) Motor unit number estimation: a technology and literature review. Muscle Nerve 50:884–893. https://doi.org/10.1002/mus.24442
26. Piasecki M, Garnés-Camarena O, Stashuk DW (2021) Near-fiber electromyography. Clin Neurophysiol 132:1089–1104. https://doi.org/10.1016/j.clinph.2021.02.008
27. Stålberg EV, Sonoo M (1994) Assessment of variability in the shape of the motor unit action potential, the "jiggle," at consecutive discharges. Muscle Nerve 17:1135–1144. https://doi.org/10.1002/mus.880171003
28. Sanders DB (2002) Clinical impact of single-fiber electromyography. Muscle Nerve 999. https://doi.org/10.1002/mus.10141
29. Stålberg E, Sanders DB, Ali S et al (2016) Reference values for jitter recorded by concentric needle electrodes in healthy controls: a multicenter study. Muscle Nerve 53. https://doi.org/10.1002/mus.24750
30. Allen MD, Stashuk DW, Kimpinski K et al (2015) Increased neuromuscular transmission instability and motor unit remodelling with diabetic neuropathy as assessed using novel near fibre motor unit potential parameters. Clin Neurophysiol. https://doi.org/10.1016/j.clinph.2014.07.018

Chapter 8

Magnetic Resonance Quantification of Muscle Phosphocreatine Resynthesis Kinetics During Exercise Recovery: An In Vivo Measure of Mitochondrial Function in Humans

Jordan J. McGing, Susan T. Francis, Sébastien Serres, Gordon W. Moran, and Paul L. Greenhaff

Abstract

Quantifying muscle phosphocreatine (PCr) resynthesis kinetics during recovery from exercise can be employed as a sensitive and noninvasive in vivo measure of mitochondrial function, or more specifically mitochondrial oxidative capacity, in humans. Here we describe a protocol employing nonlocalized ^{31}P MRS acquisition in the medial gastrocnemius muscle, which, when combined with mDIXON ^{1}H MRI scanning of muscle volume and fat fractions, ensures anatomical localization for ^{31}P MRS acquisition. The protocol involves the performance of in-bore plantar flexor exercise at a predetermined and controlled exercise intensity (50% maximal isometric force) with limb blood flow occluded, thereby ensuring that end-exercise muscle PCr, ADP, and cytosolic pH are well matched across volunteers at the point of end-exercise reinstatement of limb blood flow. This will ensure maximal stimulation of mitochondrial respiration and PCr resynthesis during subsequent resting, nonischemic recovery. Consideration is given to the view that contraction-induced muscle acidosis complicates the interpretation of ^{31}P-derived PCr resynthesis data during exercise recovery and how this can be circumvented.

Key words Exercise, Mitochondria, Energy metabolism, MRI, MRS

1 Introduction

Adenosine triphosphate (ATP) is the sole substrate that can be used directly by skeletal muscle to fuel contraction. However, the store of ATP in human skeletal muscle is relatively small (approximately 24 mmol/kg dry muscle) and, therefore, must be continually resynthesized from its breakdown products ADP, AMP, and inorganic phosphate (P_i). Resynthesis of ATP during exercise is provided by anaerobic (non-oxygen using) and aerobic (oxygen using) processes, with the contribution from each depending principally on the relative exercise intensity and duration. During

Philip J. Atherton and Daniel J. Wilkinson (eds.), *Neuromuscular Assessments of Form and Function*, Neuromethods, vol. 204, https://doi.org/10.1007/978-1-0716-3315-1_8,

high-intensity to maximal intensity exercise or lower-intensity exercise with blood flow occlusion, aerobic ATP resynthesis cannot meet the energy requirements of contraction and, therefore, anaerobic ATP resynthesis from phosphocreatine (PCr) hydrolysis and glycolysis is essential if the energy demand of contraction is to be met. However, PCr hydrolysis and glycolysis can only sustain the high ATP demand of maximal exercise for relatively short periods of time (s to min). PCr hydrolysis is catalyzed by the enzyme creatine kinase to produce creatine and P_i, which are then transferred to ADP in order to resynthesize ATP (PCr + ADP + H^+ → ATP + Cr). In the process of anaerobic glycolysis, glucose-6-phosphate, derived primarily from muscle glycogen during high-intensity to maximal exercise, is converted to lactic acid via a series of metabolic reactions, two of which (1,3-diphosphoglycerate → 3-phosphoglycerate and phosphoenolpyruvate → pyruvate) re-phosphorylate ADP to form ATP. Pyruvate that is not reduced to lactic acid enters the mitochondria via the pyruvate dehydrogenase complex (PDC), where it provides acetyl-CoA for the TCA cycle and oxidative ATP resynthesis, albeit at a much lower maximal rate of ATP resynthesis compared to PCr hydrolysis or anaerobic glycolysis. When the rate of pyruvate-derived acetyl-CoA production during exercise exceeds the rate of TCA cycle demand (at above 55% maximal oxygen consumption), the excess acetyl groups are shunted to acetylcarnitine, which accumulates.

It is well established that PCr resynthesis during recovery from exercise relies directly upon mitochondrial-derived ATP production [1, 2], and during the first seconds of exercise recovery, the most potent stimulants for mitochondrial respiration are the free [ADP] and free [creatine] [3]. In generating maximal flux through the F_0-F_1ATPase during recovery, muscle PCr stores can be depleted using high-intensity exercise with limb blood flow occluded. It has previously been shown that during recovery from such exercise, the rate of PCr resynthesis reflects the net ATP synthesis rate via the F_0-F_1ATPase [2], as both are coupled at the mitochondrial membranes by adenine nucleotide translocase and mitochondrial creatine kinase. Thus, during the period of ischemic contraction, the intra-mitochondrial accumulation of reducing equivalents [4] will increase the proton gradient, and in tightly coupled mitochondria, the energy generated by this proton gradient will be diverted to phosphorylate ADP via F_0-F_1ATPase, which will in turn determine the rate of PCr resynthesis during recovery when limb blood flow is reinstated. In short, therefore, measuring muscle PCr resynthesis kinetics during recovery from exercise performed with circulatory occlusion can be employed as a sensitive in vivo measure of mitochondrial function, or more specifically mitochondrial oxidative capacity, in humans [5, 6]. Importantly, the best time resolution will be achieved using ^{31}P MRS, rather than repeated muscle biopsy sampling, which is also minimally invasive.

2 Materials

2.1 mDIXON Scans

Anatomical localization and composition of the calf muscles is obtained in the resting state by mDIXON ^{1}H MRI scanning using a Philips 3T Achieva scanner equipped with a ^{1}H body coil (Philips Medical Systems, Best, Netherlands).

2.2 ^{31}P MR-Spectroscopy

Nonlocalized ^{31}P-MR spectra are obtained using a Philips 3T Achieva with multinuclear capability and a 14 cm ^{31}P surface coil (Philips Medical Systems, Best, Netherlands). The volume of interest is shimmed and nonlocalized pulse-acquired ^{31}P-MR spectra collected using a TR 4 s, TE 0.10 ms, spectral band with 3000 Hz, 4096 samples. Analysis of ^{31}P MR spectra is achieved using jMRUI Beta 6.0. ^{31}P-MR spectra are apodized and peaks including inorganic phosphate (Pi), PCr, and ATP subunits ($\gamma - ATP, \alpha - ATP, \beta - ATP$) fitted using the AMARES function with prior knowledge.

2.3 In-Bore Exercise

In-bore exercise is performed using a 3T MRI compatible Trispect exercise ergometer (Ergospect GmbH, Innsbruck, Austria) designed to localize the calf musculature of one leg while undertaking fixed workload plantar flexion exercise. Volunteers are secured in a supine position to undertake a plantar flexion exercise against a resistance generated by an external air pressure device, set by the operator at a predetermined resistance equivalent to 50% maximal isometric force (MVC).

3 Methods

1. Subjects report to the laboratory having abstained from strenuous physical exercise for 24 h.
2. Volunteers lay in a feet-first position on a Philips 3T Achevia scanner bed outside of the magnet with their dominant limb secured into a 3T compatible Trispect ergometer. Calf muscle isometric strength is assessed as the peak force produced during three maximum plantar flexion isometric contractions, each sustained for 2–3 s and interspersed with 60 s recovery (*see* Fig. 1a).
3. Volunteers are then moved into the MR scanner. High-resolution mDIXON scans of the calf are acquired in sagittal, coronal, and axial planes for muscle volume and fat fraction quantification (*see* Fig. 1b). This scan also familiarizes volunteers with the MRI scanner prior to performing within bore exercise.

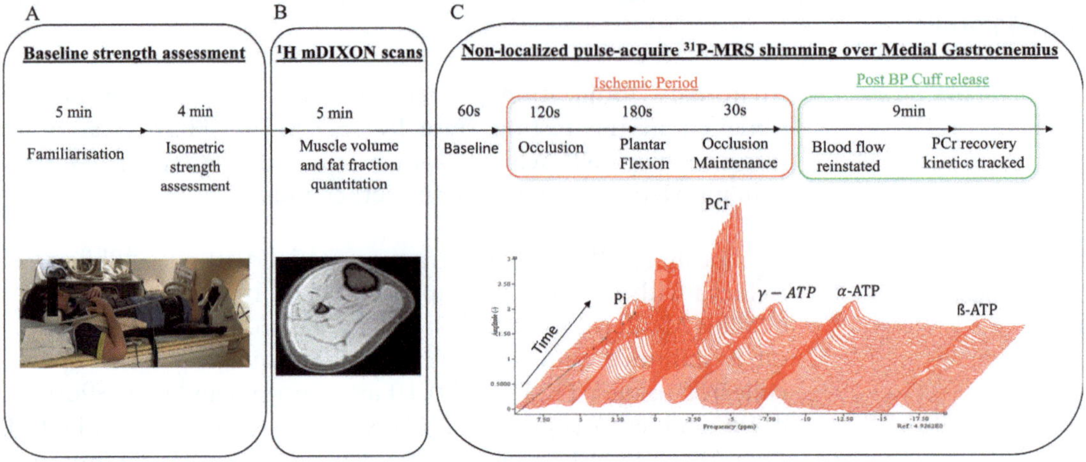

Fig 1 Protocol overview to quantify nonlocalized ^{31}P MRS assessment of gastrocnemius muscle phosphocreatine depletion and recovery following ischemic plantar flexion exercise in human volunteers

4. A 16-min nonlocalized ^{31}P acquisition of the medial gastrocnemius muscle then follows (*see* Fig. 1c, **Note 1**). Given that the volume of interest will be determined by the sensitivity profile of the ^{31}P surface coil, this has implications for the comparison of volunteers differing in lower leg tissue composition. Thus, the mDIXON scans of muscle volume and fat fractions are used to ensure medial gastrocnemius muscle was investigated.
5. Sixty seconds into nonlocalized ^{31}P acquisition, a member of the research team inside the scan room inflates a thigh blood pressure cuff to 250 mmHg to occlude lower limb blood flow. After 2 min occlusion, the volunteer is instructed to perform repeated plantar flexion exercises at 50% MVC at a frequency of one extension per second. The air-braked Trispect ergometer is controlled in the console room by the research team via a computer. The volunteer is verbally encouraged to exercise to volitional exhaustion which typically occurs ~2–3 min after the start of ischemic exercise. Importantly, the application of limb blood flow occlusion prior to exercise results in the desaturation of myoglobin ensuring that exercise is performed with minimal contribution to the oxidative process [5, 7]. Furthermore, maintenance of limb blood flow occlusion during contraction also ensures PDC flux is halted, thereby circumventing mitochondrial ATP production [4]. Limb blood flow occlusion during contraction also produces a closed muscle compartment, thereby resulting in comparable muscle lactate concentrations, acidosis, and PCr hydrolysis immediately following exercise. Muscle acidosis complicates the interpretation of ^{31}P acquisition data (*see* **Note 2**), but uniform muscle PCr

depletion during contraction is important to achieve if the comparison of PCr recovery kinetics is an objective of any study design (*see* **Note 3**). This protocol has been shown to result in almost total degradation of muscle PCr stores [5, 7].

6. Following exercise completion, cuff occlusion is maintained for 30 s to necessitate spectral averaging to determine end-exercise muscle ^{31}P metabolite concentrations. The cuff pressure is then released to immediately reinstate limb blood flow. ^{31}P acquisitions continued during resting recovery for 10 min (*see* Fig. 1c). The marked hyperemia upon reinstatement of blood flow following exercise and occlusion ensures an almost immediate increase in muscle oxygen delivery, therefore, minimizing any potential for oxygen availability to limit PCr resynthesis at the onset of exercise recovery [8].

7. Exercise kinetics for PCr can be expressed relative to the baseline (resting, nonischemic) signal amplitude. Cytosolic pH can be estimated using the chemical shift difference (δ) between Pi and PCr peaks:

$$\mathrm{pH} = pK + \log\left(\frac{\delta 1 - \delta 0}{\delta 0 - \delta 2}\right) \tag{1}$$

where $pK = 6.75$, $\delta 1 = 3.27$, and $\delta 2 = 5.63$.

Free ADP can be estimated as:

$$\mathrm{Free}\,[\mathrm{ADP}] = \left(\frac{[\mathrm{TCr}]}{[\mathrm{PCr}]} - 1\right)\left(\frac{[\mathrm{ATP}]}{(K[H+])}\right) \tag{2}$$

where $K = 1.66 \times 10^9$ L/mol and ATP and total creatine (TCr) are assumed to be 8.2 mM and 42.5 mM/L cell water.

8. The PCr resynthesis curve during recovery ($\mathrm{PCr}(t)$) following ischemic contraction can be fitted to a mono-exponential function:

$$\mathrm{PCr}(t) = \mathrm{PCr}_{\mathrm{initial}} + (\mathrm{PCr}_{\mathrm{end}} - \mathrm{PCr}_{\mathrm{initial}})\left(1 - e^{-kt}\right) \tag{3}$$

where t is the time from the start of recovery and $\mathrm{PCr}_{\mathrm{initial}}$ and $\mathrm{PCr}_{\mathrm{end}}$ are the PCr contents at the initial and end of recovery phases.

Postexercise PCr resynthesis (V_{PCr}) can be estimated as:

$$V_{\mathrm{PCr}} = k_{\mathrm{PCr}} \times \Delta\,[\mathrm{PCr}] \tag{4}$$

where k_{PCr} is the rate constant of PCr resynthesis and $\Delta[\mathrm{PCr}]$ is the end recovery PCr concentration minus the end-exercise PCr concentration in mM/l cell water.

Muscle mitochondrial oxidative capacity (Q_{max}; *ATP synthesis rate*) can be estimated as:

$$Q_{max} = k_{PCr[basal\ PCr]} \tag{5}$$

where k_{PCr} is the rate constant of PCr resynthesis and basal PCr is the resting PCr concentration in mM/l cell water.

The ADP control model of oxidative ATP synthesis ($Q_{max\ ADP}$) can be estimated as:

$$Q_{max}\text{ADP} = V_{PCr}\left(1 + \frac{K_m}{[\text{ADP}]\text{end}}\right) \tag{6}$$

where [ADP]end is the ADP at the end of the exercise, V_{PCr} is the postexercise PCr resynthesis, and $K_m = 30$ μm/L is the equilibrium constant of the creatine kinase reaction.

4 Notes

4.1 Note 1: Use of a Nonlocalized ^{31}P Sequence

While the use of gradient-based localization methods is advantageous when quantifying muscle energetics using ^{31}P MRS, the temporal resolution of localized acquisition seen with ^{31}P-image-selected in vivo spectroscopy (ISIS), and the susceptibility to motion-induced phasing issues, makes it poorly suited to quantify PCr recovery kinetics after ischemic muscle contraction, which is in the order of seconds. A nonlocalized ^{31}P sequence is, therefore, used in this protocol. Nonlocalized ^{31}P acquisition also enables the surface coil to move with the lower limb, enabling acquisitions to continue during contraction and reducing subject burden as volunteers are not required to adhere to a strict gated exercise protocol.

4.2 Note 2: Muscle Acidosis Complicates the Interpretation of ^{31}P-Derived PCr Resynthesis Data During Exercise Recovery

As outlined above, based on the premise that muscle PCr resynthesis following contraction is entirely reliant on oxidative ATP synthesis [1], the rate constant of PCr resynthesis during exercise recovery (k_{PCr}) has been utilized as an index of muscle mitochondrial oxidative capacity. However, it has been proposed that muscle acidosis perturbs PCr resynthesis in vivo [9, 10], which may be due to a direct inhibitory effect of acidosis on skeletal muscle oxidative phosphorylation [9] and/or indirectly via a pH-mediated effect on creatine kinase equilibrium [11], increased ATP consumption by cellular ion pumps [12], and inhibition of sarcoplasmic reticulum calcium release and reuptake [13]. In keeping with this, these in vivo observations are inconsistent with findings from studies utilizing isolated mitochondria where many of these cellular functions are obviously not operative [14]. These in vivo observations have led to the suggestion that the use of the rate constant of PCr resynthesis becomes unreliable when estimating muscle mitochondrial function in this scenario [15]. Indeed, a recent expert consensus recommendation endorsed limiting muscle contraction intensity such that intracellular pH does not decline by more than 0.2 units [16]. This may be difficult to achieve in practical terms but, moreover, will increase within and between volunteer reproducibility considerably adding significantly to study variance.

Even in the presence of exercise-induced acidosis, however, several ^{31}P kinetic parameters have been proposed as indices of muscle mitochondrial function. For example, the estimation of postexercise PCr resynthesis (V_{PCr}), reflecting the initial rate of oxygen utilization [17], has been shown to be independent of cytosolic pH estimation [15, 18]. Furthermore, while the reliability of Q_{max} has been disputed due to its reliance on k_{PCr} [19], $Q_{max\ ADP}$ is reported to be independent of cytosolic pH as it includes V_{PCr} [20] and remains commonly used [21]. Indeed, comparable V_{PCr} and $Q_{max\ ADP}$ values have been reported across exercise protocols evoking significant differences in end-exercise muscle pH using measured V_{PCr} [20]. Overall therefore, using V_{PCr} and $Q_{max\ ADP}$ as indices of mitochondrial function during exercise recovery minimizes any potential confounding effect of postexercise muscle acidosis, which appears to predominate in the later stages of the biphasic PCr recovery [22].

4.3 Note 3: Consistent Muscle PCr Depletion Is Important to Achieve If the Comparison of PCr Recovery Kinetics Is a Study Objective

As outlined in Subheading 4.2, the kinetic parameters specific to initial PCr recovery can reliably infer muscle mitochondrial function in the presence of end-exercise acidosis. By using limb occlusion, a closed metabolic compartment can be created such that end-exercise PCr, ADP, and cytosolic pH are tightly matched across volunteers/study visits at the point of reinstating limb blood flow, which will maximally stimulate mitochondrial respiration and PCr resynthesis. This scenario is not easily achievable during submaximal intensities used to limit acidosis [16], as muscle energy metabolism and pH changes will be highly variable between subjects [23, 24] and muscles recruited [25]. A variable end-exercise metabolic state would itself complicate the inference of muscle mitochondrial function from PCr recovery kinetics. The standardized, ischemic exercise protocol described here avoids such issues, albeit at the expense of low end-exercise muscle pH.

Acknowledgments

The work described was supported by a Crohn's and Colitis UK Medical Research Award (M2017/6). JJM was in receipt of a University of Nottingham PhD scholarship funded by the Joan Browne Legacy.

References

1. Quistorff B, Johansen L, Sahlin K (1993) Absence of phosphocreatine resynthesis in human calf muscle during ischaemic recovery. Biochem J 291:681–686
2. Harris R, Edwards R, Hultman E et al (1976) The time course of phosphorylcreatine resynthesis during recovery of the quadriceps muscle in man. Pflugers Arch 367(2):137–142
3. Tonkonogi M, Harris B, Sahlin K (1998) Mitochondrial oxidative function in human saponin-skinned muscle fibres: effects of prolonged exercise. J Physiol 510(1):279–286
4. Constantin-Teodosiu D, Cederblad G, Hultman E (1992) PDC activity and acetyl group accumulation in skeletal muscle during prolonged exercise. J Appl Physiol 73(6): 2403–2407
5. Greenhaff PL, Bodin K, Soderlund K et al (1994) Effect of oral creatine supplementation on skeletal muscle phosphocreatine resynthesis. Am J Physiol Endocrinol Metab 266(5):E725–E730
6. Hesselink MKC, Greenhaff PL, Constantin-Teodosiu D et al (2003) Increased uncoupling protein 3 content does not affect mitochondrial function in human skeletal muscle in vivo. J Clin Invest 111:479–486
7. Greenhaff PL, Söderlund K, Ren J-M et al (1993) Energy metabolism in single human muscle fibres during intermittent contraction with occluded circulation. J Physiol 460(1): 443–453
8. Hultman E, Spriet L, Söderlund K (1986) Biochemistry of muscle fatigue. Biomed Biochim Acta 45(1–2):S97–S106
9. Jubrias SA, Crowther GJ, Shankland EG et al (2003) Acidosis inhibits oxidative phosphorylation in contracting human skeletal muscle in vivo. J Physiol 553(2):589–599
10. Kemp G, Thompson C, Sanderson A et al (1994) pH control in rat skeletal muscle during exercise, recovery from exercise, and acute respiratory acidosis. Magn Reson Med 31(2): 103–109
11. Lawson J, Veech R (1979) Effects of pH and free Mg2+ on the Keq of the creatine kinase reaction and other phosphate hydrolyses and phosphate transfer reactions. J Biol Chem 254(14):6528–6537
12. Baker A, Brandes R, Schendel T et al (1994) Energy use by contractile and noncontractile processes in skeletal muscle estimated by 31P-NMR. Am J Phys Cell Phys 266(3): C825–CC31
13. Byrd S, McCutcheon L, Hodgson D et al (1989) Altered sarcoplasmic reticulum function after high-intensity exercise. J Appl Physiol 67(5):2072–2077
14. Willis WT, Jackman MR (1994) Mitochondrial function during heavy exercise. Med Sci Sports Exerc 26(11):1347–1353
15. Roussel M, Bendahan D, Mattei JP et al (2000) 31P Magnetic resonance spectroscopy study of phosphocreatine recovery kinetics in skeletal muscle: the issue of intersubject variability. Biochim Biophys Acta Bioenerg 1457:18–26
16. Meyerspeer M, Boesch C, Cameron D et al (2020) 31P magnetic resonance spectroscopy in skeletal muscle: Experts' consensus recommendations. NMR Biomed 2020:1–22
17. Kemp GJ, Ahmad RE, Nicolay K et al (2015) Quantification of skeletal muscle mitochondrial function by 31P magnetic resonance spectroscopy techniques: a quantitative review. Acta Physiol 213:107–144
18. Lodi R, Kemp GJ, Lotti S et al (1997) Influence of cytosolic pH on in vivo assessment of human muscle mitochondrial respiration by phosphorus magnetic resonance spectroscopy. MAGMA 5:165–171
19. Van Den Broek NMA, De Feyter HMML, De Graaf L et al (2007) Intersubject differences in the effect of acidosis on phosphocreatine recovery kinetics in muscle after exercise are due to differences in proton efflux rates. Am J Physiol Cell Physiol 293:228–237
20. Walter G, Vandenborne K, McCully KK et al (1997) Noninvasive measurement of phosphocreatine recovery kinetics in single human muscles. Am J Phys Cell Phys 272(2):C525–CC34
21. Heskamp L, Lebbink F, van Uden MJ et al (2021) Post-exercise intramuscular O2 supply is tightly coupled with a higher proximal-to-distal ATP synthesis rate in human tibialis anterior. J Physiol 599:1533–1550
22. McMahon S, Jenkins D (2002) Factors affecting the rate of phosphocreatine resynthesis following intense exercise. Sports Med 32(12): 761–784
23. Vandenborne K, McCully K, Kakihira H et al (1991) Metabolic heterogeneity in human calf muscle during maximal exercise. Proc Natl Acad Sci 88(13):5714–5718
24. Bendahan D, Confort-Gouny S, Kozak-Reiss G et al (1990) Heterogeneity of metabolic response to muscular exercise in humans: new criteria of invariance defined by in vivo phosphorus-31 NMR spectroscopy. FEBS Lett 272(1–2):155–158
25. Niess F, Schmid AI, Bogner W et al (2020) Interleaved 31P MRS/1H ASL for analysis of metabolic and functional heterogeneity along human lower leg muscles at 7T. Magn Reson Med 83:1909–1919

Chapter 9

Immunohistochemistry, Microscopy, and Image Analysis of Human Muscle Biopsies: Muscle Fiber Denervation as a Working Example

Casper Soendenbroe, Jesper L. Andersen, and Abigail L. Mackey

Abstract

Immunohistochemistry, fluorescent microscopy, and image analysis are key tools for visualizing and analyzing specific proteins in a variety of tissues. Together these techniques are readily available in most laboratories and in an infinite number of variations. Herein, a working example is used relating to the aging human skeletal muscle being characterized by denervated muscle fibers that undergo atrophy and eventual death. As such, proteins normally restricted to embryonic development, muscle regeneration, and the myotendinous and neuromuscular junctions of the muscle fiber are observed in denervated muscle fibers. We describe a workflow from muscle biopsy sampling to lab bench and to image analysis. Our aim is—through a concrete example—to provide the reader with knowledge and insight into how to design an effective analysis workflow, which can then be utilized in specific research contexts.

Key words Immunohistochemistry, Microscopy, Image analysis, Human skeletal muscle, Muscle biopsy, Atrophy, Denervation

Abbreviations

BSA	Bovine serum albumin
DAPI	4′,6-Diamidino-2-phenylindole
MHC	Myosin heavy chain
MHCe	Embryonic myosin heavy chain
MHCn	Neonatal myosin heavy chain
NCAM	Neural cell adhesion molecule
NMJ	Neuromuscular junction
TBS	Tris-buffered saline

Philip J. Atherton and Daniel J. Wilkinson (eds.), *Neuromuscular Assessments of Form and Function*, Neuromethods, vol. 204, https://doi.org/10.1007/978-1-0716-3315-1_9,

1 Introduction

Muscle fibers are formed during embryonic development through the fusion of single-nucleated myoblasts into primitive multinucleated myotubes [1], which, over time, mature into muscle fibers situated in elaborate networks of connective tissue, nerves, vessels, and distinctive cell types [2, 3]. Myosin heavy chain (MHC)—one of the key proteins for force production—is found in all muscle fibers; however, before the mature syncytial state is reached, muscle fibers express two developmental MHC isoforms; embryonic MHC (MHCe); and neonatal MHC (MHCn) [2, 4]. Likewise, before innervation is established, other proteins, such as Neural Cell Adhesion Molecule (NCAM), are found in the developing muscle fibers [5]. These proteins cease in expression shortly after birth and remain absent as long as the muscle fiber remains intact and innervated. Interestingly, when adult muscle fibers become denervated, developmental MHC isoforms and NCAM re-emerge [5, 6]. This chapter will provide a comprehensive description of how to analyze the denervation of muscle fibers in human biopsy tissue using immunohistochemical techniques, fluorescent microscopy, and image analysis.

1.1 Models of Muscle Denervation

Muscle fiber denervation is defined as the loss of connection between a motor neuron, situated in the spinal cord, and the muscle fibers that it previously innervated. Muscle denervation is a focus within several fields of research, such as neurology, oncology, and gerontology. The present method has primarily been used on muscle samples from healthy human subjects, with research questions aimed at understanding age-related alterations [7–11], such as muscle atrophy and fiber type switching, which occur relatively slowly in humans over decades rather than years [12]. This is in stark contrast to individuals who have suffered a traumatic spinal cord injury leading to the cessation of voluntary activation of the affected muscles; under these circumstances the musculoskeletal system undergoes a rapid and severe deterioration, with widespread disorganization, muscle atrophy, decreases in muscle stem cell numbers, and infiltration of fibrous and adipose tissue as the most central alterations [13, 14]. Furthermore, various neuromuscular diseases, such as muscular dystrophies, neurogenic atrophy, and amyotrophic lateral sclerosis, lead to disturbances in signal transmission between motor neuron endplates and muscle fibers as well as motor neuron decay, causing pronounced alterations in the musculoskeletal system [15–17]. Finally, muscle fiber denervation can also be observed in various animal models, including experimental methods such as nerve crush/sectioning [18], nerve signal blockade [6], and accelerated aging in transgenic animals [19].

1.2 Methods for Studying Muscle Denervation

Due to the altered protein composition along with morphological changes of denervated muscle fibers, immunohistochemistry is the method of choice in assessing denervation in human muscle biopsies. Immunohistochemistry is used in various fields of research and disease diagnostics, including muscle denervation in animal models [6, 18], human diseases [17], spinal cord injury [14], and aging [10, 11]. Immunohistochemistry is a broad term covering various experimental procedures (antigen retrieval, fixation, antibodies, temperature, timing, visualization system, etc.). The overall aim is to target specific antigens present in tissues using one or several primary antibodies which are subsequently visualized with antibodies labeled with enzymes of fluorescent molecules (Fig. 1). While it is beyond the scope of this chapter to outline all the procedural variations, the chapter will provide a detailed description of an indirect, two-step immunofluorescence method for visualizing denervated muscle fibers from cryosections of human biopsy material. Essentially, primary antibodies are incubated with the tissue, where they bind to the targeted antigens. Then, fluorophore-conjugated secondary antibodies bind to the primary antibodies and allow microscopic assessment using light at specific wavelengths.

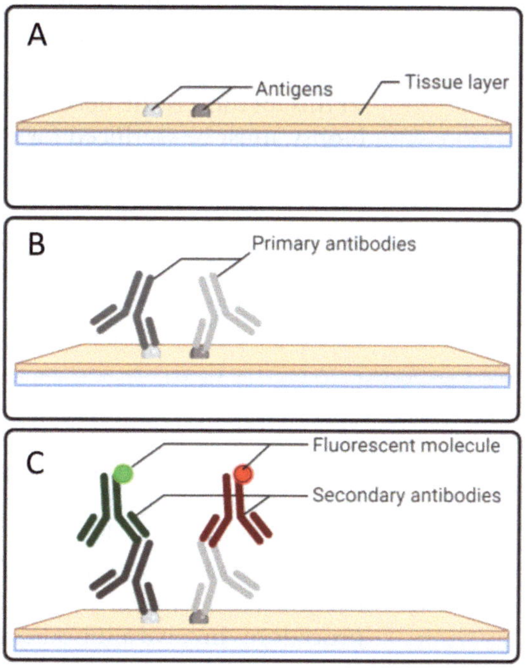

Fig. 1 The concept of indirect immunofluorescence. Immunohistochemistry is used in various fields of research and disease diagnostics, to visualize specific structures found in tissue, and can be performed in a practically unlimited number of variations. Tissue antigens (**a**) are bound by incubation with specific primary antibodies (**b**), which subsequently are bound by fluorophore-conjugated secondary antibodies (**c**), for fluorescence microscopy. Multiple antigens can be visualized in the same sample due to different fluorescent emission wavelengths

Alternative means also exist to assess characteristics of denervation, in addition to immunohistochemistry, with western blotting and qPCR analysis having also been used [20, 21]. Although both methods have value, neither provides the spatial and morphological information that is gained using immunohistochemical analysis of muscle sections. This is most exemplified by the wide range in size of denervated fibers and when serial transverse or longitudinally oriented biopsy sections are used to investigate the presence of one or more specific proteins along the length of a fiber. The latter method is particularly challenging due to the small diameter of a muscle fiber (approx. 50 μm) and since the fibers need to be aligned perfectly when being cut. Either method gives the option of following the same fiber over great distances, where uniformity of protein expression can be investigated. However, given that both methods are very time-consuming in terms of sectioning and analysis, usually only a subset of samples can be treated this way.

Immunohistochemistry is also used to directly assess aging-related structural alterations at the site of innervation, where nerve meets muscle, i.e., the neuromuscular junction (NMJ). This has been widely studied in animal models. Nevertheless, the number of human studies is limited [22–25], probably due to the difficulty in obtaining human samples containing NMJs. Recently, however, a novel biopsy method for obtaining intact NMJs from human donors has been designed [26]. A potential problem related to this method is the necessity that a large piece, perhaps the whole biopsy, must be sacrificed for a single purpose, thereby reducing the number of analytical options. The focus of this chapter is on the extra-synaptic region of the muscle fibers, where denervation induces changes in protein expression along the entire length of the muscle fiber [27], for which immunohistochemistry is an ideal method.

2 Methods and Materials

This section will outline the procedures for analyzing denervated muscle fibers from biopsy sampling to sample preparation and image analysis (see Fig. 2). We have studied healthy men and women of all ages, either at rest or following voluntary exercise, using these methods [7–11, 28, 29]. The immunohistochemical procedures involve antibodies against dystrophin, a membrane protein specific to muscle fibers, in combination with MHCe, MHCn, or NCAM. For certain purposes, it can be beneficial to enhance the fluorescent signal by combining MHCe, MHCn, and NCAM into one antibody cocktail, labeled with the same fluorophore, although this can lead to a loss of information as no discrimination between the markers is possible. A list of used materials and chemicals is provided at the end of this section.

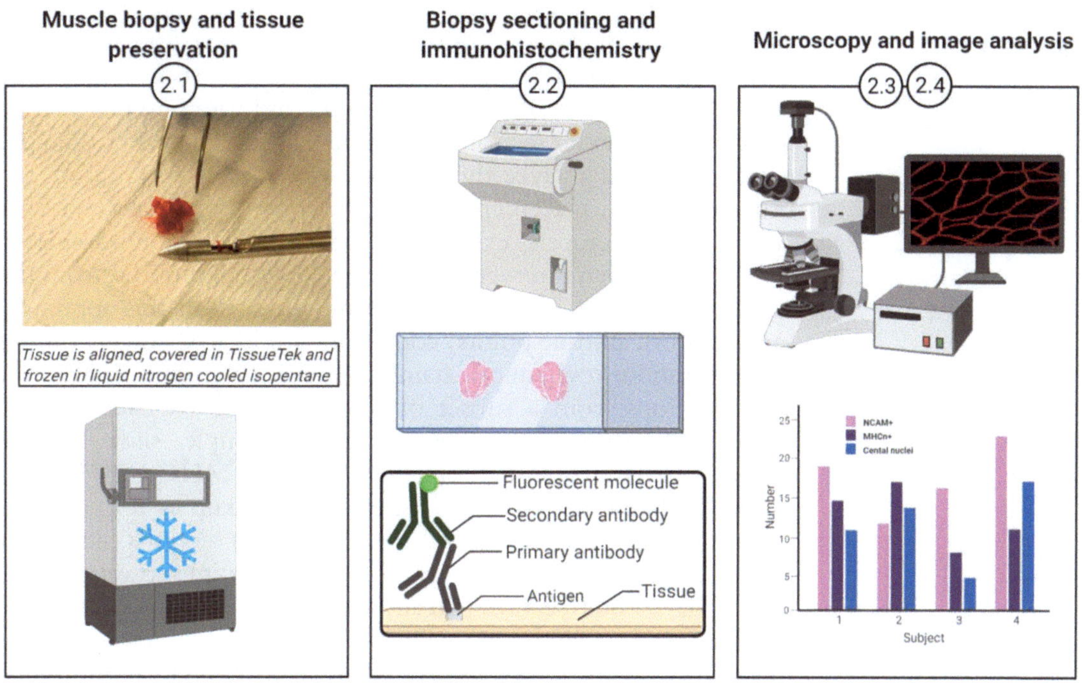

Fig. 2 Overview of muscle biopsy analysis from biopsy sampling to lab bench and to image analysis (covered in Subheadings 2.1, 2.2, 2.3, and 2.4). Subheading 2.1 will detail the muscle biopsy procedure and how muscle tissue is best preserved for histology. Subheading 2.2 deals with the immunohistochemical procedures, starting with tissue sectioning using a cryostat, placing of sections on slides, and special considerations when serial sectioning. Subheading 2.3 describes the microscopy approach and the following stitching of the acquired images. Lastly, Subheading 2.4 contains a detailed description of the image analysis procedure, including a step-by-step guide and several tips and tricks

2.1 Percutaneous Muscle Biopsy and Tissue Preservation

We obtain muscle biopsies, primarily from the vastus lateralis, the largest of the four knee extensor muscles. Following local anesthesia with lidocaine, a small incision is made in the skin and muscle fascia with a scalpel. The Bergström [30] biopsy needle (5 or 6 mm in diameter), with the addition of manual suction, is then used to obtain a tissue sample (approximately 200 mg generally expected). Sampled tissue is immediately placed on a glass slide, and the largest pieces of intact muscle, with the longest fibers set aside for histology, aligned according to the direction of the fibers and covered with Tissue-Tek (Sakura Finetek). The sample is then submerged in isopentane (J.T. Baker, Avantor Performance Materials BV) and cooled in liquid nitrogen, until frozen (a few seconds). The sample, now covered in solid white Tissue-Tek, is quickly transferred to a Cryotube (Thermo Scientific, Nunc A/S) which is then submerged in liquid nitrogen. Samples are stored at −80 °C until further processing. It is important to emphasize that while care taken in selecting and aligning the fascicles for histology will save time at the microscope later, this process should be performed as quickly as possible. Incorrect handling can cause freeze damage to the tissue, rendering the sample useless for immunohistochemistry.

2.2 Immunohistochemistry

Frozen cross sections are cut at approximately −20 °C using a cryostat, and duplicate sections are placed on a SuperFrostPlus glass slide (Thermo Scientific). In most cases, a section thickness of 7–10 μm is used. Blinding for subject and/or condition can be done at this stage, if sectioning is done by someone other than the person who will analyze the samples, for example, by altering the order of section placement on the slides or labeling slides with a code. If several biopsies exist from the same subject, for example, in relation to a given intervention, sections from all biopsies from the same individual are placed on the same slide. With this approach, samples from all time points/conditions are treated equally during the immunohistochemical staining procedures. Specifically, for serial sections from a subset of biopsy samples, a fixed pattern based on planned staining is used. For example, slide 1–3 are stained with dystrophin combined with either MHCe, MHCn, or NCAM, while slide number 4 is saved as a backup or for potential extra staining. In repeating this pattern, 40–50 sections can be cut and placed on slides from a given biopsy. It is important to keep track of any "lost" sections, to allow for the calculation of longitudinal distances (refer to an illustration of this in Fig. 3). The following procedure can be used for several different immunohistochemical stainings. However, the stainings which we are

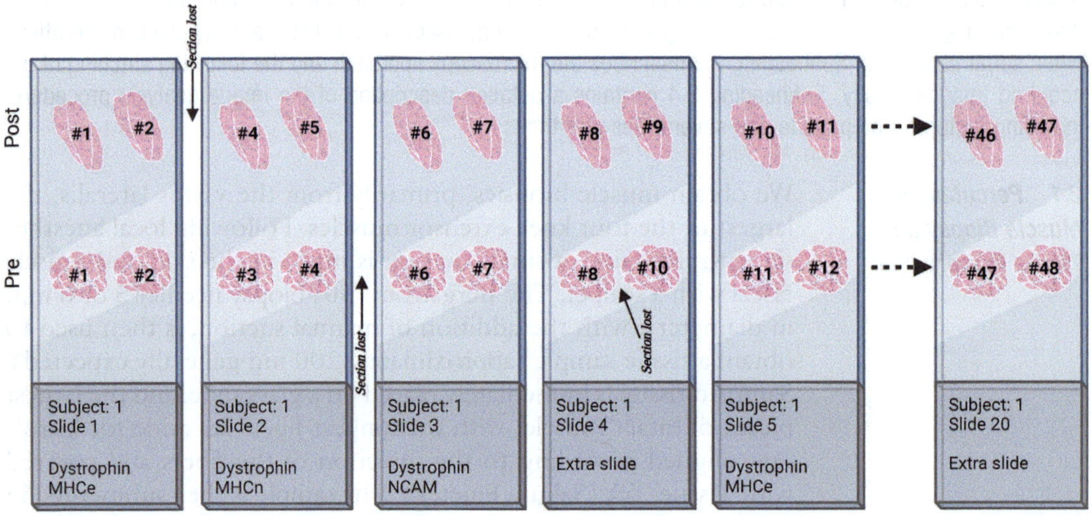

Fig. 3 Schematic showing a method for preparing serial cross sections of human muscle biopsies, which can be used to investigate uniformity of protein expression over great distances. In this example, 10 μm-thick cross sections from a pre- and post-biopsy from the same subject are placed in duplicate on each slide. Slides 1–3 are stained with dystrophin and MHCe, MHCn, or NCAM, respectively, while slide 4 is an extra slide. The extra slide can be used in case one of the prior staining being inconclusive or another antibody of interest later appears. This pattern is then repeated for about 40–50 sections in total. It is important to keep track of lost sections, for accurate calculation of longitudinal distance, as indicated with solid arrows. MHCe embryonic myosin heavy chain; MHCn neonatal myosin heavy chain; NCAM neural cell adhesion molecule

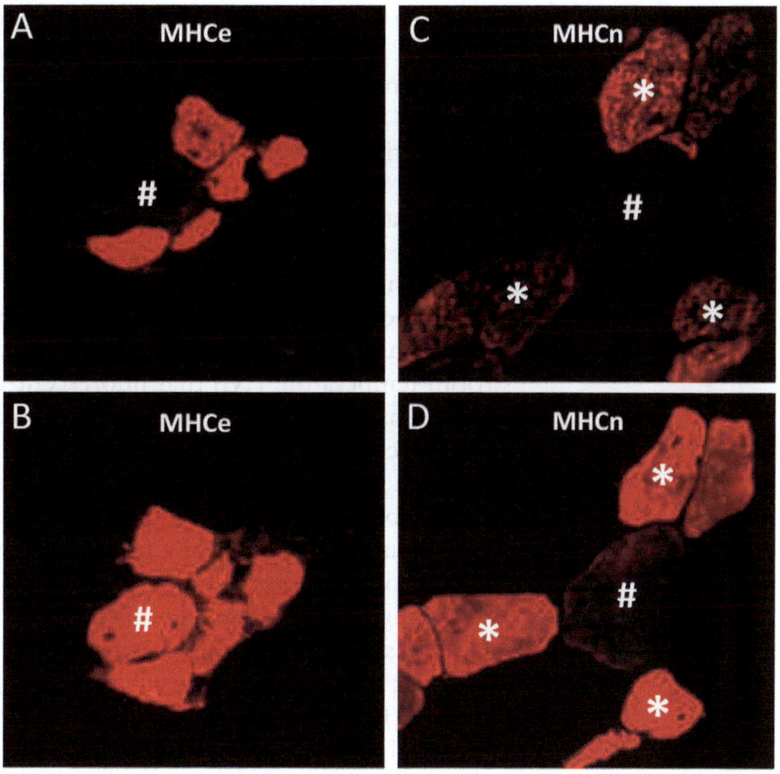

Fig. 4 Human muscle fiber regeneration as a positive control for MHCe and MHCn staining. MHCe and MHCn expressions are upregulated in regenerating human muscle fibers following muscle damage. MHCe or MHCn staining was performed on two subsequent sections from a human biopsy containing regenerating fibers, and images were acquired using identical microscope settings (filters, exposure time, etc.). The upper images (**a** + **c**) were fixed before incubation with the primary antibody (pre-fixed), while the bottom images (**b** + **d**) were not fixed. Stars indicate fibers that were only weakly positive when tissue was pre-fixed, while hashtags indicate fibers that were negative when tissue was pre-fixed. This negative effect of the fixation process, which was more pronounced for the MHCn antibody, is important to consider when using these specific antibodies. MHCe embryonic myosin heavy chain; MHCn neonatal myosin heavy chain

concerned with here are dystrophin (red channel) and 4′,6-diamidino-2-phenylindole (DAPI; blue channel) combined with NCAM, MHCn, or MHCe (green channel). Of note, positive control staining can be performed using regenerating muscle, as shown in Fig. 4 (MHCe, MHCn) and myotendinous junction (NCAM, dystrophin, and central nuclei) [31].

2.3 A: Day 1

Glass slides containing the muscle biopsy cross sections are removed from the freezer and allowed to air dry (approximately 30 min). The sections are encircled by a hydrophobic pen (Sigma-

Aldrich) to reduce the quantity of antibody solution required. The sections must not dry out at any point after the next step. Then, two 5-min washes in Tris-buffered saline (TBS) are performed. Primary antibodies are then diluted in a solution (refer to the list of materials) consisting of 1% bovine serum albumin (BSA) diluted in TBS. Sections are incubated in a humidified box overnight in a refrigerator.

2.4 B: Day 2

Following overnight incubation, two 5-min washes in TBS are performed. Secondary antibodies are then diluted in a solution consisting of 1% BSA diluted in TBS and incubated for 45 min at room temperature (NB that fluorescent-conjugated antibodies are light-sensitive and hence should be protected from prolonged light exposure). Then, another two 5-min washes in TBS are performed. Next, the sections are fixed in 5% formaldehyde (Histofix, Histolab, Gothenburg, Sweden) for 12 min. Note that Histofix is harmful and should be handled in a fume hood with appropriate handling and disposal measures. See **Note 1** for additional information regarding fixation. Then, two 5-min washes in TBS are performed. Finally, slides are mounted with coverslips using ProLong Gold Antifade (P36931; Molecular Probes/Invitrogen) containing DAPI. Hoechst (H1399, Invitrogen) can be used as an alternative to DAPI. Both fluorescent dyes are light-sensitive. Slides are then allowed to dry for 2 days in the dark before being moved to −20 °C for storage before imaging.

2.5 Fluorescent Widefield Microscopy

For microscopy, we adopted a DP71 Olympus camera (Olympus Deutschland GmbH, Hamburg, Germany), mounted on a BX51 Olympus microscope (10×/0.30NA or 20×/0.50 NA objective) with a 0.5× camera mount, operated by the Olympus cellSens software (v.1.14). Images are acquired as grayscale 2040 × 1513 or 4080 × 3072 pixel images. Importantly, any fluorescent microscope which allows for spectral discrimination of the fluorophores (358, 488, and 568 nm) used in the present protocol should suffice. Channels are imaged separately and sequentially (red channel → green channel → blue channel) and subsequently stitched together into one seamless image using the "Stitch Directory with Images (unknown configuration)" plugin [32] in Fiji (ImageJ, v.1.53) [33]. This stitched image, which can consist of up to 20 individual overlapping 10× images, is used for the analysis of muscle fiber denervation. Below we provide the code for the "stitching macro." After having acquired the images, named them, and saved them in subdivided folders, simply run the macro on a folder. When inserting the code into Fiji, remember to change the programming language to "IJ1 macro." Image names must end on "DYST.tif," "NCAM.tif," and "DAPI.tif," unless these tokens are altered in the macro code.

```
macro "Stitch [9]" {{

path = getDirectory("Choose a Directory");
filename = getFileList(path);
newDir = path + "TEMP FOLDER FOR STITCHING" + File.
separator;
File.makeDirectory(newDir);
//newDir2 = path + "Tiles With Fiber Area" + File.
separator;
//File.makeDirectory(newDir2);
newDir1 = path + "Stitched_Composite_NCAM" + File.separa-
tor;
File.makeDirectory(newDir1);
}

for (i=0; i<filename.length; i++) {
 if(endsWith(filename[i], "1_DYST.tif")) {
 open(path+filename[i]);
e=getTitle();
close();
 }}

for (i=0; i<filename.length; i++) {
 if(endsWith(filename[i], "DYST.tif")) {
 open(path+filename[i]);
g=getTitle();
g2 = replace(g,"DYST.tif","NCAM.tif");
g3 = replace(g,"DYST.tif","DAPI.tif");
open(path+g3);
open(path+g2);
g1 = replace(g,"DYST.tif","composite_RGB.tif");
run("Merge Channels...", "c1=["+g+"] c2=["+g2+"] c3=
["+g3+"] create");
run("RGB Color");
selectWindow("Composite");
close();
selectWindow("Composite (RGB)");
saveAs(newDir+g1);
close();
 }}

run("Stitch Directory with Images (unknown configura-
tion)", "image_directory=["+newDir+"] output_file_na-
me=TileConfiguration.txt rgb_order=rgb
channels_for_registration=Red fusion_method=[Linear
Blending] fusion_alpha=1.50 regression_threshold=0.30
```

```
max/avg_displacement_threshold=2.50 absolute_displace-
ment_threshold=3.50");
d1 = replace(e,"1_DYST.tif","NCAM_Stitched.tif");
selectWindow("Stitched Image");
saveAs(newDir1+d1);
close();

open(newDir1+d1);
run("Split Channels");
a1 = replace(d1,"Stitched.tif","Stitched.tif (red)");
a2 = replace(d1,"Stitched.tif","Stitched.tif (green)");
a3 = replace(d1,"Stitched.tif","Stitched.tif (blue)");
run("Merge Channels...", "c1=["+a1+"] c2=["+a2+"] c3=
["+a3+"] create");
saveAs(newDir1+d1);
close();
}
```

2.6 Quantitative and Qualitative Analyses

In the following section, we will outline a simple, semiautomated image analysis workflow using Fiji and the ObjectJ plugin. Our approach to this analysis is inherently subjective and therefore requires that samples are blinded to the operator. While more automated solutions are available, the human eye remains a valuable part of this analysis. This is especially true regarding smaller positive cells (<100 μm^2), as compared to the mean size of ~4000–5000 μm^2 of the other fibers in the sample. The percentage of positive fibers in most biopsies is below 2%, necessitating biopsies with at least ~500 fibers on a cross section (see Fig. 5). Typically, we note between 1000 and 1500 fibers.

2.7 A: Quantification of Denervated Fibers

Positively stained muscle fibers (NCAM$^+$/MHCn$^+$/MHCe$^+$) are determined as cells with a clear and complete dystrophin labeling and a clear staining of one of the three markers. Centralized nuclei can be assessed alongside the other markers. When quantifying positive fibers, instead of merely counting, we suggest also measuring the cross-sectional area of the positive fibers, as this is only slightly more time-consuming while providing additional valuable insight. All suitable fibers in a cross-section are counted before the positive fibers are assessed, and the number of positive fibers is expressed relative to the total number of fibers in the section. It is important to exclude areas with excessive space between fibers, damage from the staining procedure, and longitudinally cut fibers. Fibers at the edge of the section should also be excluded. Clusters of positive fibers should always be taken note of (*see* **Note 2**). When quantifying a marker, positive fibers are classified according to whether they also express other markers. Specifically, NCAM-

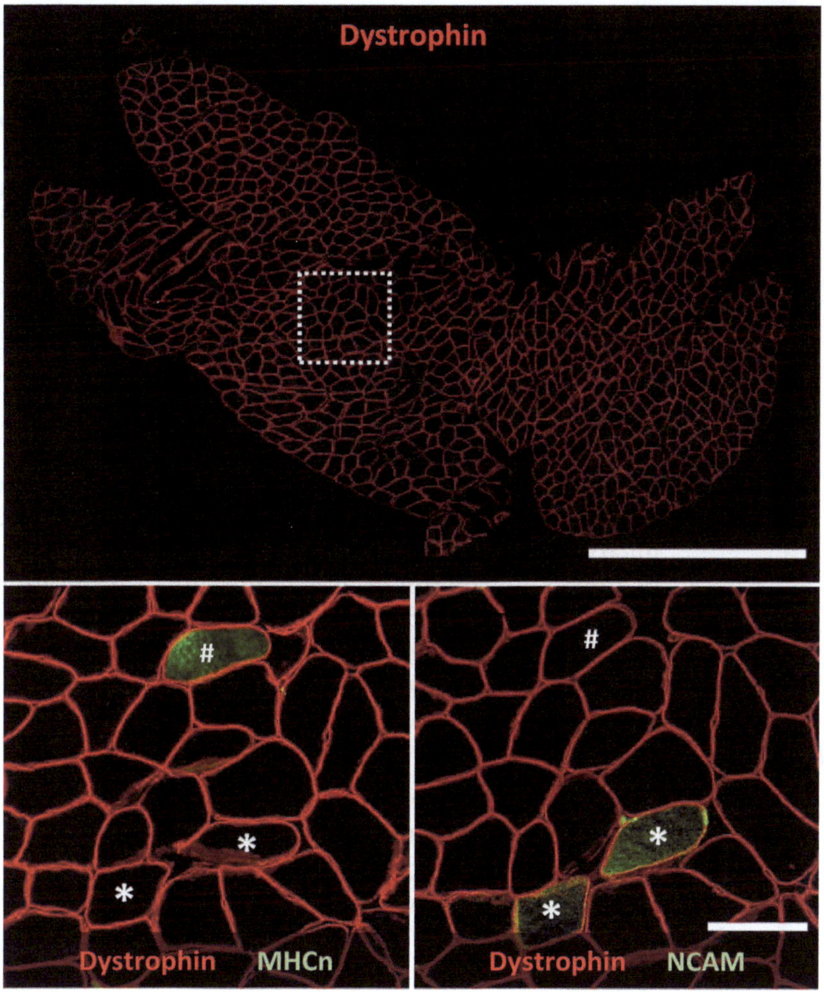

Fig. 5 Overview image of an entire cross section of a human muscle biopsy stained with dystrophin (red). A large number of muscle fibers is needed to quantify denervated fibers, as these are present in low numbers. The highlighted area is shown at higher magnification in bottom images; left stained for dystrophin (red) and MHCn (green), while the right image is stained for dystrophin (red) and NCAM (green). All fibers are not positive for both MHCn and NCAM, as shown by the stars indicating the same two $NCAM^+/MHCn^-$ fibers on both sections, while hashtags indicate the same $MHCn^+/NCAM^-$ fiber on both sections. MHCn, neonatal myosin heavy chain; NCAM neural cell adhesion molecule. Scale bar in overview image: 1000 μm. Scale bar in highlighted area: 100 μm

positive fibers are checked for central nuclei and MHCn. Afterward, MHCn-positive fibers are checked for central nuclei and NCAM. In this way co-expression will be available for both markers, since not all fibers are positive for both. It should be noted that MHCe-positive fibers in the skeletal muscle of healthy elderly individuals in the rested state are rare and, therefore, may indicate regeneration rather than denervation.

Importantly, the staining intensity of both NCAM- and MHCn-positive fibers varies between and within individuals, and even within the same fiber, thus complicating analysis. As shown in Fig. 6, Fiji contains tools which can help in establishing a lower threshold for staining intensity. A simple approach is to draw a line across a fiber and have the staining intensity shown in a plot profile, thereby enabling one to use a minimum staining intensity. Alternatively, the polygon tool can be used to outline the fiber and the staining intensity shown as a histogram. In this way, a minimum staining intensity, or a mean across the whole fiber, can be used as a threshold. While measuring signal intensity of potentially positive fibers has not been part of our routine approach, it may be useful to establish a threshold based on a subset of representative samples, which can then be used as a guide for determining if a fiber is positive or negative. However, due to the many potential artifacts related to immunofluorescence, the human eye of an experienced microscopist remains the most important factor in identifying positive fibers. Furthermore, it is important to be aware that MHCe and MHCn transiently reappear in muscle regeneration after injury [4]. However, in the basal state, regenerating fibers are rare in healthy individuals (*see* **Note 3**).

2.8 B: Step-by-Step Guide

- Setup folder and software (Windows-based)

1. Create a new folder for each sample, and place the corresponding images within. We recommend that individual images are stitched into one large image. This guide assumes that one (stitched) image per staining is used.
2. Open Fiji and load the ObjectJ plugin (must be downloaded and installed—refer to the list of materials). It can be useful to assign specific shortcuts, such as hand tool, zoom, maximize, and brightness.
3. Create a new ObjectJ project: Place the ObjectJ file in a sample folder, and name it accordingly.
4. Under the "Objects" menu, create the following elements;
   (a) "Fibrecount" as "Point"
   (b) "NCAM," "NCAM/CN," "NCAM/MHCn" and "NCAM/CN/MHCn" as "Polygon"
   (c) "MHCn," "MHCn/CN," "MHCn/NCAM" and "MHCn/CN/NCAM" as "Polygon"
5. It can be beneficial to assign specific colors to each element, as this will aid analysis.
6. Untick "Composite Objects."
7. Under the "Columns" tab, assign an "operation" to each element: Choose "Count" to "Point" and "Area" to "Polygon." Each column should be named according to the operation that it was linked to.

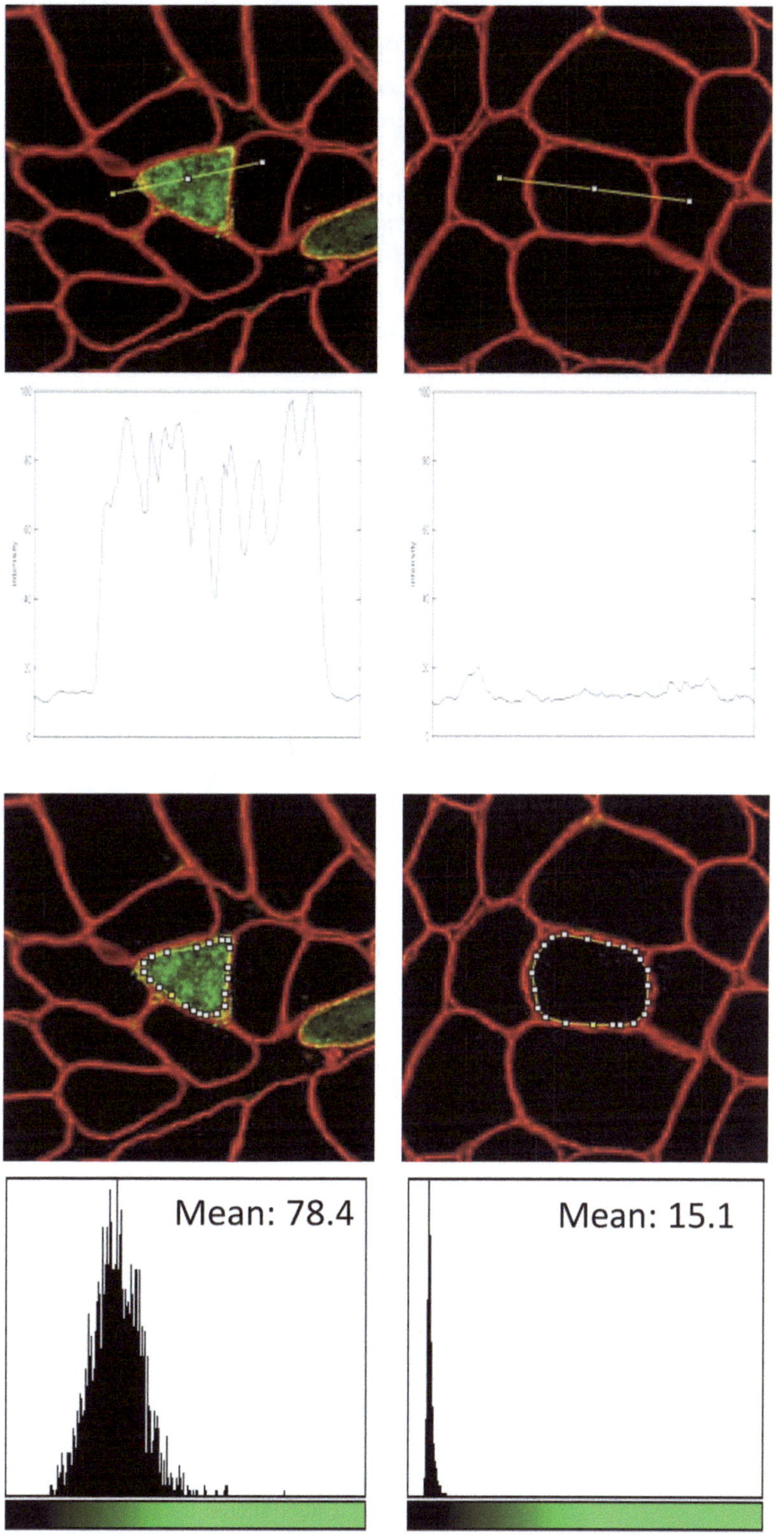

Fig. 6 Staining intensity profiling. In the top panel, the staining intensity of the green channel of two individual fibers from the same biopsy is evaluated using the line profile tool in Fiji. Results are displayed in the corresponding plot profiles

8. Once an ObjectJ project has been established, it can be copied to other folders, renamed, and used for analysis.
9. Load the stitched images into a project by dragging and dropping.
10. Calibrate images using information from your microscope using the "Set Scale" function under the "Analyze" tab in Fiji. Consider using the Global" function.

- Image Analysis

1. Open dystrophin/NCAM image, and turn off the green and blue channels.
2. Slightly overexpose the red channel to see and count all fibers.
3. Perform fiber count. Important: Leave out areas with longitudinally cut fibers or damaged areas
4. Turn on the green channel: Adjust green and red channel intensity to appropriate levels (keep constant between samples). Turn the blue channel on and off as needed for counting central nuclei.
5. Open the dystrophin/MHCn image and adjust the intensity of red and green channels appropriately. The two images should be visible side by side.
6. Working on the dystrophin/NCAM image, $NCAM^+$ fibers are now evaluated for central nuclei and MHCn, and the contours are delineated manually for the determination of fiber size.
7. Afterward, working on the dystrophin/MHCn image, $MHCn^+$ fibers are now evaluated for central nuclei and NCAM, and the contours are delineated manually.
8. Close both images and copy data from ObjectJ into a spreadsheet (ObjectJ Results → Copy/Export → Copy).

Fig. 6 (continued) below each image, where the signal intensity of each pixel along the line is recorded. In the bottom panel, the same two images are shown, but this time the staining intensity of the green channel of the individual fibers is evaluated using the polygon tool in Fiji. Results are displayed in the corresponding histograms below each image, with the mean pixel intensity provided in top-right corner. Using both tools, it is clear that the left fiber is strongly positive, while right fiber is negative. Such analyses can be used to establish a threshold based on a subset of representative samples to assist in determining if a fiber should be classified as positive or negative. Importantly, careful assessment by the investigator always precedes, as there are several potential artefacts related to immunofluorescence which can affect the staining intensity. Dystrophin (red) and NCAM (green). NCAM neural cell adhesion molecule

9. The percentage of NCAM- and MHCn-positive fibers is then available, as well as whether they express the other marker. Data on the cross-sectional area for each positive fiber is also generated.

2.9 C: Serial Cross-Sections

Collecting consecutive biopsy tissue sections in a systematic manner, termed serial sectioning, is useful for following individual fibers over a defined length, which is determined by the number and thickness of the sections. The fibers of interest can be compared to other (non-affected) fibers in the same biopsy or fibers in biopsies from control samples. Depending on the research question and immunohistochemical approach, two to three different antibodies can be combined for multiple labeling of a section, with other antibodies on the following slides, before repeating the first combination. The number of slides between the repeated sequence will depend on several factors, including how consistent the expression of the target proteins is. In addition, the ability to identify the same fibers in subsequent sections is strongly influenced by the inherent changes in muscle fiber morphology and position and artifacts related to tissue folding, missing areas, and uneven staining across the section. To exemplify the use of serial cross sections, a single muscle fiber is followed over 380 μm, where its expression of MHCe, MHCn, and NCAM is evaluated (see Fig. 7). NCAM appears to be the most stable marker, although between fiber difference in staining intensity is common. On the contrary, as also visible in Fig. 8, we have observed MHCn-positive fibers displaying variation in protein expression over relatively short distances (*see* **Note 4**). An example where 12 different antibodies in various combinations are used on serial sections of a human muscle biopsy covering a total distance of ~600 μm is found in Mackey and Kjaer (2017) [34].

2.10 List of Materials

Chemicals and Materials

- Lidocaine (443122, Xylocain, Aspen Nordic, Ballerup, Denmark).
- Tissue-Tek (4583, Sakura Finetek, Europe, AV Alphen aan den Rijn, The Netherlands).
- Isopentane/2-methylbutan (8113, J.T. Baker, Avantor Performance Materials BV, Deventer, The Netherlands).
- TBS (NaCl 154 mMol, Tris-base 50 mMol, pH 7.4–7.6). Store at 4 °C. Make fresh solution every 14th day. Phosphate-buffered saline can also be used.
- 1% BSA (A3912, Sigma-Aldrich, Merck Life Sciences A/S, Søborg, Denmark) in TBS. Store at 4 °C. Make fresh solution every 14th day.

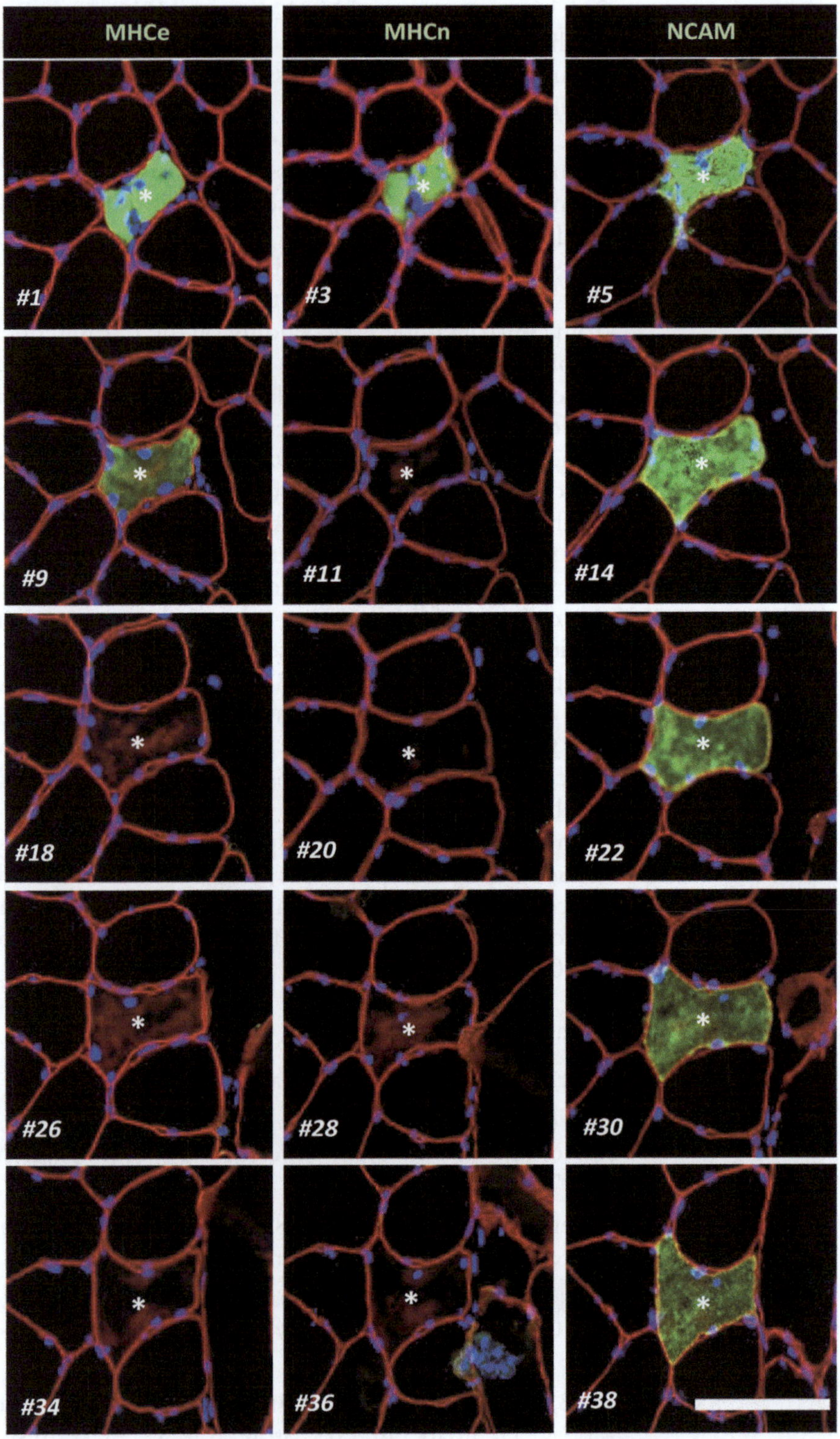

Fig. 7 Markers of muscle fiber denervation evaluated in 10 μm-thick serial cross sections from a human muscle biopsy. Microscope images of a serial cross section of a biopsy collected from an elderly individual and stained for dystrophin (red), nuclei (blue), and MHCe, MHCn, or NCAM (green). Note that while MHCe and MHCn staining is not constant across serial sections, the presence of NCAM is constant over the series of

- 5% formaldehyde (01000 Histofix, Histolab, Gothenburg, Sweden). Store at room temperature in a well-ventilated cabinet.
- ProLong Gold Antifade (P36931, Invitrogen, Molecular Probes, OR, USA).
- Cryotube 1.8 mL (Thermo Scientific, Nunc A/S, Roskilde, Denmark).
- Hydrophobic pen (Z377821; Sigma-Aldrich).
- SuperFrostPlus glass slide (J1800AMNZ, Thermo Scientific, Waltham, MA, USA).

Software

- Olympus cellSens software (v.1.14).
- Fiji (ImageJ, v.1.53, National Institutes of Health, Bethesda, Maryland, USA).
- ObjectJ plugin v. 1.04. It can be downloaded from: https://sils.fnwi.uva.nl/bcb/objectj/index.html.

Antibodies and the Dilution Used

- Mouse IgG2b dystrophin (Sigma-Aldrich D8168): 1:500.
- Mouse IgG1 MHCe (Hybridoma Bank F1.652): 1:100.
- Mouse IgG1 MHCn (Novocastra NCL-MHCn): 1:100.
- Mouse IgG1 NCAM (Becton Dickinson 347740): 1:50.
- Goat anti-mouse IgG1, 488 (Invitrogen A-21121): 1:500.
- Goat anti-mouse IgG2b, 568 (Invitrogen A-21144): 1:500.

3 Notes

1. As per manufacturer instructions (see IFU document), staining for MHCe and MHCn requires unfixed sections. A direct comparison between immunohistochemical preparations of regenerating human muscle that was fixed either before incubation with the primary antibody or after incubation with secondary antibody is shown in Fig. 4. This is most relevant for the MHCn antibody. Fixation of the tissue after the secondary antibodies provides clearer staining of the nuclei with DAPI.

Fig. 7 (continued) sections displayed. MHCe+ fibers were rare in our analyses of biopsies collected from healthy elderly individuals, so it is possible this particular fiber is regenerating rather than denervated. Centralized nuclei can also be observed in some sections. Section number is written in bottom-left corner. Asterisk indicates the same fiber on all sections. MHCe embryonic myosin heavy chain; MHCn neonatal myosin heavy chain; NCAM neural cell adhesion molecule. Scale bar: 100 μm

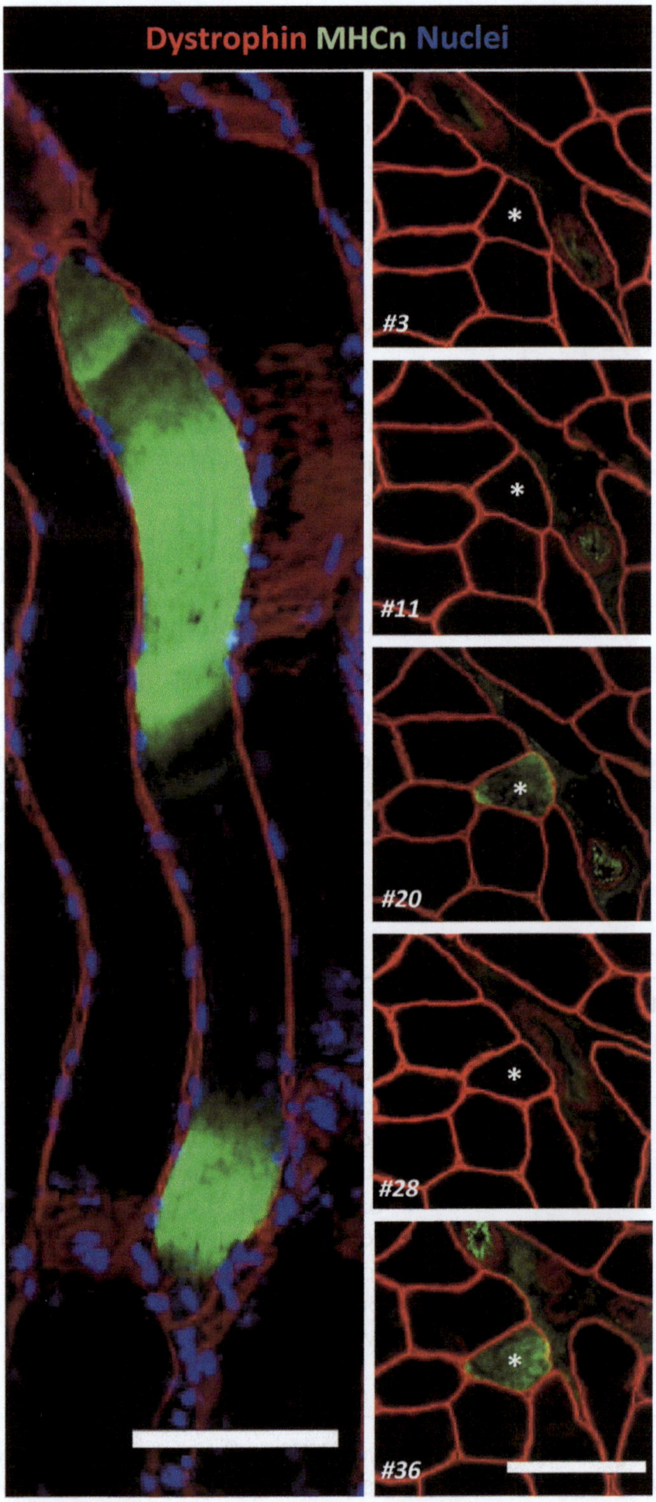

Fig. 8 Segmented staining pattern of MHCn in a longitudinally oriented fiber. Longitudinally oriented MHCn+ fibers (green) often present with a segmented staining profile, meaning that a positive signal can disappear within relatively short distances and even reappear again. This is shown in serial sections

2. In rare instances, fiber clusters of 10–20 fibers showing strong NCAM immunoreactivity are found in muscle biopsies of both young and old subjects. These fibers also show internal dystrophin and centralized nuclei and are negative for both MHCn and MHCe. We have used an antibody against collagen XXII, which labels tissue junctions [35], and confirmed these fibers to be myotendinous junction-related fibers, as all muscle fibers have been shown to express NCAM in a short band close to the tendon [31].
3. Conflicting with markers of regeneration. The reason for the similarity in markers of denervated and regenerating muscle fibers is unknown. One potential explanation is that regenerating muscle fibers are not yet innervated, supporting the view that innervation suppresses the expression of NCAM and MHCn. While it is likely that expression of these (and other) markers follows different time courses during regeneration and denervation, finding exclusive markers for each process is perhaps not realistic.
4. Longitudinally oriented MHCn-positive fibers often present with a segmented staining profile (see Fig. 8 for examples), which means that MHCn-positive fibers can transition from clearly positive to clearly negative within only a few sections. This phenomenon has been described before in animal [6] muscles and by us in human muscles [11]. It is unclear whether this segmented staining pattern is linked to the age-related occurrence of hybrid fibers [36, 37].

4 Conclusions

A well-designed workflow consisting of immunohistochemistry, fluorescent microscopy, and image analysis allows the researcher to assess specific proteins in all kinds of samples. In this chapter, we have used a qualitative and quantitative analysis of denervated muscle fibers in human biopsies as a working example of how this can be achieved. In doing so, we provide several example images and a step-by-step guide on image analysis and highlight important tips and pitfalls. It is our hope that the chapter has provided the reader with ideas and knowledge on how to tailor this approach to their research questions.

Fig. 8 (continued) following a single fiber over 330 μm, which changes repeatedly from negative to positive. The functional significance of this segmented staining pattern is unclear, but it could be related to the age-related occurrence of hybrid fibers. Section number is written in lower-left corner. Asterisk indicates the segmentally MHCn-positive fiber. MHCn neonatal myosin heavy chain. Scale bars: 100 μm

Acknowledgments

Funding from the Lundbeck Foundation (R344-2020-254) is gratefully acknowledged.

Figures 1, 2, and 3 were created with BioRender. The monoclonal antibody F1.652 (MHCe), developed by Blau, H.M., was obtained from the Developmental Studies Hybridoma Bank, created by the NICHD of the NIH, and maintained at The University of Iowa, Department of Biology, Iowa City, IA 52242.

References

1. Draeger A, Weeds AG, Fitzsimons RB (1987) Primary, secondary and tertiary myotubes in developing skeletal muscle: a new approach to the analysis of human myogenesis. J Neurol Sci 81:19–43
2. Whalen RG, Sell SM, Butler-Browne GS, Schwartz K, Bouveret P, Pinset-Härstöm I (1981) Three myosin heavy-chain isozymes appear sequentially in rat muscle development. Nature 292:805–809
3. Frontera WR, Ochala J (2015) Skeletal muscle: a brief review of structure and function. Calcif Tissue Int 96:183–195
4. Schiaffino S, Rossi AC, Smerdu V, Leinwand LA, Reggiani C (2015) Developmental myosins: expression patterns and functional significance. Skelet Muscle 5
5. Covault J, Sanes JR (1985) Neural cell adhesion molecule (N-CAM) accumulates in denervated and paralyzed skeletal muscles. Proc Natl Acad Sci 82:4544–4548
6. Schiaffino S, Gorza L, Pitton G, Saggin L, Ausoni S, Sartore S, Lømo T (1988) Embryonic and neonatal myosin heavy chain in denervated and paralyzed rat skeletal muscle. Dev Biol 127:1–11
7. Karlsen A, Bechshøft RL, Malmgaard-Clausen NM, Andersen JL, Schjerling P, Kjaer M, Mackey AL (2019) Lack of muscle fibre hypertrophy, myonuclear addition, and satellite cell pool expansion with resistance training in 83-94-year-old men and women. Acta Physiol (Oxf) 227:e13271
8. Karlsen A, Soendenbroe C, Malmgaard-Clausen NM, Wagener F, Moeller CE, Senhaji Z, Damberg K, Andersen JL, Schjerling P, Kjaer M, Mackey AL (2020) Preserved capacity for satellite cell proliferation, regeneration, and hypertrophy in the skeletal muscle of healthy elderly men. FASEB J 34: 6418–6436
9. Mackey AL, Karlsen A, Couppé C, Mikkelsen UR, Nielsen RH, Magnusson SP, Kjaer M (2014) Differential satellite cell density of type I and II fibres with lifelong endurance running in old men. Acta Physiol 210:612–627
10. Soendenbroe C, Bechshøft CJL, Heisterberg MF, Jensen SM, Bomme E, Schjerling P, Karlsen A, Kjaer M, Andersen JL, Mackey AL (2020) Key components of human Myofibre denervation and neuromuscular junction stability are modulated by age and exercise. Cell 9:893
11. Soendenbroe C, Heisterberg MF, Schjerling P, Karlsen A, Kjaer M, Andersen JL, Mackey AL (2019) Molecular indicators of denervation in aging human skeletal muscle. Muscle Nerve 60:453–463
12. Lexell J, Taylor CC, Sjöström M (1988) What is the cause of the ageing atrophy? Total number, size and proportion of different fiber types studied in whole vastus lateralis muscle from 15- to 83-year-old men. J Neurol Sci 84:275–294
13. Verdijk LB, Dirks ML, Snijders T, Prompers JJ, Beelen M, Jonkers RAM, Thijssen DHJ, Hopman MTE, Van Loon LJC (2012) Reduced satellite cell numbers with spinal cord injury and aging in humans. Med Sci Sports Exerc 44:2322–2330
14. Castro MJ, Apple DF, Staron RS, Campos GE, Dudley GA (1999) Influence of complete spinal cord injury on skeletal muscle within 6 mo of injury. J Appl Physiol (1985) 86:350–358
15. Lepore E, Casola I, Dobrowolny G, Musarò A (2019) Neuromuscular junction as an entity of nerve-muscle communication. Cell 8:906
16. Musarò A (2013) Understanding ALS: new therapeutic approaches. FEBS J 280:4315–4322
17. Winter B (1999) NCAM, vimentin and neonatal myosin heavy chain expression in human muscle diseases. Neuropathol Appl Neurobiol 25:417–424

18. Butler-Browne GS, Bugaisky LB, Cuénoud S, Schwartz K, Whalen RG (1982) Denervation of newborn rat muscle does not block the appearance of adult fast myosin heavy chain. Nature 299:830–833
19. Harkema L, Youssef SA, de Bruin A (2016) Pathology of mouse models of accelerated aging. Vet Pathol 53:366–389
20. D'Antona G, Pellegrino MA, Adami R, Rossi R, Carlizzi CN, Canepari M, Saltin B, Bottinelli R (2003) The effect of ageing and immobilization on structure and function of human skeletal muscle fibres. J Physiol Lond 552:499–511
21. Daou N, Hassani M, Matos E, De Castro GS, Galvao Figueredo Costa R, Seelaender M, Moresi V, Rocchi M, Adamo S, Li Z, Agbulut O, Coletti D (2020) Displaced myonuclei in cancer cachexia suggest altered innervation. Int J Mol Sci 21:1092
22. Arizono N, Koreto O, Iwai Y, Hidaka T, Takeoka O (1984) Morphometric analysis of human neuromuscular junction in different ages. Acta Pathol Jpn 34:1243–1249
23. Jones RA, Harrison C, Eaton SL, Llavero Hurtado M, Graham LC, Alkhammash L, Oladiran OA, Gale A, Lamont DJ, Simpson H, Simmen MW, Soeller C, Wishart TM, Gillingwater TH (2017) Cellular and molecular anatomy of the human neuromuscular junction. Cell Rep 21:2348–2356
24. Oda K (1984) Age changes of motor innervation and acetylcholine receptor distribution on human skeletal muscle fibres. J Neurol Sci 66: 327–338
25. Wokke JHJ, Jennekens FGI, van den Oord CJM, Veldman H, Smit LME, Leppink GJ (1990) Morphological changes in the human end plate with age. J Neurol Sci 95:291–310
26. Aubertin-Leheudre M, Pion CH, Vallée J, Marchand S, Morais JA, Bélanger M, Robitaille R (2020) Improved human muscle biopsy method to study neuromuscular junction structure and functions with aging. J Gerontol A Biol Sci Med Sci 75:2098–2102
27. Gundersen K, Rabben I, Klocke BJ, Merlie JP (1995) Overexpression of myogenin in muscles of transgenic mice: interaction with Id-1, negative crossregulation of myogenic factors, and induction of extrasynaptic acetylcholine receptor expression. Mol Cell Biol 15:7127–7134
28. Soendenbroe C, Heisterberg MF, Schjerling P, Kjaer M, Andersen JL, Mackey AL (2022) Human skeletal muscle acetylcholine receptor gene expression in elderly males performing heavy resistance exercise. Am J Physiol Cell Physiol 323(1):C159–C169
29. Soendenbroe C, Dahl CL, Meulengracht C, Tamáš M, Svensson RB, Schjerling P, Kjaer M, Andersen JL, Mackey AL (2022) Preserved stem cell content and innervation profile of elderly human skeletal muscle with lifelong recreational exercise. J Physiol 600(8):1969–1989
30. Bergstrom J (1975) Percutaneous needle biopsy of skeletal muscle in physiological and clinical research. Scand J Clin Lab Invest 35: 609–616
31. Jakobsen JR, Jakobsen NR, Mackey AL, Koch M, Kjaer M, Krogsgaard MR (2018) Remodeling of muscle fibers approaching the human myotendinous junction. Scand J Med Sci Sports 28:1859–1865
32. Preibisch S, Saalfeld S, Tomancak P (2009) Globally optimal stitching of tiled 3D microscopic image acquisitions. Bioinformatics 25: 1463–1465
33. Schindelin J, Arganda-Carreras I, Frise E, Kaynig V, Longair M, Pietzsch T, Preibisch S, Rueden C, Saalfeld S, Schmid B, Tinevez J-Y, White DJ, Hartenstein V, Eliceiri K, Tomancak P, Cardona A (2012) Fiji: an open-source platform for biological-image analysis. Nat Methods 9:676–682
34. Mackey AL, Kjaer M (2017) The breaking and making of healthy adult human skeletal muscle in vivo. Skelet Muscle 7:1–18
35. Koch M, Schulze J, Hansen U, Ashwodt T, Keene DR, Brunken WJ, Burgeson RE, Bruckner P, Bruckner-Tuderman L (2004) A novel marker of tissue junctions, collagen XXII. J Biol Chem 279:22514–22521
36. Andersen JL, Terzis G, Kryger A (1999) Increase in the degree of coexpression of myosin heavy chain isoforms in skeletal muscle fibers of the very old. Muscle Nerve 22:449–454
37. Murach KA, Dungan CM, Kosmac K, Voigt TB, Tourville TW, Miller MS, Bamman MM, Peterson CA, Toth MJ (2019) Fiber typing human skeletal muscle with fluorescent immunohistochemistry. J Appl Physiol (1985) 127: 1632–1639

Chapter 10

Stable Isotope Tracer Methods for the Measure of Skeletal Muscle Protein Turnover

Matthew S. Brook, Daniel J. Wilkinson, and Ken Smith

Abstract

In this chapter, we will discuss two approaches to measure protein metabolism using stable isotope tracers and the advantages and disadvantages of approaches employed to measure stable isotope labeling in saliva, plasma, urine, and tissue samples—in this case skeletal muscle. We will outline the application of the primed constant infusion approach to measure skeletal muscle protein turnover over acute periods, primarily protein synthesis. We will also provide detailed information on the re-emergence of D_2O (heavy water) as a flexible and unique stable isotope tracer to make measures of muscle protein synthesis over both acute and extended periods of time, i.e., days, weeks, and months, and highlight the advantages and disadvantages of both approaches along with the complementarity of applying both approaches to make both acute and chronic measures of protein metabolism in a single study.

Key words Skeletal muscle, Protein synthesis, Protein turnover

1 Introduction

1.1 Importance and Value of Protein Turnover Measures

Protein turnover is a dynamic process, and the amount of protein is the net result of the relative rates of protein synthesis and breakdown. When these processes are equal over a given time period, the net balance is zero, and the concentration of this protein is maintained, so simply measuring concentration will provide only a snapshot of the metabolism of any protein at any given time. More detailed mechanistic information is gathered from the measurement of the absolute rates of synthetic and proteolytic processes regulating this amount of protein and in particular the physiological regulation of these processes in both health and disease. Knowledge of the normal regulation of both synthesis and breakdown in response to feeding/nutrition, exercise/inactivity, and hormones is important to understand their impact in maintaining health and preventing disease—especially in situations where an individual protein or cell/tissue undergoes an increase or decrease in concentration or mass. This is particularly important in understanding the

Philip J. Atherton and Daniel J. Wilkinson (eds.), *Neuromuscular Assessments of Form and Function*, Neuromethods, vol. 204,
https://doi.org/10.1007/978-1-0716-3315-1_10,

growth (i.e., hypertrophy) or loss (i.e., atrophy) of muscle, for example, during the transition from childhood to adulthood (or the response to exercise, e.g., resistance exercise) or the loss of muscle mass that inevitably happens in older age and how we might accelerate or slow down those changes. It is especially important to understand the changes that result in the loss of muscle in aging, where these losses can often be accelerated with the onset of disease, e.g., cancer, cardiac cachexia, diabetes, or sedentarism/inactivity. From the robust and accurate measurement of the rates of protein synthesis and protein breakdown, we can begin to develop targeted therapeutic interventions, be they nutritional, exercise-/activity-based, or pharmacological, to attempt to minimize these losses through greater stimulation of protein synthesis or inhibition of protein breakdown or indeed a combination of the two. Over the past five or so decades, techniques for the in vivo measurement of protein turnover (synthesis and breakdown) using stable isotope tracers have been extensively developed and validated. These techniques will now be introduced in more detail, describing their implementing, key considerations and applications to our understanding of physiology.

1.2 Stable Isotopes

Stable isotopes are elements that occupy the same position in the periodic table, are essentially "chemically and functionally identical," but differ in mass due to a different number of neutrons within the atomic nucleus. This difference in mass makes these isotopes analytically distinguishable from each other—via mass spectrometry, they are inherently stable (as they do not produce ionizing radiation) and safe for use in humans [1]. Therefore, by replacing one (or more) of the ^{12}Carbon, ^{14}Nitrogen, ^{16}Oxygen, or ^{1}Hydrogen (For the purpose of this chapter, we will focus on the isotopes that are readily available, are widely used, and are metabolically useful, being constituent parts of our carbon-based/organic metabolites; however, they are not limited to these alone; *see* Table 1) within a compound of interest with their heavier less naturally abundant stable isotopic equivalents, ^{13}Carbon (^{13}C), ^{15}Nitrogen (^{15}N), ^{18}Oxygen (^{18}O), or ^{2}Hydrogen (^{2}H or Deuterium), respectively, a stable isotope tracer can be created. They can be incorporated into a range of metabolites, e.g., amino acids, carbohydrates, and fatty acids, and then introduced into a biological system of interest. The metabolic fate of these stable isotopically labeled compounds can then be "traced" or measured by sampling blood, breath, urine, or tissue. Since they are not distinguished from the endogenous compound by the metabolic processes occurring within all cells of the body, they can be used to determine rates of flux or turnover within these metabolic pathways. In most cases, the sensitivity of the analytical equipment available (e.g., mass spectrometry) ensures that these labeled compounds can be added to the system at comparatively low levels.

Table 1
Summary of metabolically useful stable isotopes and the relative abundances of the heavier stable isotopes (adapted from [1])

Element	Stable Isotopes	% Natural abundance
Hydrogen	^{1}H ^{2}H	99.985 0.015
Carbon	^{12}C ^{13}C	98.89 1.11
Nitrogen	^{14}N ^{15}N	99.63 0.37
Oxygen	^{16}O ^{17}O ^{18}O	99.76 0.037 0.204
Sulfur	^{32}S ^{33}S ^{34}S	95.00 0.76 4.22
Silicon	^{28}Si ^{29}Si ^{30}Si	92.21 4.9 3.09
Iron	^{54}Fe ^{56}Fe ^{57}Fe ^{58}Fe	5.82 91.66 2.19 0.33
Zinc	^{64}Zn ^{66}Zn ^{67}Zn ^{68}Zn ^{70}Zn	48.89 27.81 4.11 18.57 0.62
Selenium	^{74}Se ^{76}Se ^{77}Se ^{78}Se ^{80}Se	0.87 9.02 7.58 23.52 49.82

1.3 Mass Spectrometry

The primary advantage of mass spectrometry with respect to stable isotope approaches to measure turnover is that these instruments allow the detection/visualization of the heavy isotope within a compound, and the isotope can be "traced" as it moves from one body pool to another and into (and out of) protein, with appropriate sampling and preparation of blood, tissue, urine, and even breath ($^{13}CO_2$, providing estimates of oxidation, or ^{15}N in urea/ammonia, which provides estimates of protein breakdown). Gas chromatography–mass spectrometry (GC-MS) is the most accessible and widely used approach to measure the labeling of amino

Table 2
Relative performance ranges of the three most commonly used gas chromatography mass spectrometry modalities used to measure precursor and product enrichments in protein turnover studies

MS instrumentation	Concentration range	Enrichment range (APE)
GC-MS	ng–pg	~0.2–99
GC-MS-MS	ng–pg	~0.01–99
GC-C/P-IRMS	μg	~0.0005–1

C/P denotes combustion or pyrolysis modes of IRMS

acids in body fluids and intracellular water. It is extremely sensitive and flexible in terms of concentration measurement (ranging from microgram (μg) to as little as a picogram (pg) of a compound); however, it is the least sensitive approach for determining the isotope ratio or enrichment of a compound (with a range of 0.2 to 99 atoms percent excess, APE), which is not sufficiently sensitive for the routine and robust measurement of label incorporated into slowly turning over metabolic pools such as muscle protein (*see* **Note 1**). With a slower turning over pool (e.g., muscle, bone, tendon), the MS method of choice is isotope ratio mass spectrometry (IRMS), which can measure down to 0.01–0.005 APE; however, a caveat of this approach is that much more material (i.e., μM of amino acid) is required, as the amino acid is combusted back to a simple gas (e.g., CO_2) for measurement. Depending on the duration of the experiment and the model systems involved (i.e., how rapidly they turnover), GC-MS-MS can also be used. A summary of these different instrumentations is provided in Table 2 below; however, for a more detailed comparison of the mass spectrometric instruments and systems available for measuring stable isotopically labeled amino acids and their relative sensitivity in terms of enrichment and amount of material required, readers are directed to Wilkinson [1]. The power of these techniques for uncovering key aspects of the regulation and dysregulation of protein turnover in response to nutrition, healthy aging, and disease and the importance of muscle mass in their regulation has been clear over the years. Although beyond the scope of this chapter, interested readers are directed toward the following articles for more information [2–10]. The remainder of this article will discuss key aspects of the methods and considerations required to set up and analyze stable isotope tracers for the analyses of human muscle protein turnover, ranging from the traditional substrate specific, i.e., tracer amino acid, approaches to the recent reintroduction of deuterium oxide (D_2O) tracer methods.

2 Material and Reagents

2.1 Stable Isotope Tracers—Storage and Preparation—Substrate Specific and D20

Most tracers can be stored at room temperature for many years in powder form, sealed, and placed in a dry and cool environment. In the past, it was possible to prepare the tracer infusion in the laboratory under aseptic conditions (i.e., in a Class II cabinet, using a sterile aseptic technique); however, most studies now require that tracers are QC tested and manufactured by pharmacies and that infusions are fully certified, sterile, and pyrogenicity tested (pre- and post-manufacture). Current shelf life limits on both Phe and Leu tracers of 18 months to 2 years provide a degree of reassurance and flexibility—in that infusion can be prepared on the day and at short-notice, under aseptic conditions, making their use more cost-effective, efficient and manageable. The tracer is prepared in 0.9% sterile saline bags (typically 250 mL), the volume of tracer required is calculated, and this volume is removed from the bag and replaced, via a 0.2 μM sterile filter, by the pre-prepared tracer solution. The infusate is then connected to an infusion pump and the lines are primed; a priming dose is given (*see* Fig. 1) and then a constant infusion is given over the duration of the study (*see* **Note 2**). The storage and preparation of D_2O are simpler due principally to the fact that it is given orally rather than intravenously. For storage and prior to human consumption, D_2O should be passed through a 0.2 μM sterile filter. This can be easily achieved with single-use filter units.

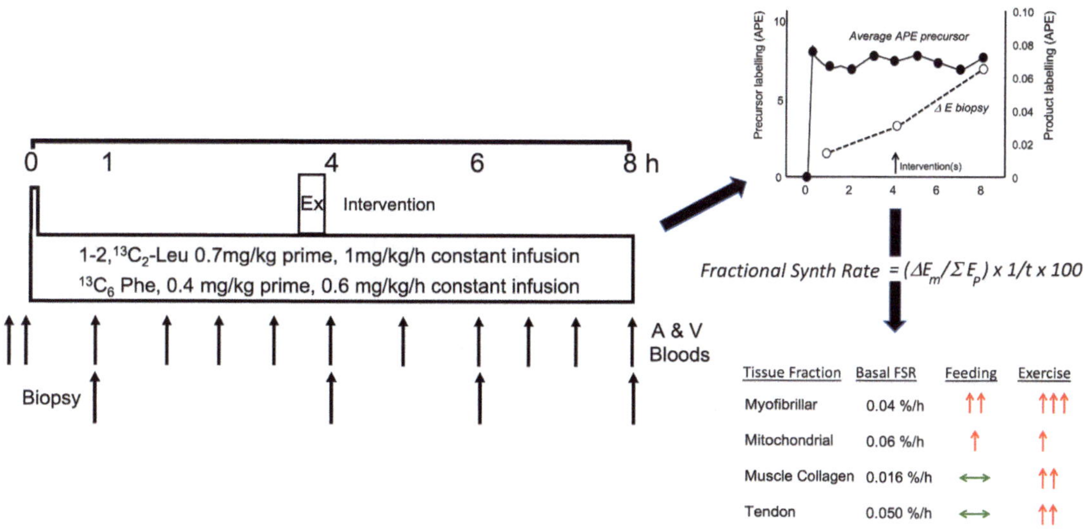

Fig. 1 A typical protocol for the acute determination of muscle protein synthesis (MPS), using a primed constant infusion approach in a human volunteer, the precursor and product labeling measurements undertaken and the basal turnover rates of muscle fractions and their responses to both feeding (a mixed meal or protein) and exercise (resistance type). Where E_m is the change in enrichment of labeled AA between biopsies, E_p is the average enrichment of labeled AA in the precursor pool of choice, and t is the time in hours

2.2 Reagents Required for Preparation of Amino Acid and Keto Acid Derivatives—Precursor Enrichment and Concentration

As with most derivatization processes (*see* **Note 3**), this will depend upon the quality of reagents, and these should be appropriately sealed and stored. Moisture is often the main cause of deterioration of reagents, as is contamination from poor pipetting practices; always decant the volume you require from the source bottle into clean and dry glassware for use. Many of the reagents come in sealed ampoules; although more expensive, these are stable for longer. However, good housekeeping and appropriate storage and handling of reagents can extend their shelf life and help reduce costs, as buying in bulk is more cost-effective.

Chemical	Source	Cat No	Safety
Ethanol 100%	Fisher Scientific	64-17-5	Flammable
Ortho-phenylenediamine (OPD)	Sigma-Aldrich	P1516	Toxic and dangerous to the environment
Hydrochloric acid (4N, 167 mL in 500 mL)	BDH/VWR	20252.335	Corrosive
Ethyl acetate	BDH	101086J	Flammable and irritant
Pyridine	Sigma	360570	Harmful and flammable
Acetonitrile	Sigma	00709	Highly flammable and harmful
N-methyl-tert-butyl dimethyl-silyl-trifluoro acetamide (MTBSTFA)	Sigma	375934	Irritant

2.3 Reagents Required for Derivatization of Amino Acid Hydrolysates for IRMS Analyses

As n-acetyl, n-propyl ester (NAP)—*For product enrichment, i.e., incorporation of labeled AA into muscle protein*

Chemical	Source	Cat No	Safety
Propyl acetate	Sigma-Aldrich	133108	Flammable and irritant
BF_3:propanol	Sigma-Aldrich	15718	Toxic, causes burns, and reacts with H_2O
Acetonitrile	Sigma-Aldrich	00709	Flammable and irritant

(continued)

Chemical	Source	Cat No	Safety
1,4 dioxan	Sigma-Aldrich	296309	Flammable, irritant, and carcinogenic
Triethylamine	Sigma-Aldrich	39428	Flammable and harmful
Acetic anhydride	Sigma-Aldrich	33085	Harmful and flammable
Chloroform	Sigma-Aldrich	288306	Harmful, irritant, and carcinogenic
$NaHCO_3$	BDH	27778.260	Not classified dangerous
Molecular sieve	Sigma-Aldrich	334316	Irritant

2.4 Reagents Required for Derivatization of Amino Acid Hydrolysates as Methoxy-Carbonyl-Methyl Ester (MCME)—An Alternative Derivative for Product Enrichment Analysis

Chemical	Source	Cat No	Safety
Methanol 100%	Fisher Scientific	M/4056/17	Flammable and toxic
Methyl chloroformate	Sigma-Aldrich	M35304	Very flammable, toxic, – corrosive, and reacts violently with water to form toxic gas
Pyridine anhydrous	Sigma-Aldrich	270970	Flammable, toxic, and toxic to the environment
Chloroform	Sigma-Aldrich	C2432	Harmful, irritant, and carcinogenic
Sodium bicarbonate	BDH	27778.260	Not classified dangerous
Molecular sieve	Sigma-Aldrich	334316	Irritant

The methoxy-carbonyl-methyl-ester (MCME) derivative has a number of advantages over the NAP derivative, primarily that it adds fewer carbons to the amino acid and, therefore, dilutes the ^{13}C tracer less. In addition, it can be used in an aqueous environment, removing the need to ensure that there is no water in the final hydrolysate, and the derivatization is faster, has fewer steps, and is more efficient, resulting in greater sensitivity of the IRMS measurement.

2.5 Reagents for Muscle Fractionation and Processing

Chemicals required for homogenization buffer composition—this should be at a final pH of 7.5 prior to use. This can be made in large volumes and stored for several weeks at room temperature. Once sodium orthovanadate and the inhibitor have been added, use them immediately.

Chemical	Source	Cat No	Safety	Final concentration
Tris-HCl	Sigma-Aldrich	T3253	Not classified dangerous	50 mM
EDTA	Sigma-Aldrich	E16421	Irritant	1 mM
EGTA	Sigma-Aldrich	E4378	Irritant	1 mM
B-Glycerophosphate	Sigma-Aldrich	G6251	Not classified dangerous	10 mM
Sodium fluoride	Sigma-Aldrich	57920	Toxic	50 mM
Sodium orthovanadate	Sigma-Aldrich	56508	Irritant	0.5 mM (add only before use)
Protease inhibitor tablet	Roche	05892791001	Corrosive	1 tablet per 10 mL

If isolation of mitochondrial fraction is required during processing, the following chemicals are required for the mitochondrial extraction buffer—this should be at a final pH of 7.5 and can be stored at room temperature for several weeks.

Chemical	Source	Cat No	Safety	Final concentration
MOPS	Fisher Scientific	BP308-100	Irritant	20 mM
KCl	Sigma-Aldrich	P3911	Not classified dangerous	110 mM
EGTA	Sigma-Aldrich	E4378	Irritant	1 mM

Other general chemicals for muscle processing are listed below:

Chemical	Source	Cat No	Safety
100% ethanol	SLS	Che1938	Flammable
Perchloric acid	VWR	20583.26	Flammable/corrosive
Sodium hydroxide	Sigma-Aldrich	30620	Corrosive
Ammonium hydroxide	Fisher Scientific	A3240/PB17	Corrosive
Hydrochloric acid	VWR	20252.335	Corrosive

3 Methods for Measuring Protein Synthesis

3.1 Precursor-Product Relationship—Measurement of Protein Synthesis—A Typical Labeled AA Constant Infusion Approach

In its simplest design, the measurement of protein turnover requires only two measurements: determination of the amino acid enrichment of the precursor pool for protein synthesis over the duration of the measurement and determination of the change in enrichment of the amino acid in the protein product over the same time—the precursor—product approach. The simplest and most widely adopted approach involves a primed constant infusion of a labeled amino acid over several hours with a biopsy at the beginning and again at the end of the measurement period.

The study design is crucial to answering your research and must be taken in line with the limitations of your tracer choice, your sampling capabilities, and the analytical instrumentation available to make the measurements. A typical human protocol is outlined in Fig. 1. The prime and infusion rate are chosen to achieve a steady state quickly and to limit the labeling to around 10 atoms percent excess (APE). Typically, when using leucine tracer, the priming dose is 0.7 mg/kg body weight, and the infusion is given as 1 mg/kg/h; this will generate a steady-state labeling between 6 and 10 APE. As the Phe pool is smaller, the prime is 0.4 mg/g, with an infusion of 0.6 mg/kg, which results in a similar range of labeling (6–10 APE: *see* **Notes 4** and **5**).

Fasting (or baseline) muscle protein synthesis (MPS) can be determined easily and robustly over 2–3 h (with multiple labeled Phe/Leu tracers and IRMS), prior to an intervention, e.g., feeding and exercise, so this approach allows the volunteer to act as their own control, and a biopsy is taken before the intervention and then 2–4 h following the intervention. Typically, we acquire a biopsy 2–3 h post-feeding and 3–4 h post-exercise due to the temporal response of human muscle to these interventions; i.e., the feeding response is generally rapid and short-lasting (~90 min), whereas the exercise response is delayed (~60–120 min), likely due to the energy demands of the exercise and tissue remodeling, but last longer, up to 4–5 h [2, 4]. By employing a unilateral exercise model, where only one leg is exercised, both the feeding and feeding plus exercise effects can be studied (*see* **Note 6**).

3.2 Choice of Precursor Pool

The choice of precursor pool represents a potential source of error; under ideal conditions, the aminoacyl-tRNA pool (within the tissue/cells under study) represents the most accurate estimate of the precursor labeling—it represents the point at which the AA is committed to protein synthesis. However, this measurement requires a tissue biopsy; the aminoacyl pool is extremely small and extremely labile, so MS determination is very challenging and is generally avoided—though given advances in protecting these volatile molecules and increased MS sensitivity, this should be achievable. Previous work has also demonstrated that the labeling of the intracellular AA pool is very similar (±10%) to that of the AA-tRNA pool, and so most commonly, researchers select the labeling in the intracellular pool as their surrogate precursor [11]. A minor disadvantage of choosing intracellular labeling is that you may only have 1 or 2 measures representing the precursor pool over time; therefore, some have chosen to measure venous AA labeling as an alternative surrogate precursor. However, with certain amino acids (such as Phe), this can overestimate the precursor pool and underestimate the true rate, as being representative of the intracellular pool [12]. There is one caveat in this when using leucine as an AA tracer; venous α-KIC has been shown to accurately represent the intracellular leucine precursor pool as it is formed in muscle via transamination, through the branched-chain amino transferase (BCAT) enzyme found in muscle [13]—this also has the advantage that sampling venous blood regularly will provide a more complete temporal profile of the labeling over the duration of the measurement.

The flooding dose approach was conceived to overcome these limitations discussed above, wherein a large bolus of highly labeled AA (~3 g, 20 AP) is administered over a few minutes, with the aim of labeling all the possible precursor pools to the same extent; hence, sampling the blood AA labeling would represent the AA-tRNA labeling over the duration of the experiment (~90 min; [14]). Unfortunately, this approach, when using essential amino acids (EAA), i.e., Leu or Phe, broke one of the tenets of stable isotope tracer methodologies, which is they do not perturb the system they are tracing. Many studies have since demonstrated that providing a large bolus of EAA in fact stimulates muscle protein synthesis (MPS) to the same extent as a mixed meal or large protein feed ([9, 10]; *see* **Note 7**).

There are several factors to bear in mind when choosing the best AA for measuring protein turnover. EAA are usually the AA of choice as they are only obtained from the diet (i.e., accounting for de novo synthesis or intermediary metabolism, which could dilute the tracer being infused, is not necessary); they also tend to have smaller metabolic pools in blood and intracellularly in tissue, and so less material is required to label the precursor pools readily. The exchange between the extracellular and intracellular pool is also rapid, meaning the steady state is achieved rapidly. In addition, choosing an EAA that is abundant in protein (e.g., muscle comprises ~50% EAA, of which 8–10% is leucine; Phe is around 3%) makes measurement of incorporation easier. Most human studies involve one of either Leu or Phe; Phe has an added advantage as it is not extensively metabolized in muscle (*see* **Note 8**).

3.3 Sample Collection (Fluid/Tissue) Preparation and Storage

Sample collection requires the siting of at least two cannulae, one for the infusion of the tracer on one arm and another on the contralateral arm for blood sampling. The blood sampling cannula is kept patent, via a slow saline drip, to prevent clotting and blockage. Blood samples are removed via a 3-way extension set via a syringe and stored on ice in either Lithium Heparin (LiHep) or Ethylenediaminetetraacetic acid (EDTA)-treated tubes, prior to centrifugation at 4 °C, 2000 g for 20 min, to separate red cells from plasma. The plasma is aliquoted and stored at −20 °C or below until analyses (*see* **Note 9**).

Muscle tissue is obtained usually from the vastus lateralis (although the anterior tibialis, gastrocnemius, and biceps muscles have all been sampled). Firstly, the skin is anesthetized with 1% lidocaine, and this continues down to the level of the muscle fascia. A 1 cm incision is then cut in the skin with a scalpel and then the fascia. The sterile conchotome biopsy forceps are then placed through the opening in the fascia, opened to allow muscle to fill the space, then closed to cut the muscle, rotated slightly, and then removed (*see* **Note 10**). Muscle tissue is quickly blotted free of excess blood; any obvious fat or collagen is removed prior to freezing in liquid nitrogen and stored at −80 °C till analyses. Washing with ice-cold saline, prior to blotting, is also helpful in removing excess blood from the biopsy and has no deleterious effect upon the stable isotope measurements that are subsequently made and likely reduces contamination from red cell protein and extracellular free amino acids.

3.4 Preparation of Plasma for Measurement of the Plasma Precursor Enrichment and Amino Acid Concentration

The choice of amino acid pool or surrogate pool to represent the precursor enrichment for protein synthesis has been discussed above and will depend upon the study design. Assuming we have chosen to use venous plasma as the surrogate pool, the following protocol would determine the labeling; it can also be used to provide information re changes in concentration of the tracer and other plasma AA, for example, in situations of nutritional intervention.

Plasma is allowed to defrost and then centrifuged at 10,000 g (4 °C) for 4 min to pellet any fibrin clots or red cell debris. An aliquot (100–200 μL) is then removed, and 10 μL of an internal standard mix is added for determination of concentration (*see* **Note 11**), and the plasma proteins are precipitated with ice-cold ethanol (1 mL). Following 20 min on ice, samples are centrifuged at 10,000 × *g* (4 °C) for 4 min. The supernatant containing the free amino acids (and keto acids if using leucine) is then transferred to a glass culture tube and completely dried under nitrogen at 90 °C. The dried material is reconstituted in 0.5 M HCl (300 μL) to dissolve the AA, and the fatty acids are removed by liquid:liquid extraction with ethyl acetate (2 mL) (*see* **Note 12**). The remaining aqueous layer is then dried down, and the sample is then derivatized in preparation for GC-MS analyses. The preferred method of derivatization involves the addition of pyridine (100 μL) (or other cosolvents include acetonitrile and or dimethylformamide) and a silylating reagent (MTBSTFA, 100 μL, other silylating, acylating reagents are available that may better suit your GC-MS setup; *see* also MCME derivatization protocol). The samples are heated at 90 °C for 60–90 min (simply covering the tubes with a suitable lid prevents excessive evaporation of the reagents), allowed to cool, and then transferred to autosampler vials ready for mass spectrometric analyses.

3.5 Preparation of Muscle for Measurement of the Intracellular Precursor Amino Acid Enrichment

The intracellular fraction can be obtained in several ways, as a by-product of the muscle preparation (*see* Subheading 3.6) or by direct liquid extraction of muscle tissue. Approximately 25 mg (wet weight) of muscle tissue is required; the extraction is facilitated by grinding the frozen tissue to a powder in liquid nitrogen using a mortar and pestle (and appropriate personal protective equipment (PPE), mask, and freezer gloves) and transferring the powder to a tube containing ice-cold 0.2 M perchloric acid (PCA). The sample is vortexed, then centrifuged, and the supernatant containing free amino acids and metabolites is removed and neutralized with 2 M KOH (universal indicator can be added to visualize the pH, or pH paper can be used); this precipitates the perchlorate and the supernatant can then be further purified using a Dowex cation-exchange resin (*see* below). The eluate is dried and then derivatized as above for plasma. Internal standards can be added to enable quantitation of the intracellular free amino acid pool, after correction for the extracellular space contribution [15].

3.6 Preparation of Muscle Fractions for the Measurement of Tracer AA Incorporation, i.e., the Product

The muscle tissue can be easily subfractionated into four protein pools to provide greater detail as to the impact of an intervention upon these pools; it also permits the use of these fractions for other analyses, e.g., immunoblotting to investigate underlying cell signaling mechanisms and activity. The mixed muscle can be separated into myofibrillar (contractile protein), sarcoplasmic, mitochondrial, and collagen fractions using properties of solubility and differential centrifugation to enable separation.

Anywhere between 20 and 100 mg of muscle tissue should be processed, depending on analyses of importance (*see* note on muscle weight). Muscle is chipped to the required weight over liquid nitrogen and then homogenized on ice in 10 μL/mg of ice-cold homogenization buffer (*see* Subheading 2.5 for composition) using clean scissors, followed by vortex mixing at 1500 rpm for 10 min. The sarcoplasmic fraction can be separated from the myofibrillar, collagen, and mitochondrial fractions following high-speed centrifugation (11,000 g at 4 °C for 15 min). The supernatant containing sarcoplasmic proteins is removed, and the myofibrillar/collagen pellet is washed and centrifuged to minimize contamination. The mitochondrial fraction is then isolated from the myofibrillar/collagen following homogenization of the pellet using a glass homogenizer in the mitochondrial extraction buffer, following a slow-speed centrifugation (1000 g at 4 °C for 5 min) to pellet the remaining myofibrillar/collagen fraction; the mitochondrial fraction is retained in the supernatant and can be used for further downstream analyses if needed. The myofibrillar/collagen fraction is then separated using 0.3 M NaOH (300 μL) to solubilize the myofibrils, the collagen is pelleted by centrifugation, the alkali-soluble myofibrillar supernatant is removed, the pellet is washed with 0.3 M NaOH and centrifuged, and the supernatants are combined (*see* **Note 13**). Following precipitation of the solubilized myofibrillar proteins with PCA, this can then be hydrolyzed back to its constituent AAs using equal volumes of 0.1 M HCl and Dowex H^+ cation-exchange resin slurry, incubated at 110 °C overnight.

3.7 Use of Dowex Resin to Purify Free Amino Acids or to Hydrolyze Protein Fractions

Cation-exchange resins are traditionally used to isolate amino acid fraction from complex matrices, due to the ability of amino acids to act as zwitterions, wherein at extremes of acid and basic pH, the AA exists as positively charged species (accepting protons, H^+, in strong acid, NH_3^+) or negatively charged ($-O^-$, in strong base). These properties allow the amino acids to interact with the Dowex H^+ resin in acid, sticking to the resin, and then by changing the pH to strongly basic, causing the amino acids to be repelled from the resin and eluted in a basic solution.

The amino acids are added to the resin in an acidified solution (0.5 M HCl); other compounds are eluted with excess acid (1 M HCl) and then washed with distilled deionized water (ddW) to neutralize the pH and wash away the HCl (it will interact with the ammonium hydroxide to form NH_4Cl, an insoluble salt that interferes with subsequent derivatization steps); the amino acids are eluted in 2 M NH_4OH, which has the added advantage of being easily evaporated under nitrogen at 90 °C.

3.8 Determination of Enrichment of the Product by Isotope Ratio Mass Spectrometry

Following purification, the next step involves derivatization of the amino acids prior to mass spectrometric analyses; the choice of derivative will depend upon the MS technique used to perform this final analysis. Given that muscle turns over at a relatively slow rate and infusions are performed acutely over 4–8 h, invariably the MS method of choice is IRMS. This technique involves the separation of amino acids by gas chromatography, followed by combustion (at 960 °C) of the amino acid derivative, which generates CO_2 and NO_2 as fixed gases, the NO_2 undergoes catalytic reduction to N_2, and these gases are first separated by chromatography and then are directed to the isotope ratio mass spectrometer where they are ionized, and the relative amount of ^{13}C, ^{18}O (in CO_2), and ^{15}N (in N_2) is determined. Therefore, it is important that any dilution of label is minimized by using derivatives that add as few C, O, and N as possible to improve the sensitivity.

We have traditionally used the n-acetyl, propyl ester (NAP) derivative, which adds five carbons to most AA's, thereby diluting the ^{13}C label in the amino acid by a further five carbons (*see* **Note 14**). Briefly, samples (eluates from Dowex) and standards (*see* **Note 15**) are dried completely in 2 mL autosampler vials, under nitrogen at 70 °C in a TurboVap (Biotage UK, Hengoed, UK)). Propyl acetate (200 μL) and boron trifluoride (BF_3):propanol (14%) (100 μL) are added to the vial, vortex mixed, capped, and then heated at 110 °C for 30 min. They are allowed to cool to room temperature and then dried in a TurboVap (at 70 °C, N_2 flow 1.2 mL/min, for 12 min). Additional BF_3:propanol (14%, 100 μL) is added, mixed, and then evaporated again as above. Following this acetonitrile (50 μL), 1,4, dioxane (25 μL), triethylamine (37.5 μL), and acetic anhydride (22.5 μL) are added; the vials are capped, vortex mixed, and heated at 55 °C for 15 min. Once cool, the sample is transferred to a 1.5 mL microcentrifuge tube, and chloroform (100 μL) and 0.001 M $NaHCO_3$ (150 μL) are added in 3 × 50 μL aliquots to prevent overheating, and the tubes are tapped vigorously to mix. The upper layer is removed completely and discarded, and a molecular sieve is added to the lower chloroform layer to remove remaining water. The sample is transferred to a 2 mL autosampler vial containing a low-volume insert and capped ready for analyses by GC-C-IRMS. More recently, we have used an alternative methoxy-carbonyl-methyl-ester (MCME) derivative; *see* details in D_2O Subheading 3.9 below, which can be performed in aqueous solution, at room temperature, more rapidly, with improved AA GC separation. For details on this please, *see* [16, 17].

3.9 Mass Spectrometric Analyses

Once both the precursor AA and the product protein-bound AA fractions have been extracted and derivatized, they can be analyzed using the relevant forms of GC-MS. There are many different types and vendors that provide GC-MS instrumentation, and it is always best to optimize your samples to your instrumentation when setting up these measurements. However, we have provided in Table 3 below a summary of the GC conditions that we have optimized for the separation of specific derivatives and sample types as an exemplar.

4 Measurement of Protein Synthesis—A D_2O Approach

Deuterium oxide (D_2O) was one of the first stable isotope tracers to make a breakthrough into physiology studies, after Harold Urey was able to provide experimental evidence of a hydrogen isotope with a mass of 2 in the 1930s [18]. Building on this finding, Urey's

Table 3
Typical columns and conditions to enable baseline separation of amino acids of interest, with the three main derivatives. Dimensions relate to column length, internal diameter (ID), and film thickness (FT). GC ramp rates can be altered (slowed or increased) to optimize peak separations. Amino acid retention times reflect approximate elution times and can vary. Setup and conditions for the measurement of labeled $^{13}C_2$-α-KIC quinoxalinol-*t*-BDMS derivative are very similar to those for the AA- *t*-BDMS derivative

Derivative	*AA-tBDMS*	*AA-NAP*	*AA-MCME*
Sample type	Precursor AA	Product AA	Product AA
MS instrument	GC-MS or MS-MS	GC-C-IRMS	GC-C/P-IRMS
Column type	Rtx -5MS	DB-1701	StabiloWax/DB-1701
Dimensions Length (m), ID (mm), and FT (μm)	30 × 0.25 × 0.25	30 × 0.25 × 1.0	30 × 0.25 × 0.25/ 30 × 0.25 × 1.0
Column flow (ml/min)	1.4	1.5	1.5
Injector temp (°C)	250	250	240
Ramp rate	100 °C (1 min) 15 °C/min–280 °C (2 min)	100 °C (1 min) 15 °C/min–280 °C (2 min)	50 °C (1 min) 10 °C/min–240 °C (15 min)
Retention time (min)	Leu (~7), Phe (~9)	Leu (~11), Phe (~14)	Ala (~13)

colleagues David Rittenberg and Rudolf Schoenheimer were able to develop some of the first stable isotope tracers by incorporating ^2H into compounds such as lipids and amino acids [19]. Using these deuterated tracers, Schoenheimer and Rittenberg were able to show that metabolism was highly active and not static, leading to a rapid publication of many impactful papers in the 1930s. These developed techniques provided a substantial amount of our knowledge regarding dynamic mammalian metabolism (*see* [19, 20] for details). Yet during the following years, there was a decline in the use of stable isotope tracers and an increase in the use of radioisotopes, due to greater availability and ease of measurement. However, due to concerns of ionizing radiation in humans and the development of highly sensitive GC-IRMS equipment, the use of ^2H and D_2O and the use of D_2O would make a resurgence in the 1980s. There has been great progress in realizing the potential benefits of D_2O as a stable isotope tracer, primarily due to the work of Marc Hellerstein [21–23] and Stephen Previs [24–27] independently of each other. D_2O is now frequently used throughout metabolic research, providing researchers an alternate tool to stable isotope-labeled compounds [1].

A primary benefit in the use of D_2O is that it can be easily administered via oral consumption, negating the need for infusions. Consumption of a single, small 100–150 mL bolus of 70% AP D2O

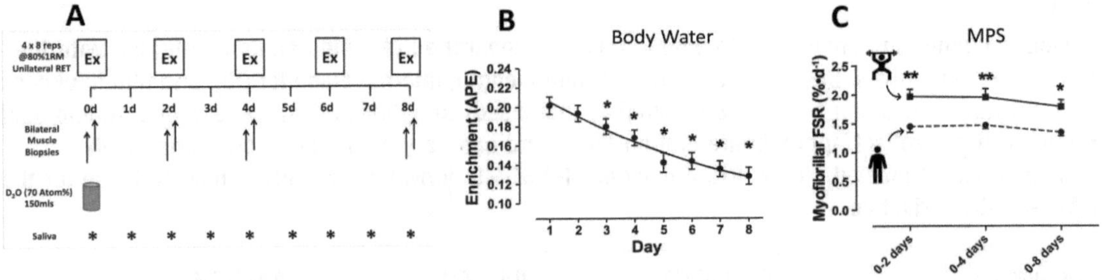

Fig. 2 An example of how D_2O can be used to measure muscle protein synthesis (MPS) over days. (**a**) Study schematic highlighting the timings for the collection of muscle biopsies and saliva samples over 1 week of unilateral resistance exercise. (**b**) Results demonstrating body water enrichment over 8 days after consumption of 150 mL 70 Atom% deuterium oxide (D_2O). (**c**) Results demonstrating rested and exercise muscle protein synthesis over 8 days (adapted from Wilkinson et al. [17])

can be used to measure MPS, both temporally and continuously, over 24 h to 8 days [17, 28] (*see* **Note 16**). Similarly, continual dosing strategies have also been used to measure MPS over a period of days [29]. A critical step with both protocols is the collection baseline saliva and muscle to enable accurate determination of increases in ^{2}H (*see* **Note 17**). Subsequent biopsy timings should be designed to assess the research question most appropriately. For instance, multiple bilateral biopsies over 1 week demonstrated the temporal nature of MPS in response to RET [17] and immobilization [30]. A typical protocol for the determination of daily MPS using a single-bolus approach in human volunteer is shown in Fig. 2. The number and frequency of saliva samples should be kept high (at least weekly but preferably more frequently), especially in a bolus design where body water enrichment changes over time. At a minimum, it is essential that a measure of the precursor (saliva or plasma) be made in conjunction with every muscle biopsy collection; however, the collection of precursor samples daily is optimal over short (daily) time periods.

The measurement of MPS over weeks can be achieved using either highly sensitive GC-pyrolysis-IRMS or GC-MS. The primary difference in the study design is that GC-MS requires high levels of body water enrichment to achieve detectable ^{2}H increase in muscle. For instance, body water enrichment is maintained at ~1.5–2% by implementing a 1-week priming stage of 50 mL 3 × /d = 150 mL/d followed by 5 weeks of 50 mL 2 × d = 100 mL/d enriched muscle to 1.2% APE after 4 weeks in young sedentary individuals. Although this is above the 0.5% detection limit of GC-MS equipment, consuming regular high doses of D_2O may become burdensome for participants. Whereas consumption of a single 150 mL bolus, followed by a 50 mL "top up" per week, maintained body water enrichment to ~0.15–2%, this resulted in a muscle enrichment of ~0.012% after 3 weeks. This 2H enrichment is much higher than the 0.0005% detection limit of GC-pyrolysis-IRMS equipment and reduces the invasiveness of the participant. As such study designs using GC-pyrolysis-IRMS equipment are preferable for the measurement of MPS over weeks, yet this can be achieved using GC-MS. Again, doing strategies of continuous daily dosing or weekly "top ups" have been successfully used to capture weekly rates of MPS. Regular "top ups" provide a practical way to maintain body water enrichment in a pseudo steady state with minimal burden on the participant.

Capturing acute, hourly, measures of MPS over a period of hours requires the consumption of larger quantities of D_2O (400 mL of 70% D_2O) compared to chronic measures. This is to increase the amount of ^2H incorporated into bound muscle protein per hour, resulting in a greater and therefore more easily detected, δ increase in sequential biopsies. To avoid incurring adverse events (*see* below), the larger volume of D_2O should be consumed the day before the study due to the longer time to reach a stable plateau enrichment. Prior to D_2O consumption, a baseline saliva sample should always be collected to determine current ^2H levels. Additionally, it is important to monitor the stability of the body water enrichment throughout the study, as any substantial change in enrichment will change precursor enrichment levels. As such, regular saliva samples (every 30 min) should be collected throughout the study period to monitor the stability of body water enrichment. Due to the prior consumption of D_2O, baseline muscle biopsies must be collected to set the basal level of ^2H. To accurately determine changes in muscle δ, ~2–3 h is ideally required between muscle biopsies. This was shown by Wilkinson et al.; using 400 mL of 70% AP, body water enrichment increased ~33,000 δ (~0.5% APE). Muscle biopsies collected at 0 and 3 h showed an increase of ~16 δ (~0.0007% APE) in bound enrichment representing a MPS rate of 0.05%.h^{-1}. Further, after consumption of 20 g of EAA, bound muscle protein increased by 27 δ (~0.0012% APE), representing an MPS rate of 0.088%.h^{-1} [17].

Below is a summary of considerations associated with the use of D_2O as a tracer and associated analytical analyses.

4.1 Saliva Sample Collection and Analysis by TCEA-IRMS for Precursor Enrichment—D_2O Only

Having an accurate representation of precursor enrichment is essential in any precursor-product calculation, and considerable work has been undertaken to establish ideal precursors for the use of D_2O. In muscle, the incorporation of bound deuterated alanine into proteins has become the amino acid of choice due to multiple factors. The passive diffusion of D_2O through tissues means that no active transport is required and, therefore, no dilution of tracer across a membrane occurs. Alanine becomes labeled intracellularly through rapid transamination reactions with pyruvate, incorporating ^2H from labeled body water. The rate of these reactions results in alanine enrichment having a strong relationship with that of body water and is not easily perturbed by feeding or exercise. Further, alanine incorporates ^2H onto an average of 3.7 out of 4 available hydrogens, amplifying the ^2H signal and increasing the ability to detect enrichment in the product. As such, it has been demonstrated on multiple occasions that monitoring body water enrichment through saliva can be reliably used to represent intracellular alanine enrichment. Importantly, studies using either saliva or intracellular alanine have produced similar results.

Precursor enrichment using D_2O can simply and accurately be performed through the collection of saliva samples, yet care must be taken to ensure correct sampling. Saliva can be sampled through the use of cotton swabs or expelling saliva into plastic tubes (*see* **Note 18**). Further, after consumption of D_2O, it takes at least 2 h for the body water pool to equilibrate and for saliva enrichment to be representative of plasma alanine. Upon receipt of saliva samples, they should be cold centrifuged at 16,000 g to remove any debris, aliquoted into appropriate storage, such as 2 mL glass vials, and frozen at −20 °C until analysis. Direct measurement of saliva enrichment can be determined through liquid injection of saliva (0.1 μL volume) into a high-temperature conversion elemental analyzer (TC/EA) connected to an IRMS. Prior to injection, samples should be purified to reduce contamination. This can be achieved by placing 50 to 100 μL of saliva into the cap of inverted autosampler vials. These are then placed onto a heating block set at 90 °C for at least 2 h, with the water distillate collected by rapidly cooling the vials on ice for 10 min and transferred to fresh vials for direct liquid injection. To minimize the effect of carryover between samples, each should be performed a minimum of four times. In the absence of a TC/EA, indirect saliva enrichment can be measured using GC-MS techniques. The exchange of deuterium from water with hydrogen positions on acetone can be achieved by incubating 100 μL of saliva with 2 μL of 10 M NaOH and 1 μL of acetone for 24 h at room temperature. Following incubation, the acetone is extracted into 200 μL of n-heptane, and 0.5 μL of the heptane phase is injected into the GC for analysis [17]. A standard curve of known D_2O enrichment should run alongside the saliva samples to demonstrate the linearity of the enrichment measure, over the range of enrichments expected in the muscle-derived alanine. In addition, pooled QC samples at low and high enrichments should be prepared with each batch to determine inter- and intra-batch variability, with CV's of more than 5% to be used as a cutoff value.

The same protocols (3.4 and 3.5) are used for extracting and hydrolyzing proteins for amino acid analyses, to determine the incorporation rate of labeled alanine (*the Product*) are used for the D_2O approach, where the analysis differs is in the final derivatization step, using the MCME derivative and the determination of the isotope ratio by GC-pyrolysis-IRMS. The use of the MCME derivatization approach limits the dilution of the alanine deuterium enrichment, by hydrogens added during derivatization, and using pyrolysis, at 1400 °C, produces H_2 gas, permitting the accurate and very sensitive measurement of the D:H ratio. Using the D:H ratio, rates of protein synthesis can be calculated. Typically, this ratio is expressed as δ2H, which can be converted to atom% using the following equation:

$$\text{Atom\%} = 100 \times \text{AR} \times (\delta H2 \times 0.001 + 1)\, 1 + \text{AR} \times (\delta H2 \times 0.001 + 1)$$

where AR represents the absolute ratio constant for deuterium based on the Vienna Standard Mean Ocean Water (VSMOW) standard and equates to 0.00015595. This is then converted to APE by correcting for baseline sample, i.e., background enrichment. The Fractional Synthetic Rate (FSR) can then be determined using precursor-product calculation (Fig. 1) where the body water enrichment (corrected for the mean number of deuterium moieties incorporated per alanine, 3.7) can be used as the surrogate precursor labeling between subsequent biopsies. During long-term labeling studies, isotope incorporation follows a rise to plateau, and therefore, nonlinear calculations are preferentially used.

$$\text{FSR}\left(\%\text{day}^{-1}\right) = -\text{In}\left[1 - \left(\text{APE}_{\text{Ala}}/\text{APE}_{\text{P}}\right)/t\right]$$

where APE_{Ala} is deuterium enrichment of protein-bound alanine, APE_{P} is mean precursor enrichment of the body water over the period, and t is the time between biopsies.

4.2 Mass Spectrometry Analytical Requirements for D_2O-Based Tracer Studies

GC-MS and GC-MS-MS are primarily used to measure precursor labeling where enrichments are relatively high (>0.2 APE up to around 10 APE). However, the hybrid-IRMS systems are used to accurately measure the low levels of enrichment found in the product amino acid, due to the slow rate of incorporation of AA into muscle protein in humans. Clearly in the case of the D_2O approach, longer or higher dosing protocols may permit the measurement of product enrichment by GC-MS or GC-MS-MS, when the level of labeling reaches the enrichment range of those instruments (*see* Table 1). Similarly, the measurement of smaller protein pools, i.e., the mitochondrial pool, is also achievable in this circumstance by GC-MS-MS or GC-MS where sample concentration is a limitation for IRMS. In preclinical or cell-based models where turnover is 10–50 times more rapid then it may be possible to use only GC-MS instrument over shorter measurement periods; the main limitation of these approaches is getting enough tracer/label into the analytical window (enrichment range) of the instrument at your disposal. This incorporation is a function of the amount of tracer given, the duration of the experiment/intervention, and the rate of turnover of the protein or tissue of interest, all subject to the limitations described in this chapter, i.e., not perturbing the pool being measured and cost.

5 Summary and Comparison of Both Primed Constant Infusion and D_2O Approaches

The choice of which approach to adopt will be driven primarily by your research question, the information you require, whether or not you are looking at acute or chronic changes, and the population under investigation (D_2O lends itself to studies in a clinical population, given the relative simplicity of the approach and relatively limited invasiveness, i.e., oral dosing and a single biopsy). Additional considerations will obviously be the cost/funding available and the mass spectrometric instrumentation at your disposal. For a study using D_2O, the primary costs that need to be taken into consideration are of the tracer itself, whereas a primed constant infusion approach will require sterile preparation. Figure 3 summarizes the advantages and disadvantages of both approaches. Ultimately, stable isotope-labeled AAs have excellent resolution over hours and are the tracer of choice in acute studies. D_2O is ideally suited to measure integrative MPS over days to weeks and provides an alternate research tool if tracer infusions are not permitted. However, if funding permits, then the two approaches can and have been used simultaneously, providing complementary measures of acutes response and the chronic effect of any given intervention [30].

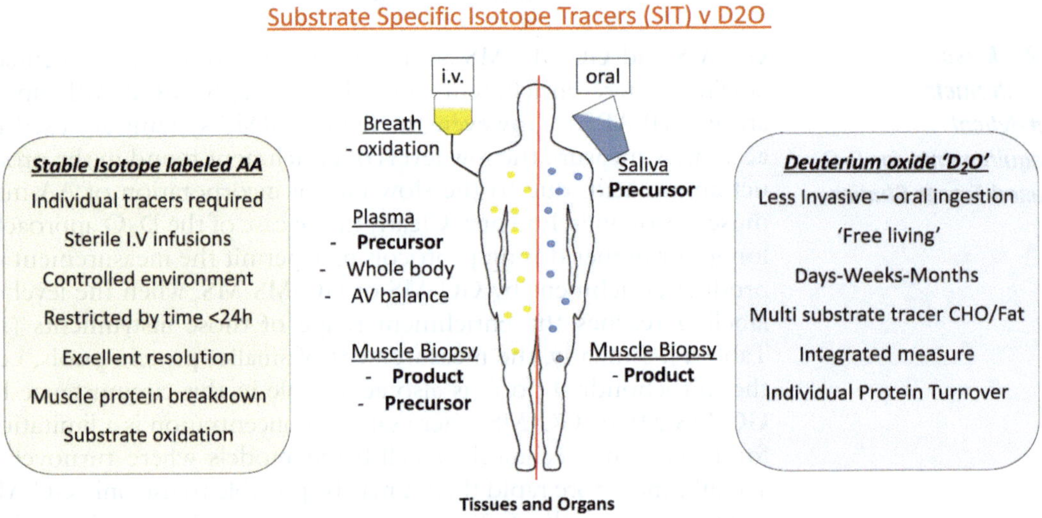

Fig. 3 A comparison of the traditional acute tracer infusion and the deuterium oxide (D_2O) oral tracer approaches—detailing the perceived advantages and disadvantages of the two approaches. In reality, both approaches can be used concurrently to investigate different aspects, e.g., whole body, tissue/limb, and muscle protein turnover (synthesis, oxidation, breakdown) and investigate the impact of both acute and chronic long-term interventions

6 Notes

1. Over recent decades, the rapid development of liquid chromatography and its coupling to high-resolution Fourier transform mass spectrometers has introduced these as alternatives to IRMS in certain situations for analyses of accurate low-level enrichment.
2. It is necessary to account for the infusate needed to prime the lines and standard to prepare an extra volume to allow for any delays that may occur during the study, e.g., time taken to put in cannula, resite sampling lines, and biopsy delays.
3. Derivatization is often a necessary precursor to measuring polar metabolites (amino acids, sugars, organic acids, etc.) from human matrices, i.e., plasma, saliva, urine, and extra- and intracellular fluids by gas chromatography. This is performed to make the compounds more volatile and nonpolar. The GC then uses temperature to put energy into these compounds, which become more volatile and then move into the gas phase of the GC column and are transported to the detection system, in this case a mass spectrometer for determination of the enrichment or labeling of the compound or fragment of interest.
4. Enrichment levels higher than this have shown that the increase in concentration of AA can exert effects upon the system under study. Indeed, it is well known that some individual EAA can robustly stimulate muscle protein synthesis, when administered as a flooding dose (i.e., 3 g of EAA of 20 atoms percent [9, 10, 31]). It is also common to take the first biopsy around 60–90 min after the infusion has started, i.e., once a steady state has been attained—see Fig. 1.
5. As an alternative to a baseline biopsy, plasma or whole blood protein sampled prior to the initiation of the infusion can be used as a measure of baseline (pre-infusion) labeling—this slightly overestimates the precursor, therefore reducing calculated MPS—the use of a plasma protein baseline is more common in longer duration studies, i.e., over days and weeks.
6. Uncovering the temporality and on-off nature of MPS to feeding and exercise was only possible due to the availability of multiple labeled AA and advances in gas chromatography-combustion-isotope ratio mass spectrometry (GC-C-IRMS), which increased both the number of heavy labeled atoms and the sensitivity and precision of the isotope ratio measurement, permitting the measurement of the incorporation of tracer over as little as 45–60 min.

7. Studies have shown that using nonessential AA in this way does not impact MPS, and so AA such as proline can be easily and robustly applied to study MPS (Smith et al.) with the added advantage that it can be utilized to study collagen turnover in all ECM-rich tissues, including bone, tendon, skin, and ligament. Since proline is converted posttranslationally to hydroxyproline (OHPro), the measurement of OHPro labeling provides a clean measure of collagen synthesis [32].
8. Phenylalanine is the tracer of choice when using arteriovenous (AV) balance approaches to measure protein breakdown; you can estimate the rate of appearance of Phe from protein breakdown, from the dilution of the Phe tracer and the concentration balance across the limb or organ. When using leucine in an AV model, the tracer dilution and concentration of (-KIC and CO_2 must also be measured, which is not trivial.
9. When adopting an AV balance approach to measure protein breakdown, in the absence of an arterial cannulation, the blood sampling cannula will be placed retrogradely into a hand vein, and the hand is heated to 55 °C for 20 min prior to sampling. The warming of the hand increases blood flow and shunting through the hand, which limits metabolism in the hand, effectively providing arterial blood (this can be checked using a blood gas analyzer, with O_2 saturation being above 95%, as proof of arterialization); venous samples would be drawn simultaneously from the femoral vein (ensuring the tip of the cannula lies below the venous drainage from the liver).
10. Other biopsy procedures are available such as the Bergstrom needle and microbiopsy procedures.
11. An internal standard mix of all stable isotopically labeled AAs is typically used to generate concentration measures for AAs. ^{13}Carbon, ^{15}Nitrogen, and Deuterium labels can be used; however, care must be taken to not use labels that positionally interfere with the label provided by the tracer; otherwise, issues can arise in analyses.
12. If venous (-KIC is the precursor choice, an additional step involves preparation of the quinoxalinol derivative of the keto acids. The dried ethanol extract residue is reconstituted in 0.15% *ortho*-phenylenediamine (OPD), and 4N HCl is added, vortex mixed, and then incubated at 90 °C for 90 min. Following extraction with ethyl acetate, the BCKA-quinoxalinol derivative is retained in this layer and can be dried and derivatized using MTBSTFA (50 μL) and pyridine (50 μL) at 90 °C for 30 min and run for (-KIC enrichment/concentration.
13. The relative purity of each of the fractions and cross contamination can be assessed using immunoblotting for proteins that should only be present in that particular fraction, e.g., myosin

in the myofibrillar fraction and cytochrome c in the mitochondria. This can be performed when the extraction protocol is being optimized in your lab.

14. In theory, any reaction capable of forming a stable AA derivative can be used for this; you just need to consider the dilution of your label. The NAP derivative is a good one for carbon tracers as it doesn't add too many additional carbons during formation. MCME is an even smaller addition and is useful for both carbon- and hydrogen-based tracers due to low-level dilution. Others have used silylating reagents; however, they add additional carbons, and silicon quickly contaminates the oxidation and pyrolysis reactors, so in our opinion, they are best to be avoided.
15. We highly recommend users to prepare 0.01, 0.05, and 0.1 APE isotopically labeled standards, along with baseline and enriched muscle QC's, to ensure linearity of the measurement and batch-to-batch variation.
16. Although a bolus of this size is well tolerated in healthy young individuals, caution must be taken in elderly or compromised individuals.
17. In tracer naïve individuals, the number of biopsies can be minimized by collecting a single biopsy from one leg to represent the baseline deuterium enrichment of both legs or indeed plasma or blood protein. In animal studies, control animals not exposed to the tracer can be used to provide baseline unenriched samples, e.g., blood or tissue.
18. Importantly, saliva samples must be collected at least ≥60 min after their last meal or drink for the samples to reflect that of body water and not the recent contents of the mouth.

References

1. Wilkinson D (2018) Historical and contemporary stable isotope tracer approaches to studying mammalian protein metabolism. Mass Spectrom Rev 37:57–80
2. Atherton PJ, Smith K (2012) Muscle protein synthesis in response to nutrition and exercise. J Physiol 590:1049–1057
3. Brook M et al (2016) Skeletal muscle homeostasis and plasticity in youth and ageing: impact of nutrition and exercise. Acta Physiol 216:15–41
4. Atherton PJ, Wilkinson DJ, Smith K (2016) Chapter 9 – Feeding modulation of amino acid utilization: role of insulin and amino acids in skeletal muscle. In: The molecular nutrition of amino acids and proteins, pp. 109–124. https://doi.org/10.1016/B978-0-12-802167-5.00009-8
5. Mitchell W et al (2016) Human skeletal muscle protein metabolism responses to amino acid nutrition. Adv Nutr 7:828S–838S
6. Mitchell WK et al (2012) Sarcopenia, dynapenia, and the impact of advancing age on human skeletal muscle size and strength; a quantitative review. Front Physiol 3:260
7. Gharahdaghi N et al (2019) Testosterone therapy induces molecular programming augmenting physiological adaptations to resistance exercise in older men. J Cachexia Sarcopenia Muscle 10:1276–1294
8. Brook MS et al (2016) Synchronous deficits in cumulative muscle protein synthesis and

ribosomal biogenesis underlie age-related anabolic resistance to exercise in humans. J Physiol 594

9. Smith K, Reynolds N, Downie S, Patel A, Rennie MJ (1998) Effects of flooding amino acids on incorporation of labeled amino acids into human muscle protein. Am J Phys 275: E73–E78
10. Smith K, Barua JM, Watt PW, Scrimgeour CM, Rennie MJ (1992) Flooding with L-[1-13C] leucine stimulates human muscle protein incorporation of continuously infused L-[1-13C] valine. Am J Phys 262:E372–E376
11. Rennie M, Smith K, Watt P (1995) Measurement of human tissue protein synthesis: an optimal approach. Am J Physiol 266:E298–E307
12. Witard OC et al (2014) Myofibrillar muscle protein synthesis rates subsequent to a meal in response to increasing doses of whey protein at rest and after resistance exercise. Am J Clin Nutr 99:86–95
13. Watt PW et al (1991) Isolation of aminoacyl-tRNA and its labeling with stable-isotope tracers: use in studies of human tissue protein synthesis. Proc Natl Acad Sci U S A 88:5892–5896
14. McNurlan M et al (1991) Measurement of protein synthesis in human skeletal muscle: further investigation of the flooding technique. Clin Sci 81:557–564
15. Bergstrom J, Furst P, Noree LO, Vinnars E (1974) Intracellular free amino acid concentration in human muscle tissue. J Appl Physiol 36: 693–697
16. Husek P, Simek P (2006) Alkyl Chloroformates in sample derivatization strategies for GC analysis. Review on a decade use of the reagents as esterifying agents. Curr Pharm Anal 2:23–43
17. Wilkinson DJ et al (2014) A validation of the application of D(2)O stable isotope tracer techniques for monitoring day-to-day changes in muscle protein subfraction synthesis in humans. Am J Physiol Endocrinol Metab 306: E571–E579
18. Urey HC, Brickwedde F, Murphy GM (1932) A hydrogen isotope of mass 2 and its concentration. Phys Rev 40:1–15
19. Schoenheimer R (1942) The dynamic state of body constituents. Harvard University Press
20. Guggenheim K, Rudolph Y (1991) Schoenheimer and the concept of the dynamic state of body constituents. J Nutr 121:1701–1704
21. Busch R, Neese RA, Awada M, Hayes GM, Hellerstein MK (2007) Measurement of cell proliferation by heavy water labeling. Nat Protoc 2:3045–3057
22. Gardner JL et al (2007) Measurement of liver collagen synthesis by heavy water labeling: effects of profibrotic toxicants and antifibrotic interventions. Am J Physiol Gastrointest Liver Physiol 292:G1695–G1705
23. Li KW, Turner SM, Emson CL, Hellerstein MK, Dale DC (2011) Deuterium and neutrophil kinetics. Blood 117:6052–6053; author reply 6053–4
24. Previs SF et al (2011) Quantifying cholesterol synthesis in vivo using (2)H(2)O: enabling back-to-back studies in the same subject. J Lipid Res 52:1420–1428
25. Herath K et al (2011) Equilibration of (2)H labeling between body water and free amino acids: enabling studies of proteome synthesis. Anal Biochem 415:197–199
26. Kasumov T et al (2013) Assessment of cardiac proteome dynamics with heavy water: slower protein synthesis rates in interfibrillar than subsarcolemmal mitochondria. Am J Physiol Heart Circ Physiol. https://doi.org/10.1152/ajpheart.00933.2012
27. Mahsut A et al (2011) Headspace analyses of ^{2}H labeling of acetone: enabling studies of fatty acid oxidation in vivo. Anal Biochem 408:351–353
28. Damas F et al (2016) Resistance training-induced changes in integrated myofibrillar protein synthesis are related to hypertrophy only after attenuation of muscle damage. J Physiol. https://doi.org/10.1113/JP272472
29. Holwerda AM et al (2018) Daily resistance-type exercise stimulates muscle protein synthesis in vivo in young men. J Appl Physiol (1985) 124:66–75
30. Brook MS et al (2022) Declines in muscle protein synthesis account for short-term muscle disuse atrophy in humans in the absence of increased muscle protein breakdown. J Cachexia Sarcopenia Muscle 13:2005–2016
31. Wilkinson DJ et al (2013) Effects of leucine and its metabolite β-hydroxy-β-methylbutyrate on human skeletal muscle protein metabolism. J Physiol 591: 2911–2923
32. Babraj JA et al (2005) Collagen synthesis in human musculoskeletal tissues and skin. Am J Physiol Endocrinol Metab 289:E864–E869

Chapter 11

Ex Vivo Human Single Muscle Fibers: An Insightful Approach to Skeletal Muscle Function

Carlo Reggiani

Abstract

Skeletal muscle, a highly heterogeneous tissue, is composed of a variety of cells and different muscle fiber types, mostly classified as slow and fast; the latter is divided into two subgroups, fast 2A and fast 2X. The functional properties of a muscle are determined not only by muscle fibers with their contractile and metabolic features but also by all cellular components. To appreciate the contribution of muscle fibers and, additionally, the specific contribution of each fiber type, single-fiber studies offer the most fruitful avenue. Single-fiber dissection from human muscle biopsy samples started 50 years ago in the 1970s and has generated a wealth of data on molecular, metabolic, and contractile characteristics of each fiber type in a variety of physiological conditions such as aging, and adaptive responses to training and disuse, and in pathological conditions. This chapter is aimed at presenting the analytical power of single-fiber studies and its impact on our knowledge of skeletal muscle in normal and pathological conditions. Special attention is given to the description and discussion of the methodologies and associated considerations for accurate analyses.

Key words Human skeletal muscle, Muscle biopsy, Single fiber dissection, Ex vivo single-fiber physiology, Single-fiber biochemistry

1 Three Main Fiber Types in Human Skeletal Muscles

Human skeletal muscle is a heterogeneous tissue composed of many different cell types. The classical observation of Schmalbruch and Hellhammer [1] that only 50% of the nuclei belong to muscle fibers has been recently confirmed with a novel updated approach based on single nuclei transcriptomics [2]. In close contact with muscle fibers, many other cell types are present in the skeletal muscles: satellite cells, fibroblasts, adipocytes, endothelial cells, and blood cells. Due to their mass, however, myofibers represent the greatest component, and, importantly, they are the major determinants of contractile performance, force, and energy consumption. To add further complexity, myofibers exhibit a high degree of diversity and, for this reason, are categorized into two

Philip J. Atherton and Daniel J. Wilkinson (eds.), *Neuromuscular Assessments of Form and Function*, Neuromethods, vol. 204, https://doi.org/10.1007/978-1-0716-3315-1_11,

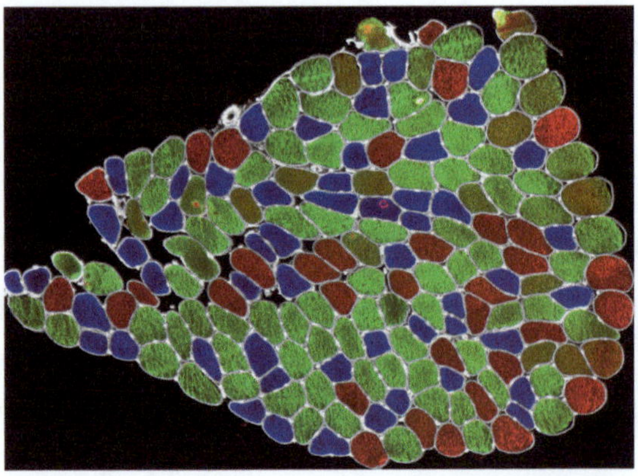

Fig. 1 Cryosection of a bundle/fascicle obtained from a biopsy of M. vastus lateralis. Fibers have been immunostained with antibodies against MyHC isoforms (slow green, fast 2A blue, fast 2X red) and laminin (white). Note that some hybrid fibers with intermediate staining are present in the section

major types, slow and fast, with the fast type further categorized into two subsets of fast 2A and fast 2X, this categorization being dependent upon the expression of the myosin heavy chain isoform (MyHC, *see* Fig. 1). Nonetheless, myosin heavy chain isoform expression is just one of the many molecular, metabolic, and functional features that distinguish fiber types. Thus, analysis of whole muscle samples (or fibre bundles) can only provide a rather general overview, and important information can be easily overlooked. The study of single muscle fibers can, at least partially, account for this.

The three main fiber types differ in their protein complement, metabolic activity, and contractile performance (*see* for a review [3]) (Fig. 2). The difference is especially obvious when type 1/slow and type 2/fast fibers are compared, whereas a more continuous distribution of values is often observed with the different subsets of type 2 fibers. The mechanical power output, i.e., the ability to move a load with a given velocity, is possibly the most relevant expression of skeletal muscle contractile function and shows clear-cut difference between slow and fast fibers. The main determinant of mechanical power is the isoform of myosin, which is the molecular motor able to transfer the chemical energy contained in the phosphate bond of Adenosine TriPhosphate (ATP) into mechanical and thermal energy. Slow fiber contraction generates less tension (force per cross-sectional area, CSA) and exhibits a slower shortening velocity, thus lower mechanical power compared to fast fibers—but the ATP consumption is also lower (*see* Fig. 2a, b; Refs. [4, 7]). The energy economy associated with the greater ability to produce ATP via mitochondrial respiration makes the slow fibers able to resist fatigue and to perform long-lasting, low-intensity contractile activity. In contrast, fast fibers are suited to motor tasks that require greater power, either higher force or higher speed, and are duration

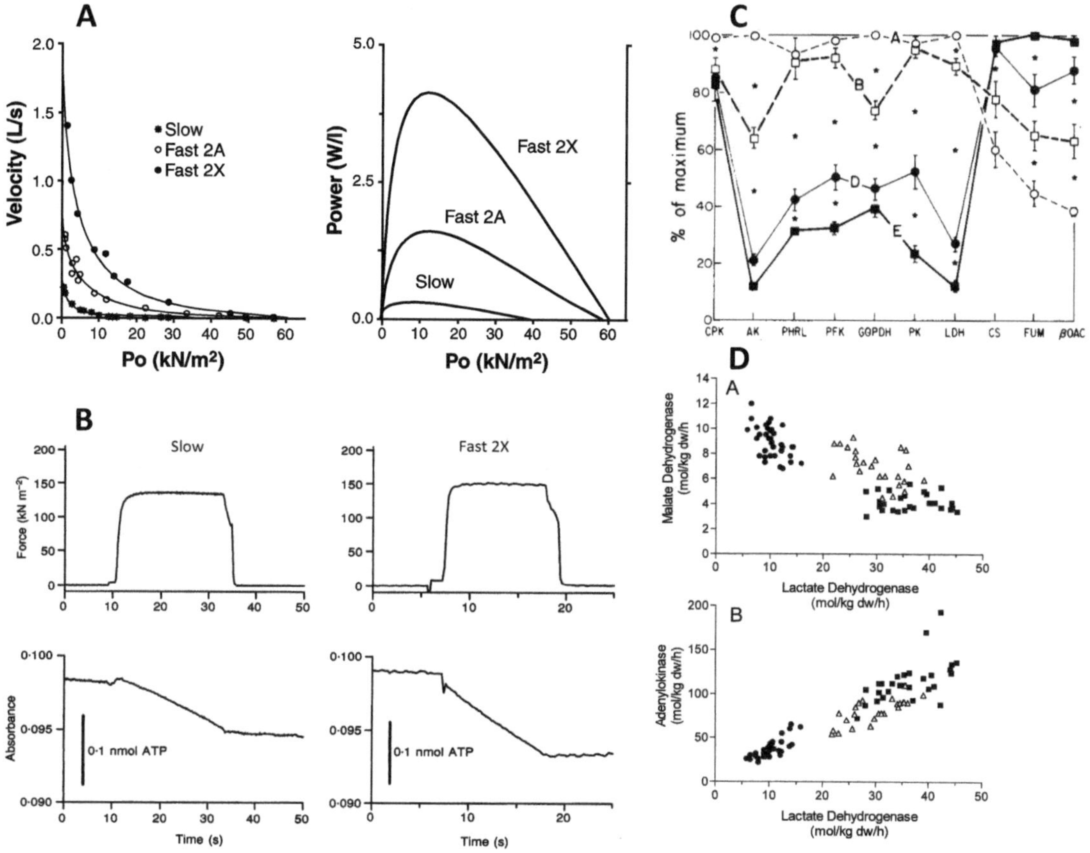

Fig. 2 Human fiber diversity in mechanical, energetic, and metabolic properties. (**a**) Force-velocity (left) and power-force (right) curves of the three main fiber types from vastus lateralis biopsy. (From Schiaffino and Reggiani [3]). (**b**) ATP consumption rate during isometric contractions of slow and fast 2X fibers from vastus lateralis biopsy. (From Stienen et al. [4]). (**c**) Metabolic diversity among single muscle fibers. Four distinct groups are formed on the basis of their lactate dehydrogenase activity, high in Group A and B, low in Group D and E. Activities of ten enzymes are expressed as the percentage of the highest enzyme in that specific group. The average of these percentages are shown together with error bars (±S.E.). Statistically significant differences ($p < 0.05$) from one group to the next are indicated by stars. Abbreviations: AK adenylokinase, bOAC b-hydroxiacylCoA, CPK creatine kinase, CS citrate synthase, FUM fumarase, GOPDH glycero-P dehydrogenase, PFK phoshofructokinase, PHRL phosphorylase, PK pyruvate kinase. (From Lowry et al. [5]). (**d**) Enzymatic activities of malate dehydrogenase (enzyme of the citric acid cycle), lactate dehydrogenase (enzyme that reduces pyruvate to lactate), and adenylokinase (enzyme that produces one ATP molecule from two ADP molecules, also called myokinase) determined in human single fibers. In each individual fiber, two activities were determined, and each fiber was classified on the basis of myofibrillar ATPase reaction as slow or type 1 (circles), fast 2A (empty triangles), and fast 2X (squares). Note the inverse relation between the activities of malate dehydrogenase and lactate dehydrogenase and the direct relation between adenylokinase and lactate dehydrogenase. (Modified from Bottinelli and Reggiani [6])

limited. Their metabolism is characterized by a more developed glycolytic pathway and a relatively lesser mitochondrial content (Fig. 2 c, d; Refs. [5, 6, 8]). The difference between the two subgroups of fast fibers is more subtle and depends for a large part upon the motor properties of the myosin isoform, with myosin 2X able to produce four times more power than myosin 2A at the expense of a proportional ATP consumption rate. Since mitochondrial enzyme abundance is approximately two times lower in 2X compared to 2A fibers and glycolytic enzyme abundance is only 20% higher in 2X fibers, the net energy balance is much more favorable in 2A compared to 2X. Thus, the two fast fiber subtypes exhibit distinct specializations: 2A fibers being suitable for less intense and more prolonged activity and 2X fibers for very intense but very short performance. Another aspect of fiber-type diversity relates to the regulation of cytosolic calcium, the trigger linking sarcolemma action potential and myofibrillar contractile activity. For instance, the sarcoplasmic reticulum (SR) is much more developed in fast than in slow fibers reflected in a greater storage capacity [9]. Intracellular calcium dynamics also shows differences, where the cytosolic calcium transient induced by 200 ms voltage clamp has a greater amplitude and a faster decline in fast compared to slow human fibers [10] in concert with the faster calcium uptake rate by sarcoplasmic reticulum observed in human fast fibers [11].

2 Single-Fiber Historical Overview

The discovery that skeletal muscles are composed of different types of fibers was made long ago. At the end of the nineteenth century, the microscopy observation allowed the identification of "translucent" and "opaque" fibers later characterized by "Felderstruktur" and "Fibrillenstruktur" [12, 13]. By 1926, so much information was available on the issue of muscle fiber diversity that Dorothy Needham could collate enough data to publish a seminal paper as a Physiological Review [14]. In those years, although Lindell and Sherrington [15] had made clear that the motor unit was the irreducible output element in motor control, i.e., the true functional unit of the muscles, the idea was spreading that whole muscle functional studies could only provide a general and approximate picture and that it was necessary to focus in on analyzing single muscle fibers. The first paper showing functional recordings of contraction of single muscle fibers from frog sartorius was published in 1930, in Science, by Brown and Sichel from New York University [16]. It was followed shortly after by a paper by Erling Asmussen [17] from Copenhagen, demonstrating force-frequency and force-length curve of single fibers from frog semitendinosus. Frog fibers were selected because of their large size; their loose connective tissue, which makes manual dissection possible, even if not easy; and their survival in good condition at low temperatures without extra oxygen supply. Single frog fibers became thus the preferred model for these studies in the 1940s; *see*, for example, the detailed length-tension analysis published by Ramsey and Street in 1940 [18] from Rochester University (NY) and the virtually contemporary comprehensive work by Fritz Buchthal in Copenhagen [19–21].

Amphibian muscle fibers remained the model of choice for the development of more and more refined mechanical studies in the middle of twentieth century, as those carried out on single fibers by Andrew F. Huxley [22, 23] and by Paul Edman [24]. Frog single fibers were also the model for high-resolution X-ray diffraction studies, which allowed Vincenzo Lombardi and coworkers to link refined mechanics and crossbridge structure [25, 26]. Not only the contractile mechanical properties but also the electrical properties of single muscle fibers and intracellular calcium transients [27] were investigated in frog or toad fibers.

In these earlier times, single mammalian fibers, and thus also human fibers, remained out of reach because the manual dissection procedure with forceps and scissors, which was the basic step to prepare single amphibian fibers, was not compatible with the preservation of an intact sarcolemma. The successful way to the study of single fibers of mammalian muscles was opened by the implementation of methods to remove the sarcolemma and preserve the intracellular organelles and their functions with the immersion of the fiber in a solution simulating the intracellular environment. The permeabilization of the sarcolemma and suppression of excitability were achieved with prolonged treatment with glycerol [28] or with mechanical skinning using fine needles to roll back the surface membrane [29]. Later on, detergents such as Triton X-100 and saponin were also used to permeabilize the sarcolemma. The methods to study frog single fibers after sarcolemmal permeabilization or removal were improved in the 1970s with the design of appropriate solutions and of a setup for quick solution exchange [30–33]. Soon after, in 1975, this experimental approach was applied to study single fibers from biopsy of human muscles [34]. At a similar time, methods for biochemical studies on single fibers were designed by Bengt Saltin [35] and further refined by Oliver Lowry and coworkers [5] by applying the recently developed micromethods for enzyme activity. A fragment of muscle biopsy was immediately frozen in liquid nitrogen and further dried under a vacuum at −40 °C. Single fibers were dissected with a hair loop and a knife blade. From each fiber, several short segments could then be obtained, allowing detailed characterization of various enzymatic activities within the same fiber.

On the basis of the achievements of the 1970s, the study of human single muscle fibers progressed by further refinement of experimental methods and characterizing more physiological and pathophysiological conditions. Measures of peak tension, maximum shortening velocity, peak power [7, 35, 36], ATP consumption during contraction [4], and efficiency of chemo-mechanical transduction [37], i.e., the contractile properties at maximal activation determined in human single fibers, were directly linked to the molecular composition of the myofibrillar motors, in particular MyHC and MyLC (Myosin Light Chain) isoforms' expression. Sarcoplasmic reticulum function and intracellular calcium dynamics were studied [9, 10, 38] as well as the response of myofibrils to calcium [39]. Mitochondrial respiration of single human muscle fibers has been also recently recorded [40]. Recently, omics approaches were applied at the single muscle fiber level [41, 42]. This allowed to extend the knowledge of the protein complement of each fiber type and to describe changes of thousands of proteins in relation to aging, training, and mitochondrial diseases.

3 Methods for Experiments on Human Single Muscle Fibers

3.1 The Biopsy

Muscle biopsy is the first step to have access to human muscle fibers. Only a few muscles are generally considered for biopsy: biceps and triceps in the upper limbs, vastus lateralis, gastrocnemius, and tibialis anterior in the lower limbs. To minimize the invasive nature of the procedure, the closed or needle biopsy is generally adopted as an alternative to open surgical biopsy. The latter is however considered an unavoidable requirement for some specific diagnostic questions, for example, diagnosis of malignant hyperthermia (MH) susceptibility or biochemical investigation in mitochondrial myopathies. The needle biopsy is a rapid procedure, generally performed under local anesthesia in an aseptic clinical environment, and leaves only a small scar. A needle for muscle biopsy was first proposed by Charriere and Duchenne 150 years ago [43] but was not taken into consideration until later, and the technique was largely ignored until Bergstrom in 1963 [44] designed a specialized needle to sample muscle tissue for his studies on electrolytes. The needle was composed of a sharp, cannula needle with a small side opening near the tip; a punch with a sharp edge, which fits tightly into the cannula; and a stylet to push samples from the punch. The sample is sucked into the opening of the cannula when the punch is removed and a negative pressure is created. Later, several variants of needles have been designed, and the classical Bergstrom needle with a 5–6 mm diameter has been largely replaced by smaller Bergstrom's needles based on the same principle but with diameters of 3–4 mm. A substantial improvement was the introduction of suction, which allows a greater amount of tissue to be recovered [45]. More recently, microbiopsy approaches based on 14G–16G needles, originally designed for liver biopsy, have been adopted when smaller amounts of tissue are sufficient for molecular analysis, for example, polymerase chain reaction (PCR) or Western blot, although some small histological preparation is also possible [46, 47]. On the other hand, when larger amounts of tissue are required, a conchotome with alligator forceps has been designed [48]. The semi-open biopsy with conchotome requires a somewhat wider skin and fascia incision but allows sampling of larger fragments of muscle, albeit not to a level reached by an open surgical biopsy.

3.2 Sarcolemma Permeabilization and Sarcolemma Removal

To prepare suitable fiber segments of a few millimeters long, as representative of a fiber with a length of several centimeters, it is essential to remove—partly or completely—the sarcolemma to avoid depolarization and any increase in calcium concentration and to protect intracellular structures from contact with the extracellular medium or solutions with a similar composition (Ringer, Krebs, or other). In this way, the structure and function of the intracellular organelles are preserved. As mentioned in the historical overview in Subheading 2, chemical permeabilization based on prolonged immersion in glycerol-containing solution and the mechanical skinning achieved by manual removal of the sarcolemma were proposed 70 years ago and later improved and updated. The chemical permeabilization can now be achieved either with refined glycerol treatment [49] or with detergents such as Triton X-100 and Brij-35 or saponin; the latter being selective for sarcolemma with minimal damage to endoplasmic reticulum and mitochondrial membranes [50, 51]. The mechanical, manual removal of the sarcolemma allowed an important breakthrough when resealing of the T-tubules was reported, thus allowing the preservation of electrical activity, although restricted to T-tubule membranes [52]. An alternative approach is represented by the freeze-drying method proposed by Stienen and coworkers [53]. With this method, bundles of fibers are frozen in liquid nitrogen and dried in a ventilated cryostat at −20 °C. Not only sarcolemmal but also endoplasmic reticulum membranes are thus permeabilized, whereas myofibrils are well preserved. Importantly, immediate freezing in liquid nitrogen is also the initial step of single-fiber preparation for enzyme activity as proposed by Lowry et al. [5].

3.3 Methods for Sample Storage

If single fibers are not immediately dissected and studied, they require storage until the day of the experiment. However, the manual, mechanical removal of the sarcolemma can only be performed on fresh samples, blotted dry to remove extracellular medium, and immersed quickly in paraffin oil to start dissection (*see* [54]). In contrast, chemical permeabilization and freeze-drying allow storage and use of the sample for several days after biopsy.

Different storage procedures aimed at specific requirements can be adopted. The choice of the procedure for storage is relevant for the preservation of functional properties as shown by Frontera and Larsson [55]. Thus, comparative studies should be restricted to fibers permeabilized and stored in the same way after biopsy sampling. The freeze-drying procedure allows storage at low temperature (−20 °C) for several weeks. Then, before dissection and functional experiments, the preparation is allowed to equilibrate for at least 5 min in a relaxing solution. Samples permeabilized with glycerol can be stored in skinning solution with 50% glycerol at −20 °C for some weeks. Alternatively, samples can be preserved in skinning solution containing 2 mol/L of sucrose to protect against freezing artifacts, even at −80 °C. Importantly, such a high concentration of sucrose needs to be removed very gradually [56]. For biochemical or molecular studies on single fibers, deep freezing in liquid nitrogen or freeze-drying procedures allow storage at −20 °C [5] for several weeks. For RNA extraction, bundles from freshly collected biopsy samples can be placed in a petri dish filled with RNA*later* where dissection of single-fiber segments can be manually performed. After dissection, segments can be transferred to RNA extraction reagent, for example, TRIzol [57].

3.4 Single Muscle Fiber Dissection

Carried out at the time as the biopsy or, later, on the day of the experiment, dissection is the most critical, time-consuming step. Alternative methods for biochemical and molecular studies can be chosen, for example, cross sections can be obtained with a cryostat, and individual fibers in each section can be etched or removed by laser capture [58]. To this end, artificial intelligence can help to recognize the same fiber in subsequent sections [59, 60] and collect its fragments in the same vessel. However, the manual dissection under a stereomicroscope remains the approach of choice if fiber segments must be isolated carefully and without damage. The work is carried out in a small petri dish or in a similar chamber filled with relaxing or skinning solution, with the bottom covered with Sylgard® (Dow, Midland, MI, USA), on which fiber segments can be pinned, and generally kept at low temperature, under a stereomicroscope with the help of fine forceps, iris scissors, needles, and razor blades. Fiber segments of the suitable length are gently pulled out of the fiber bundle, fixed at the ends with insect pins of 50–100 μm diameters or with knots made with thin thread (braided silk n 10). If required to mount the fiber segment in the setup, aluminum clips can be attached at the ends of the fiber segment. A drop of glutaraldehyde applied with care on the edge of the clip where the fiber segments disappear between the two aluminum layers can significantly contribute to mechanically stabilize the preparation.

3.5 Setup, Solutions, and Protocols for Mechanical Experiments

The fragment of muscle biopsy and then the segments of fiber isolated during the dissection need to be immersed or bathed in solutions that reproduce the intracellular environment and its variations, as the sarcolemma has been removed or permeabilized and thus cannot fulfill its task to separate intra- and extracellular environment. The "solutions" must meet the following requirements:

(a) K^+ is the largely predominant cation accompanied by organic anions as propionate, acetate, EGTA, or HDTA, whereas the concentration of Cl^- is kept very low. Mg^{2+} concentration is generally around 1 mM, with some variations to account for its binding with ATP (Mg ATP is the substrate for myosin ATPase) and to its influence on Ca^{2+} release from sarcoplasmic reticulum. Na^+ enters in the composition as Na_2ATP. Free Ca^{2+} is strictly controlled and kept at very low levels, *see* below.

(b) ATP is present at physiological (4–8 mM) concentrations, as it is essential for fueling the various ionic pumps (plasma membrane Na-K-ATPase and SR Ca-ATPase) still active in skinned and even in permeabilized fibers [61] and, of course, for supporting force production by the contractile apparatus. Lack of ATP brings the fiber to a rigor state.

(c) 10–25 mM creatine phosphate (CP) is added to the solution. Although skinned mammalian muscle fibers may retain some endogenous creatine phosphokinase (CPK) activity, CPK is generally added to the solutions. The CP-CPK system can regenerate ATP from CP and ADP and, importantly, keeps the ADP concentrations low. For experiments aimed to measure the rate of ATP hydrolysis by skinned or permeabilized fibers, the ATP-regenerating system based on CP and CPK is replaced by an enzyme cascade system based on the enzymes pyruvate kinase (PK) and lactate dehydrogenase (LDH). PK catalyzes the regeneration of ATP from ADP and phosphoenol pyruvate (PEP), whereas LDH catalyzes the reduction of pyruvate to lactate with oxidation of reduced nicotinamide adenine dinucleotide (NADH) to NAD^+. The oxidation of NADH can be then monitored by measuring its concentration by absorbance at 340 nm or in fluorescence at 470 nm.

(d) pH buffering capacity is very important and aims to maintain the pH stable around 7 even when intense ATP hydrolysis occurs. Buffering is generally based on HEPES, TES, or imidazole.

The osmotic concentrations and ionic strength used are generally near the physiological levels (290–300 mOsm and 220 mM, respectively). In skinned and permeabilized fibers, an increase in ionic strength is accompanied by a decrease in isometric force and stiffness, which is caused by alterations in crossbridge dynamics [62, 63]. When the sarcolemma is removed or permeabilized, fibers swell as first described by Matsubara and Elliot [64] and Godt and Maughan [65]. This is attributed to repulsion between filaments and thus increase of lattice spacing (the distance between actin and myosin filaments). Fiber swelling corresponds to an increase of 20% in diameter and, thus, a 40% increase in cross sectional area. To counteract swelling, addition of high molecular weight polysaccharides as dextran 500 at 4% has been proposed and often adopted [66].

Anti-proteolytic factors, for example, E64 10 μM and leupeptin 40 μM, are frequently added to the solutions to protect proteins during the storage but also during experiments.

The concentration of free calcium and its control are of paramount importance in designing the solution for skinned or permeabilized muscle fibers. A change in free calcium concentration from 10 nM to 25 μM (i.e., from pCa = 8 to pCa = 4.6) abolishes troponin-tropomyosin inhibition and allows acto-myosin interaction, force generation, filament sliding, and ATP hydrolysis. The control of free calcium concentration is mostly based on chelating molecules, among them EGTA being the most used. Manipulating the proportions of EGTA and $CaCl_2$ makes it possible to obtain very low free calcium concentrations as required for the relaxing and skinning solutions (pCa 8) or, alternatively, relatively high free calcium concentrations (pCa between 5 and 4) as required for the activating solution able to produce maximal activation. In most experimental protocols where fibers need to be activated to contract, a pre-activating solution is also prepared. Pre-activating solutions contain a low free calcium concentration, like the relaxing solution, but alongside a low EGTA concentration, which allows a faster contractile response when fibers move from pre-activating to activating solution. Once the relaxing and the activating solutions are prepared, a complete set of solutions with intermediate free calcium concentrations can be easily obtained by mixing relaxing and activating solutions in appropriate proportions. With these solutions, contractile response at submaximal activation can be obtained.

Several variants of the "setup" for functional experiments on single muscle fibers have been developed (*see* Figs. 3 and 4). Traditionally, they were designed and built in the lab where they are then used, but currently, devices are also commercially available, for example, the 1400A Permeabilized Fiber System by Aurora Scientific (Dublin, Ireland). The main components of a setup for single fibers are:

(a) An array of chambers containing various solutions, organized in such a way that a fiber segment can be quickly (2–3 s) moved from one to the other, either manually or by a motor. The array is thermoregulated to keep the temperature of the solutions constant. The size of the chambers is in the order of

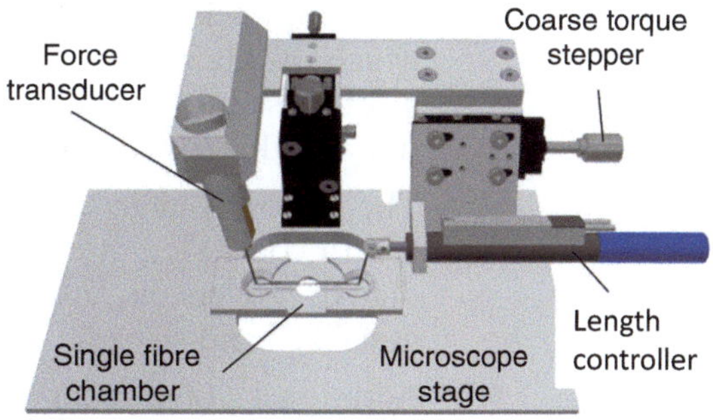

Fig. 3 Setup for single-fiber mechanics. The main components can be easily identified: the chamber in the center, the force transducer from which a shaft to fix the fiber emerges, and the length controller with a shaft to fix the fiber. (Modified from Schneidereit et al. Ref. [67])

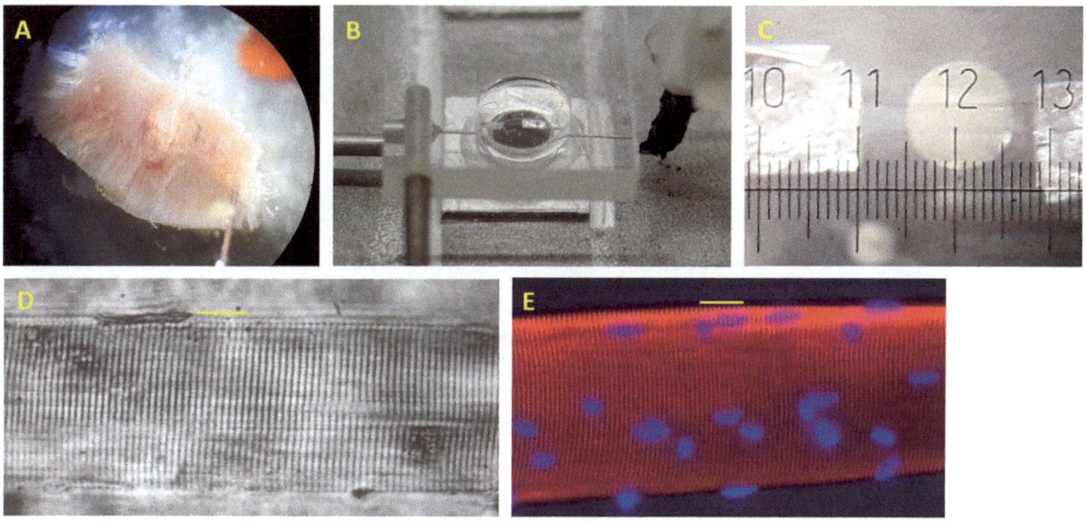

Fig. 4 Details of a single-fiber experiment. (**a**) Fragment of biopsy immersed in skinning solution, bundles, and fibers can be easily recognized. (**b**) Single fiber equipped with two aluminum clips at the ends placed in the setup between the length controller (left) and the transducer (right) and immersed in a drop of relaxing solution. (**c**) Enlarged image of single fibers with clips, small divisions of the micrometric scale correspond to 50 μm; thus the fiber segment is approximately 900 μm long and 100 μm thick. (**d**) High magnification view of a fiber segment mounted in the setup. The scale segment is 20 μm. Sarcomere length approximately 2.7 μm. (**e**) High magnification of a fiber segment after the mechanical experiment stained with anti-alpha-actinin antibody (Z line, sarcomere pattern) and Hoechst (nuclei). Sarcomere length approximately 2.2 μm

tens of μL. Importantly, the bottom of the chamber is transparent to allow viewing of the segment from below with an inverted microscope under a high magnification. A stereomicroscope for fiber mounting and manipulation is placed above the chamber array.

(b) The myograph or tensiometer is composed of a force transducer and motor or length controller attached to micromanipulators, which permit exact alignment of the fiber segment, with both being equipped with a small hook or a thin rod on which the fiber segments can be attached. Aluminum clips are mostly used to connect the fiber segments to the setup, but thin silk threads can be also used as a ligature.

(c) The analog voltage signals from the force transducer and the length controller, after appropriate amplification and analog-digital conversion, are sent to a computer where they are visualized and stored.

(d) Additional components are video cameras to view, record, and, eventually, analyze the images from the stereomicroscope and from the high magnification inverted microscope. The latter images provide a way also to measure sarcomere length and its variations. Alternatively, sarcomere length can be measured by the diffraction of a beam of laser light with a wavelength of 635 nm.

(e) For ATP hydrolysis rate measurement, a UV light beam crosses the fiber segment or the solution in which it is immersed to collect variation of fluorescence or absorbance of NADH.

Various experimental "protocols" are designed according to the scientific question that the experiment is expected to answer; the most popular protocol to measure contractile parameters follows this sequence:

(a) Dissection of the fiber segment and transfer to the setup.

(b) Mounting of the fiber segment between the force transducer and the length controller while under immersion in the relaxing solution.

(c) Alignment of the segment and adjustment of sarcomere length. In a human fiber experiment, sarcomere length is generally set at 2.7–2.8 μm, which is at the upper or right corner of the plateau of the length-tension curve [68, 69]. This implies a stretching of approximately 20% from the slack length, which is about 2.4 μm, virtually without any development of resting tension. The choice for the upper right corner considers that an internal shortening of up to 5% can occur during maximal contractions. Diameters are mostly measured at this point on the image obtained with high magnification inverted microscopy and thereafter are used to calculate cross sectional area (CSA) according to a circular or elliptical assumption of the transversal section of the fiber segment.

(d) Transfer of the fiber segment from the chamber containing relaxing solution to the chamber containing pre-activating solution and then approximately after 1 min to the chamber containing activating solution. Force develops rather quickly

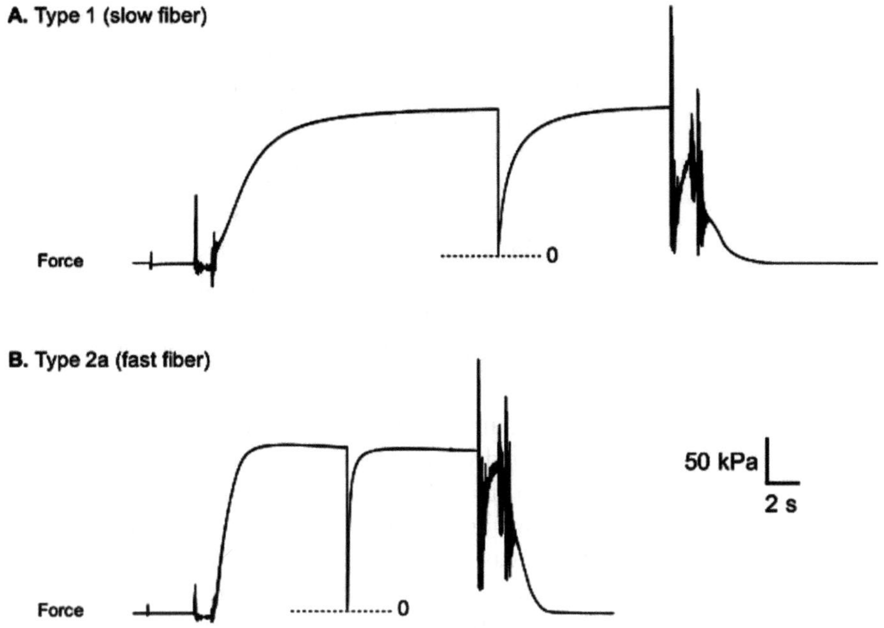

Fig. 5 Representative force traces from healthy human vastus lateralis muscle fibers. (**a**) Type 1 fiber (CSA: 5710 μm^2, maximum isometric force (Fo): 0.89 mN and Fo/CSA (Po): 156 kPa). (**b**) Type 2a fiber (CSA: 9510 μm^2, Fo: 1.66 mN and Po: 174 kPa). Fiber type was determined with electrophoretic separation of MyHC isoforms. (From Roche et al. Ref. [70])

and reaches a plateau, which remains stable for a long time (Fig. 5). Relaxation is achieved by going quickly back to a chamber containing relaxing solution. During the force plateau, several maneuvers are possible:

(i) Quick shortening to completely unload the fiber segment to zero force, thus allowing reliable measurement of force, followed within 3–5 ms by a stretch back to initial length. The isometric maximal force (Fo), which is in the order of 0.5–1 mN, can be then normalized to CSA, and an isometric specific force or tension (Po) of 100–200 kPa is obtained. The redevelopment of tension follows an approximately monoexponential time course and allows the determination of a rate constant indicated as K_{TR}.

(ii) A series of quick shortening of different amplitudes (ΔL), all sufficient to slacken the fiber. The times (Δt) required to take up the slack and redevelop tension can be plotted against ΔL, and the slope of the straight line provides a measure of the velocity of shortening at zero load, Vo (slack test method, ref. [24]).

(iii) Shortening or elongation at moderate velocity allows the collection of pairs of values, velocity, and force, which are utilized to obtain a force-velocity curve, which is interpolated with the hyperbolic Hill equation, characterized by three parameters, namely, isometric tension or Po, maximum shortening velocity or Vmax, and curvature coefficient or "a." The force-velocity curve also allows the calculation of the power, each power value being the product of velocity × force (*see* Fig. 2a). The resulting power-force and power-velocity curves reach their peak at the Wmax: the maximal mechanical power during a maximal activation, which is the product of optimal velocity or Vopt and optimal force or Fopt. The two latter values identify the conditions at which the contractile apparatus can attain the optimal contractile performance.

(iv) If the suitable experimental conditions are set (*see* above), ATP hydrolysis rate can be measured during maximal activation from NADH oxidation in isometric conditions or during shortening at moderate velocity (*see* Fig. 2b). From the measured values of ATP hydrolysis, the cost of force generation in isometric conditions and the efficiency during shortening, i.e., when external work is generated by fiber shortening, can be calculated. Importantly, all the contractile parameters measurable during maximal activation are mainly dependent on the myosin motors. Thus, they vary in relation to different myosin isoforms (slow and fast), myosin mutations, pharmacological interventions acting on myosin, etc.

(e) If the fiber segment, after immersion in pre-activating solution, is placed in a series of chambers containing free calcium concentrations intermediate between that of the relaxing and activating solutions, the sensitivity or responsiveness of the fiber to free calcium can be determined. The fiber segment is moved through a sequence of solutions with increasing free calcium (or decreasing pCa) dwelling in each solution for 20–60 s, i.e., a time sufficient to reach a plateau of tension. The tension and pCa values are utilized to generate a tension-pCa curve, which is then interpolated with the sigmoid Hill equation, characterized by two parameters: the pCa value at which 50% of the maximum tension is generated (EC50) and the Hill coefficient; n.

(f) If the experiment is aimed at studying the function of the sarcoplasmic reticulum (SR), the fiber will undergo a different sequence of steps. The fiber is mounted in relaxing solution and treated as described above and is then moved to a loading solution where the sarcoplasmic reticulum is loaded with calcium via SR Ca-pump (SERCA) in the presence of free calcium (pCa 6.5–6.7) and ATP. The duration of the loading step is variable to regulate the level of filling of sarcoplasmic reticulum. The fiber is then transferred to a washing solution,

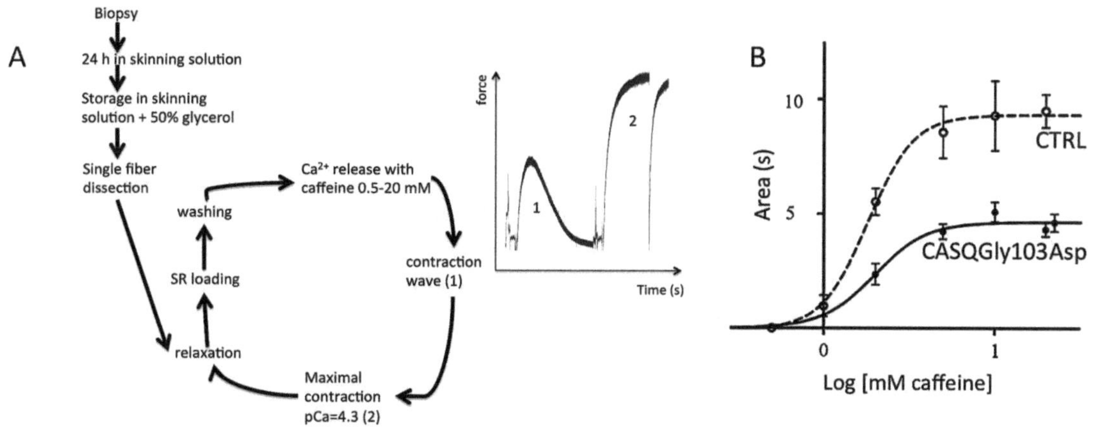

Fig. 6 Sarcoplasmic reticulum functional tests using caffeine to induce calcium release. (**a**) Schematic representation of the loading-release cycle designed to study sarcoplasmic reticulum (SR) function in single skinned fiber experiments. In the inset, the response to caffeine is indicated with 1 and the maximal isometric activation with 2. (**b**) Caffeine dose–response curves of single human skinned fibers from a proband carrying calsequestrin mutation Gly103Asp and from controls (CTRL) (Barone et al., Ref. [72])

followed by a calcium release solution, where calcium release via ryanodine receptor (RyR) channel can be induced in a variety of ways:

(i) By applying caffeine (at mM concentrations) or other pharmacological stimulants (e.g., 4-chloro-m-cresol).

(ii) By lowering free Mg^{2+}concentration from its normal level of 1 mM to 0.05 Mm.

(iii) In mechanically skinned fibers, where T-tubules are still present and functional, by triggering voltage-sensor activation of the RyRs, with appropriate electrical stimulation.

In many protocols, after the calcium release solution, the segment is transferred to the activating solution to test its maximal contractile performance. The cycle relaxing-loading-washing-releasing-activating can be repeated several times with each fiber segment and characterizes the function of the SR (*see* [71] and Fig. 6).

3.6 Biochemical and Molecular Analysis on Single Fibers

Single-fiber segments dissected from biopsy samples are also an important source for a number of biochemical and molecular analyses. The most popular molecular analysis is the determination of the fiber type. As mentioned above, MyHC isoforms are considered the molecular markers of fiber type. Thus, their separation via gel electrophoresis, possibly followed by Western blot with anti-MyHC antibodies, or the identification via mATPase with preincubation on a fragment of fiber, is associated to virtually all biochemical or mechanical experiments. A small fragment of the fiber or, in some case, the whole fiber segment at the end of the experiment is used for this purpose.

The manual dissection of fiber segments from a freeze-dried biopsy sample [8] can allow the determination of several biochemical parameters in segments of single fibers, for example, the concentrations of glycogen, lipids, high-energy phosphates, and enzyme activities based on the micromethods first developed by Lowry and Passonneau [73]. According to the original protocol [5], fiber segments are dissected from the freeze-dried samples under a stereomicroscope in a room at 20 °C and 35% humidity to allow a partial rehydration. Fiber segments are then placed on a glass slide, end fragments are removed for determination of fiber type (*see* above), and the remaining part of the segment is weighed and used for biochemical or enzymatic determinations either as single-fiber segment or after pooling fibers of the same type. The application of the micromethods allows for multiple enzymatic determinations on the same fiber segments, cut after dissection in small fragments with a length of 30–50 μm and dry weight of 20–40 ng [5].

Polymerase chain reaction (PCR) amplification of specific genes from segments of individual fibers has been developed by Jemiolo and Trappe [57]. Bundles from freshly collected biopsy samples are placed in a petri dish filled with RNA*later* solution, and the dissection of single-fiber segments is quickly performed manually. Segments are then transferred to the RNA extraction reagent. Global transcriptomics of single muscle fibers has been first developed in mice using enzymatically dissociated Flexor Digitorum Brevis (FDB) fibers by Chemello et al. [74] while implementing a specific amplification protocol. In humans, single muscle fiber transcriptomics has only been performed in pools of muscle fibers [75]. In this case, manual muscle fiber isolation from samples collected from the vastus lateralis is followed by muscle fiber clipping to allow for fiber-type identification via electrophoretic MyHC separation and by pooling of fibers of the same type. This is then followed by RNA extraction, quantification, and RNA-seq protocols [75].

More recently, methods starting from manual dissection of human single fibers from biopsy samples have been implemented for mass spectrometry (MS)-based proteomics [41]. For each fiber segment (2–4 mm long), the whole enzymatic processing and peptide purification steps are performed in a single vessel, thus minimizing sample loss. The liquid chromatography (LC)-MS analysis of individual fibers is then performed in single-run format to increase sensitivity. A linear quadrupole ultra-high-field Orbitrap mass analyzer (Q Exactive HF), characterized by high sensitivity, sequencing speed, and mass accuracy, is used to maximize the number of peptides detected, although other mass spectrometer configurations are also valid for proteomics. This protocol has allowed the identification of more than 60,000 peptides and more than 5400 proteins, in single fibers, at a false discovery rate (FDR) of 1%, with an average of 2100 proteins per single muscle fiber segment.

4 What Can Experiments on Human Single Muscle Fibers Tell Us?

The study of human single fibers isolated from biopsy samples, dating back to 50 years ago, has greatly contributed to the understanding of the contractile function of skeletal muscles in health and disease as well as muscles' plasticity in terms of ability to adapt to new functional demands or environmental signals.

As briefly reviewed in Subheading 2, single-fiber experiments have revealed how much the presence of a specific myosin isoform (MyHC and MyLC isoform combination) determines the contractile performance and thus explains the diversity between the power generated by human slow and fast fibers and among the two subgroups of fast fibers expressing 2A and 2X myosin, respectively ([7, 35] *see* Fig. 2). Those experiments also allowed the quantification of the energetic cost of such contraction [4, 37] and the understanding of the specific contribution given by slow or fast fibers to human motor performance. This has been further expanded through the analysis of the ATP regeneration enzymatic pathways with the quantification of the specific contribution of glycolysis and mitochondrial respiration to this [5]. Furthermore, proteomics have recently provided a robust molecular background for the contractile and biochemical diversity of single fibers [41, 42].

A second field where single-fiber studies have greatly contributed to current knowledge is represented by the adaptive response of skeletal muscles to various physiological and pathological conditions. Without any claim to be complete or exhaustive, some examples are given here below. The decline of skeletal muscles in terms of mass and contractile performance with aging is due to a multiplicity of causes and mechanisms, among them the motor neuron degeneration and death, increase in intramuscular fat, and alteration of connective tissues in the extracellular matrix as well as in the tendons (*see* for review [76]). Thus, the question arises as how to identify the specific contribution of changes occurring in single fibers. Comparison between older adults (50–70 years old) and young adults shows reduction in fiber thickness accompanied by a decrease in maximum isometric tension (Po) both in slow and fast 2A fibers and a decrease in unloaded shortening velocity (Vo) only in slow fibers [77, 78]. The decrease in Vo has been attributed to myosin glycation [79]. The study of single muscle fibers in oldest old (around 90 years old) people compared to young people shows a surprising preservation of fiber thickness and contractile performance [80, 81]. This finding suggests that the extreme decline of muscle mass and strength in this particular age group is due to loss of muscle fibers more than to atrophy and weakness of the fiber themselves.

Disuse and inactivity have a marked negative impact on skeletal muscle structure and function. Studies on healthy subjects exposed to microgravity in space flight or to simulated microgravity (bed rest, unilateral limb suspension) on the ground have shown that muscle mass decreases at the pace of 0.5–1%/day of unloading and muscle isometric force declines at an even faster rate (*see* for a review Narici and de Boer, [82]). Single-fiber studies offer a unique possibility to identify the specific contribution of the muscle fibers to such adaptive response. The analysis of the adaptation of single muscle fiber to inactivity has revealed that maximal isometric specific force (Po), i.e., force/cross sectional area, is virtually unchanged [83] with force and fiber thickness changing in close proportion. The maximum shortening velocity undergoes a transitory increase during the first days of disuse. This surprising variation has been explained with a structural reorganization of the sarcomere due the different degradation rates of thick and thin filaments [84]. Alterations of intracellular calcium signaling are likely responsible for the reported decrease of the specific isometric force in submaximal contractions [85, 86].

Functional and structural studies on single muscle fibers have fuelled a live debate on whether force increases in proportion to hypertrophy development during resistance training with data supporting a proportional increase in fiber performance and thickness [87] and data showing that the increase in isometric force does not follow the increase in muscle fiber thickness [88, 89].

Moving to pathophysiological conditions, studies on single muscle fibers have provided much information on pathogenesis of muscle diseases and disorders. As a first example, the analysis of single fibers from patients affected by Duchenne muscular dystrophy (DMD), which is arguably the most severe primary muscle disease, did not show any significant contractile impairment and confirmed that myofibrils assemble and function normally in DMD. Thus, the absence of dystrophin leads to clinical weakness by causing the breakdown of muscle fibers, which are in principle capable of normal contractile performance [90]. Accordingly, also in patients affected by facioscapulohumeral dystrophy (FSHD), single muscle fiber specific force is preserved even in severely altered tissues with marked replacement of muscle fibers with adipocytes [91]. The early data showing a reduced specific force in fast type 2 fibers from muscles of FSHD patients have not been confirmed by more recent studies [92, 93]. In contrast, specific force developed during maximal isometric contraction has been found to be generally reduced in myopathies due to mutations of structural components of the sarcomere involved in force generation, such as nemaline myopathy caused by mutations in NEB, ACTA1, TPM2, and TPM3 [94]. A reduction in maximal isometric fiber specific force has been also found in single fibers from patients affected by myotonic dystrophy, related to repeat expansion of the DMPK gene [95].

Interestingly, single muscle fibers sampled from patients carrying mutations of sarcoplasmic reticulum (SR) proteins have been found to show altered calcium storage and/or release. Such alterations are likely responsible for weakness, which is a major complaint of those patients. Altered calcium handling by the SR has been recently described in vacuolar and tubular myopathies associated to calsequestrin mutations [72, 96]. Single muscle fibers of patients carrying mutations of the RyR channel and with susceptibility to malignant hyperthermia (MH) episodes show a virtually normal contractile performance during maximal activation but also a subtle difference in calcium handling in the triads even in the absence of agents triggering MH crisis [97, 98].

Clinically, muscle biopsy accompanied by contractile or biochemical tests is considered essential for the diagnosis and the prognosis of various types of muscle dystrophy and myopathy, for MH susceptibility with the specific in vitro contracture test (IVCT, [99]), and for mitochondrial myopathies, which require detailed biochemical investigations. In all these conditions, however, the dissection and the analysis of single fibers may not add additional value.

5 Conclusions

Single muscle fibers represent a powerful model to study human muscle function and its molecular basis. The same or very similar preparations of single fibers can be used to determine contractile properties, metabolic enzyme activity, and gene expression at transcript and protein level. Single muscle fiber studies have given, over 50 years, important contributions to the knowledge of the fundamental mechanisms of force production, ATP regeneration, and chemo-mechanical transduction and of their diversity in various fiber types. Moreover, they have allowed the identification of the role of muscle fibers to adaptive responses, which involve the muscle-tendon complex. Finally, at the pathophysiological level, single muscle fibers have provided useful insights into the mechanisms of the muscle disorders and diseases.

However, it is necessary to recall at least two major limitations inherent in the single muscle studies. The first limitation is that physiological, biochemical, or molecular data are collected from a segment of fiber of a few millimeters in length. Even when few hundreds of such segments are studied, they represent only a small fraction of the thousands of fibers present in a muscle and, inside each fiber, only a small fraction of the fiber length, which is in the order of several centimeters. Variations in structural and functional parameters along the length of each fiber cannot be excluded. The second limitation is that all functional and biochemical experiments are performed in conditions that simulate the physiological conditions in vivo but are unavoidably different. At least two major differences are worth underlining: first, temperature as most experiments are carried out at 12–20 °C, well below the muscle temperature in the body, which varies between 35 and 39 °C depending on contractile activity, and second, the removal or permeabilization of the sarcolemma creates a condition useful for experiments as the intracellular organelles are directly accessible but at the expense of potential structural (swelling, loss of soluble proteins) alterations.

The last few years have opened a new and almost unexpected perspective to single-fiber studies as the "miniaturization" of omics, transcriptomics, and proteomics approaches has allowed global studies at the level of single cells and, in our specific case, of the segments of single fibers or even of single nuclei inside a fiber. The identification of thousands of transcripts or proteins seems promising both further advancement in the understanding the molecular basis of fiber specialization and adaptation and progress in better identifying the mechanisms, and thus the potential interventions, of muscle disorders and diseases.

References

1. Schmalbruch H, Hellhammer U (1977) The number of nuclei in adult rat muscles with special reference to satellite cells. Anat Rec 189:169–175
2. Dos Santos M, Backer S, Saintpierre B, Izac B, Andrieu M, Letourneur F, Relaix F, Sotiropoulos A, Maire P (2020) Single-nucleus RNA-seq and FISH identify coordinated transcriptional activity in mammalian myofibers. Nat Commun 11:5102
3. Schiaffino S, Reggiani C (2011) Fiber types in mammalian skeletal muscles. Physiol Rev 91: 1447–1531
4. Stienen GJM, Kiers J, Bottinelli R, Reggiani C (1996) Myofibrillar ATPase activity in skinned human skeletal muscle fibres: fibre type and temperature dependence. J Physiol 493:299–307
5. Lowry CV, Kimmey JS, Felder S, Chi MMY, Kaiser KK, Passoneau PN, Kirk KA, Lowry OH (1978) Enzyme patterns in single human muscle fibres. J Biol Chem 253:8269–8277
6. Bottinelli R, Reggiani C (2000) Human skeletal muscle fibres: molecular and functional diversity. Prog Biophys Mol Biol 73:195–262
7. Bottinelli R, Canepari M, Pellegrino MA, Reggiani C (1996) Force-velocity properties of human skeletal muscle fibres: myosin heavy chain isoform and temperature dependence. J Physiol 495:573–586
8. Essen B, Jansson E, Henriksson J, Taylor AW, Saltin B (1975) Metabolic characteristics of fibre types in human skeletal muscle. Acta Physiol Scand 95:153–165
9. Lamboley CR, Murphy RM, McKenna MJ, Lamb GD (2013) Endogenous and maximal sarcoplasmic reticulum calcium content and calsequestrin expression in type I and type II human skeletal muscle fibres. J Physiol 591: 6053–6068
10. Delbono O, O'Rourke KS, Ettinger WH (1995) Excitation-calcium release uncoupling in aged single human skeletal muscle fibers. J Membr Biol 148:211–222
11. Salviati G, Betto R, Danieli-Betto D, Zeviani M (1984) Myofibrillar protein isoforms and sarcoplasmic reticulum calcium-transport activity of single human muscle fibres. Biochem J 224: 215–225
12. Gruetzner P (1883) Zur Physiologie und Histologie der Skelettmuskeln. Breslauer Aerztl Z 5:257–258
13. Knoll P (1891) Ueber protoplasmaarme und protoplasmareiche Musculatur. Denkschr Kais Akad Wiss Wien, math-naturwiss Cl 58:633–700
14. Needham D (1926) Red and white muscles. Physiol Rev 6:1–27
15. Liddell EGT, Sherrington CS (1925) Recruitment and some other factors of reflex inhibition. Proc R Soc Lond Ser B 97:488–518

16. Brown DES, Sichel FJM (1930) The myogram of the isolated skeletal muscle cell. Science 72: 17–18
17. Asmussen E (1934) Untersuchungen über die mechanische Reaktion der Skelettmuskelfaser. Skand Arch f Physiol 70:233–272
18. Ramsey RW, Street SF (1940) The isometric length-tension diagram of isolated skeletal muscle fibers of the frog. J Cell Comp Physiol 15:11–34
19. Buchthal F, Lindhard J (1939) The physiology of striated muscle. Biol Med Kgl Danske Vid Selskab 14(6)
20. Buchthal F, Kaiser E, Knappeis GG (1944) Elasticity, viscosity and plasticity in the cross striated muscle fibre. Acta Physiol 8:16–37
21. Buchthal F, Kaiser E (1944) Factors determining tension development in skeletal muscle. Acta Physiol 8:38–74
22. Gordon AM, Huxley AF, Julian FJ (1966) The variation in isometric tension with sarcomere length in vertebrate muscle fibres. J Physiol 184:170–192
23. Ford LE, Huxley AF, Simmons RM (1977) Tension responses to sudden length change in stimulated frog muscle fibres near slack length. J Physiol 269:441–515
24. Edman KAP (1979) The velocity of unloaded shortening and its relation to sarcomere length and isometric force in vertebrate muscle fibres. J Physiol 291:143–159
25. Piazzesi G, Reconditi M, Linari M, Lucii L, Bianco P, Brunello E, Decostre V, Stewart A, Gore DB, Irving TC, Irving M, Lombardi V (2007) Skeletal muscle performance determined by modulation of number of myosin motors rather than motor force or stroke size. Cell 131:784–795
26. Reconditi M, Linari M, Lucii L, Stewart A, Sun YB, Boesecke P, Narayanan T, Fischetti RF, Irving T, Piazzesi G, Irving M, Lombardi V (2004) The myosin motor in muscle generates a smaller and slower working stroke at higher load. Nature 428:578–581
27. Blinks JR, Rüdel R, Taylor SR (1978) Calcium transients in isolated amphibian skeletal muscle fibres: detection with aequorin. J Physiol 1978: 291–323
28. Szent-Gyorgyi A (1949) Free-energy relations and contraction of actomyosin. Biol Bull 96: 140–161
29. Natori R (1954) The property and contraction process of isolated myofibers. Jikeikai Med J 1: 177–192
30. Hellam DC, Podolsky RJ (1969) Force measurements in skinned muscle fibres. JPhysiol 209:807–819
31. Julian FJ (1971) The effect of calcium on the force-velocity relation of briefly glycerinated frog muscle fibres. J Physiol 218:117–145
32. Ashley CC, Moisescu DG (1973) Tension changes in isolated bundles of frog and barnacle myofibrils in response to sudden changes in the external free calcium concentration. J Physiol 233:8P–9P
33. Moisescu DG (1976) Kinetics of reaction in calcium-activated skinned muscle fibres. Nature 262:610–613
34. Wood DS, Zollman JR, Reuben JP, Brandt PW (1975) Human skeletal muscle: properties of the "chemically skinned" fiber. Science 187: 1075–1076
35. Larsson L, Moss RL (1993) Maximum velocity of shortening in relation to myosin isoform composition in single fibres from human skeletal muscles. J Physiol 472:595–614
36. Linari M, Bottinelli R, Pellegrino MA, Reconditi M, Reggiani C, Lombardi V (2004) The mechanism of the force response to stretch in human skinned muscle fibres with different myosin isoforms. J Physiol 554:335–352
37. He ZH, Bottinelli R, Pellegrino MA, Ferenczi MA, Reggiani C (2000) ATP consumption and efficiency of human single muscle fibers with different myosin isoform composition. Biophys J 79:945–961
38. Lamboley CR, Wyckelsma VL, Dutka TL, McKenna MJ, Murphy RM, Lamb GD (2015) Contractile properties and sarcoplasmic reticulum calcium content in type I and type II skeletal muscle fibres in active aged humans. J Physiol 593:2499–2514
39. Fink RH, Stephenson DG, Williams DA (1990) Physiological properties of skinned fibres from normal and dystrophic (Duchenne) human muscle activated by Ca^{2+} and Sr^{2+}. J Physiol 420:337–353
40. Dollerup OL, Chubanava S, Agerholm M, Søndergård SD, Altıntaş A, Møller AB, Høyer KF, Ringgaard S, Stødkilde-Jørgensen H, Lavery GG, Barrès R, Larsen S, Prats C, Jessen N, Treebak JT (2019) Nicotinamide riboside does not alter mitochondrial respiration, content or morphology in skeletal muscle from obese and insulin-resistant men. J Physiol 598:731–754
41. Murgia M, Toniolo L, Nagaraj N, Ciciliot S, Vindigni V, Schiaffino S, Reggiani C, Mann M (2017) Single muscle fiber proteomics reveals fiber-type-specific features of human muscle aging. Cell Rep 19:2396–2409
42. Deshmukh AS, Steenberg DE, Hostrup M, Birk JB, Larsen JK, Santos A, Kjøbsted R, Hingst JR, Schéele CC, Murgia M, Kiens B,

Richter EA, Mann M, Wojtaszewski JFP (2021) Deep muscle-proteomic analysis of freeze-dried human muscle biopsies reveals fiber type-specific adaptations to exercise training. Nat Commun 12:304
43. Charriere M, Duchenne GB (1865) Emporte piece histologique. Bull Acad Natl Med 30: 1050–1051
44. Bergstrom J (1963) Clinical studies on cell electrolytes. Scand J Clin Lab Invest 15(Suppl 76):16–18
45. Evans WJ, Phinney SD, Young VR (1982) Suction applied to a muscle biopsy maximizes sample size. Med Sci Sports Exerc 14:101–102
46. Hayot M, Michaud A, Koechlin C, Caron MA, LeBlanc PC, Maltais F (2005) Skeletal muscle microbiopsy: a validation study of a minimally invasive technique. Eur Respir J 25:431–440
47. Paoli A, Pacelli QF, Toniolo L, Miotti D, Reggiani C (2010) Latissimus dorsi fine needle muscle biopsy: a novel and efficient approach to study proximal muscles of upper limbs. J Surg Res 164:e257–e263
48. Henriksson KG (1979) "Semi-open" muscle biopsy technique. A simpleoutpatient procedure. Acta Neurol Scand 59:317–323
49. Salviati G, Sorenson MM, Eastwood AB (1982) Calcium accumulation by the sarcoplasmic reticulum in two populations of chemically skinned human muscle fibers. Effects of calcium and cyclic AMP. J Gen Physiol 79:603–632
50. Herland JS, Julian FJ, Stephenson DG (1990) Halothane increases Ca^{2+} efflux via Ca^{2+} channels of sarcoplasmic reticu- lum in chemically skinned rat myocardium. J Physiol 426:1–18
51. Launikonis BS, Stephenson DG (1997) Effect of saponin treatment on the sarcoplasmic reticulum of rat, cane toad and crus- tacean (yabby) skeletal muscle. J Physiol 504:425–437
52. Posterino GS, Lamb GD, Stephenson DG (2000) Twitch and tetanic force responses and longitudinal propagation of action potentials in skinned skeletal muscle fibres of the rat. J Physiol 527:131–137
53. Stienen GJM, Guth K, Ruegg JC (1983) Force and force transients in skeletal muscle fibres of the frog skinned by freeze-drying Pflugers. Archiv 397:272–276
54. Lamb GD, Stephenson DG (2018) CORP: measurement of force and calcium release using mechanically skinned fibers from mammalian skeletal muscle. J Appl Physiol 125: 1105–1127
55. Frontera WR, Larsson L (1997) Contractile studies of single human skeletal muscle fibers: a comparison of different muscles, permeabilization procedures, and storage techniques. Muscle Nerve 20:948–952
56. Kraft T, Messerli M, Rothen-Rutishauser B, Perriard JC, Wallimann T, Brenner B (1995) Equilibration and exchange of fluorescently labeled molecules in skinned skeletal muscle fibers visualized by confocal microscopy. Biophys J 69:1246–1258
57. Jemiolo B, Trappe S (2004) Single muscle fiber gene expression in human skeletal muscle: validation of internal control with exercise. Biochem Biophys Res Commun 320:1043–1050
58. Emmert-Buck MR, Bonner RF, Smith PD, Chuaqui RF, Zhuang Z, Goldstein SR, Weiss RA, Liotta LA (1996) Laser capture microdissection. Science 274:998–1001
59. Koob AO, Bruns L, Prassler C, Masliah E, Klopstock T, Bender A (2012) Protein analysis through Western blot of cells excised individually from human brain and muscle tissue. Anal Biochem 425:120–124
60. Murgia M, Tan J, Geyer PE, Doll S, Mann M, Klopstock T (2019) Proteomics of cytochrome c oxidase-negative versus -positive muscle fiber sections in mitochondrial myopathy. Cell Rep 29:3825–3834
61. Szentesi P, Zaremba R, van Mechelen W, Stienen GJ (2001) ATP utilization for calcium uptake and force production in different types of human skeletal muscle fibres. J Physiol 531: 393–403
62. Seow CY, Ford LE (1993) High ionic strength and low pH detain activated skinned rabbit skeletal muscle crossbridges in a low force state. J Gen Physiol 101:487–511
63. Wang L, Bahadir A, Kawai M (2015) High ionic strength depresses muscle contractility by decreasing both force per cross-bridge and the number of strongly attached cross-bridges. J Muscle Res Cell Motil 36:227–241
64. Matsubara I, Elliott GF (1972) X-ray diffraction studies on skinned single fibres of frog skeletal muscle. Mol Biol 72:657–669
65. Godt RE, Maughan DW (1977) Swelling of skinned muscle fibers of the frog. Experimental observations. Biophys J 19:103–116
66. Percario V, Boncompagni S, Protasi F, Pertici I, Pinzauti F, Caremani M (2018) Mechanical parameters of the molecular motor myosin II determined in permeabilised fibres from slow and fast skeletal muscles of the rabbit. J Physiol 596:1243–1257
67. Schneidereit D, Nübler S, Prölß G, Reischl B, Schürmann S, Müller OJ, Friedrich O (2018) Optical prediction of single muscle fiber force production using a combined biomechatronics

and second harmonic generation imaging approach. Light Sci Appl 24(7):79

68. Walker SM, Schrodt GR (1974) I segment lengths and thin filament periods in skeletal muscle fibers of the rhesus monkey and the human. Anat Rec 178:63–81
69. Ottenheijm CAC, Witt CC, Stienen GJ, Labeit S, Beggs AH, Granzier H (2009) Thin filament length dysregulation contributes to muscle weakness in nemaline myopathy patients with nebulin deficiency. Hum Mol Genet 18:2359–2369
70. Roche SM, Gumucio JP, Brooks SV, Mendias CL, Claflin DR (2015) Measurement of maximum isometric force generated by permeabilized skeletal muscle fibers. J Vis Exp (100): e52695
71. Reggiani C (2020) Caffeine as a tool to investigate sarcoplasmic reticulum and intracellular calcium dynamics in human skeletal muscles. J Muscle Res Cell Motil 42:281
72. Barone V, Del Re V, Gamberucci A, Polverino V, Galli L, Rossi D, Costanzi E, Toniolo L, Berti G, Malandrini A, Ricci G, Siciliano G, Vattemi G, Tomelleri G, Pierantozzi E, Spinozzi S, Volpi N, Fulceri R, Battistutta R, Reggiani C, Sorrentino V (2017) Identification and characterization of three novel mutations in the CASQ1 gene in four patients with tubular aggregate myopathy. Hum Mutat 38:1761–1773
73. Lowry OH, Passonneau JV (1972) A flexible system of enzymatic analysis. Academic, New York
74. Chemello F, Bean C, Cancellara P, Laveder P, Reggiani C, Lanfranchi G (2011) Microgenomic analysis in skeletal muscle: expression signatures of individual fast and slow myofibers. PLoS One 6:e16807
75. Rubenstein AB, Smith GR, Raue U, Begue G, Minchev K, Ruf-Zamojski F, Nair VD, Wang X, Zhou L, Zaslavsky E, Trappe TA, Trappe S, Sealfon SC (2020) Single-cell transcriptional profiles in human skeletal muscle. Sci Rep 10:229
76. Mitchell WK, Williams J, Atherton P, Larvin M, Lund J, Narici M (2012) Sarcopenia, dynapenia, and the impact of advancing age on human skeletal muscle size and strength: a quantitative review. Front Physiol 3:260
77. Frontera WR, Suh D, Krivickas LS, Hughes VA, Goldstein R, Roubenoff R (2000) Skeletal muscle fiber quality in older men and women. Am J Physiol Cell Physiol 279:C611–C618
78. D'Antona G, Pellegrino MA, Adami R, Rossi R, Carlizzi CN, Canepari M, Saltin B, Bottinelli R (2003) The effect of ageing and immobilization on structure and function of human skeletal muscle fibres. J Physiol 552: 499–511
79. Ramamurthy B, Höök P, Jones AD, Larsson L (2001) Changes in myosin structure and function in response to glycation. FASEB J 15: 2415–2422
80. Grosicki GJ, Standley RA, Murach KA, Raue U, Minchev K, Coen PM, Newman AB, Cummings S, Harris T, Kritchevsky S, Goodpaster BH, Trappe S for The Health ABC Study (2016) Improved single muscle fiber quality in the oldest-old. J Appl Physiol 121: 878–884
81. Venturelli M, Saggin P, Muti E, Naro F, Cancellara L, Toniolo L, Tarperi C, Calabria E, Richardson RS, Reggiani C, Schena F (2015) In vivo and in vitro evidence that intrinsic upper- and lower-limb skeletal muscle function is unaffected by ageing and disuse in oldest-old humans. Acta Physiol (Oxf) 215: 58–71
82. Narici MV, de Boer MD (2011) Disuse of the musculo-skeletal system in space and on earth. Eur J Appl Physiol 111:403–420
83. Fitts RH, Trappe SW, Costill DL, Gallagher PM, Creer AC, Colloton PA, Peters JR, Romatowski JG, Bain JL, Riley DA (2010) Prolonged space flight-induced alterations in the structure and function of human skeletal muscle fibres. J Physiol 588:3567–3592
84. Widrick JJ, Knuth ST, Norenberg KM, Romatowski JG, Bain JLW, Riley DA, Karhanek M, Trappe SW, Trappe TA, Costill DL, Fitts RH (1999) Effect of a 17 day spaceflight on contractile properties of human soleus muscle fibres. J Physiol 516:915–930
85. Lamboley CR, Wyckelsma VL, Perry BD, McKenna MJ, Lamb GD (2016) Effect of 23-day muscle disuse on sarcoplasmic reticulum Ca2+ properties and contractility in human type I and type II skeletal muscle fibers. J Appl Physiol 1985(121):483–492
86. Monti E, Reggiani C, Franchi MV, Toniolo L, Sandri M, Armani A, Zampieri S, Giacomello E, Sarto F, Sirago G, Murgia M, Nogara L, Marcucci L, Ciciliot S, Šimunic B, Pišot R, Narici MV (2021a) Neuromuscular junction instability and altered intracellular calcium handling as early determinants of force loss during unloading in humans. J Physiol 599:3037–3061
87. Monti E, Toniolo L, Marcucci L, Bondì M, Martellato I, Šimunič B, Toninello P, Franchi MV, Narici MV, Reggiani C (2021b) Are muscle fibres of body builders intrinsically weaker? A comparison with single fibres of aged-

matched controls. Acta Physiol (Oxf) 231: e13557

88. D'Antona G, Lanfranconi F, Pellegrino MA et al (2006) Skeletal muscle hypertrophy and structure and function of skeletal muscle fibres in male body builders. J Physiol 570:611–627
89. Meijer JP, Jaspers RT, Rittweger J et al (2015) Single muscle fibre contractile properties differ between body-builders, power athletes and control subjects. Exp Physiol 100:1331–1341
90. Horowits R, Dalakas MC, Podolsky RJ (1990) Single skinned muscle fibers in Duchenne muscular dystrophy generate normal force. Ann Neurol 27:636–641
91. Lassche S, Voermans NC, Schreuder T, Heerschap A, Küsters B, Ottenheijm CA, Hopman MT, van Engelen BG (2021) Reduced specific force in patients with mild and severe facioscapulohumeral muscular dystrophy. Muscle Nerve 63:60–67
92. Lassche S, Voermans NC, van der Pijl R, van den Berg M, Heerschap A, van Hees H, Kusters B, van der Maarel SM, Ottenheijm CAC, van Engelen BGM (2020) Preserved single muscle fiber specific force in facioscapulohumeral muscular dystrophy. Neurology 94: e1157–e1170
93. Lassche S, Stienen GJ, Irving TC, van der Maarel SM, Voermans NC, Padberg GW, Granzier H, van Engelen BG, Ottenheijm CA (2013) Sarcomeric dysfunction contributes to muscle weakness in facioscapulohumeral muscular dystrophy. Neurology 80:733–737
94. de Winter JM, Ottenheijm CAC (2017) Sarcomere dysfunction in nemaline myopathy. J Neuromuscul Dis 4:99–113
95. Krivickas LS, Ansved T, Suh D, Frontera WR (2000) Contractile properties of single musclefibers in myotonic dystrophy. Muscle Nerve 23:529–537
96. Rossi D, Vezzani B, Galli L, Paolini C, Toniolo L, Pierantozzi E, Spinozzi S, Barone V, Pegoraro E, Bello L, Cenacchi G, Vattemi G, Tomelleri G, Ricci G, Siciliano G, Protasi F, Reggiani C, Sorrentino V (2014) A mutation in the CASQ1 gene causes a vacuolar myopathy with accumulation of sarcoplasmic reticulum protein aggregates. Hum Mutat 35: 1163–1170
97. Cully TR, Choi RH, Bjorksten AR, Stephenson DG, Murphy RM, Launikonis BS (2018) Junctional membrane Ca2+ dynamics in human muscle fibers are altered by malignant hyperthermia causative RyR mutation. Proc Natl Acad Sci U S A 115:8215–8220
98. Duke AM, Hopkins PM, Halsal JP, Steele DS (2004) Mg2+dependence of halothane-induced Ca2+release from the sarcoplasmic reticulum in skeletal muscle from humans susceptible to malignant hyperthermia. Anesthesiology 101:1339–1346
99. Ording H, Brancadoro V, Cozzolino S, Ellis FR, Glauber V, Gonano EF, Halsall PJ, Hartung E, Heffron JJ, Heytens L, Kozak-Rib-bens G, Kress H, Krivosic-Horber R, Lehmann-Horn F, Mortier W, Nivoche Y, Ranklev-Twetman E, Sigurdsson S, Snoeck M, Stieglitz P, Tegazzin V, Urwyler A, Wappler F (1997) In vitro contracture test for diagnosis of malignant hyperthermia following the protocol of the European MH Group: results of testing patients surviving fulminant MH and unrelated low-risk subjects. The European Malignant Hyperthermia Group. Acta Anaesthesiol Scand 41:955–966

Chapter 12

Myokines, Measurement, and Technical Considerations

Craig R. G. Willis, Colleen S. Deane, and Timothy Etheridge

Abstract

Skeletal muscle has long been established as a highly multifunctional organ, playing a vital role in locomotion, whole-body metabolic and energy homeostasis, and thermoregulation. More recently, emergent evidence has highlighted a potent secretory role for muscle, producing and releasing "myokine" molecules that act in autocrine, paracrine, or endocrine fashion to govern muscle physiology and regulate whole-body homeostasis via multi-tissue cross talk mechanisms. Myokines represent promising therapeutic targets in health and disease, with their discovery, measurement, and functional importance being a hotbed of research across numerous physiological contexts. Here, we provide an overview of myokines and summarize current understanding of their biological role(s). We also outline primary approaches for myokine analysis, including detailed methodology for performing omics-driven myokine prediction, while further appraising both method-specific and general technical considerations to provide an evidence-based approach for designing and conducting myokine experiments.

Key words Skeletal muscle, Secretome, Myokine, Measurement, Analysis, Omics, Technical considerations

1 Introduction

Skeletal muscle represents the largest tissue constituent of the human body, comprising approximately half the total body mass of healthy individuals [1]. In occupying such a profound mass, skeletal muscle contributes significantly to multiple bodily functions essential for physical function and health [2]. The most notable is the muscles' mechanical role, generating the force necessary to maintain posture and enable movement [2]. Muscle tissue also facilitates thermoregulation and respiration [3, 4] and, as the body's largest substrate (protein, fat, and carbohydrate) store, plays a vital role in whole-body energy metabolism [2]. Beyond these well-established functions, however, recent evidence points toward skeletal muscle as a prominent secretory organ, responsible for the production and release of small molecules termed "myokines" that perform autocrine/paracrine roles to directly influence the form and function of muscle and/or endocrine roles to facilitate multi-

Philip J. Atherton and Daniel J. Wilkinson (eds.), *Neuromuscular Assessments of Form and Function*, Neuromethods, vol. 204, https://doi.org/10.1007/978-1-0716-3315-1_12, © Springer Science+Business Media, LLC, part of Springer Nature 2023

tissue cross talk and regulate whole-body physiology [5–8]. Myokines are sensitive to a variety of different (patho)physiological stimuli, including nutrients, stress, hormonal alterations, environmental factors, metabolic dysfunction, and aging [9–11]. Also a potent stimulator of myokine expression/secretion is muscle contraction itself [6, 7, 12], with myokines holding promise as biomarkers for positive exercise-related health adaptations, as well as pharmacological "exercise mimetic" targets in clinical populations [8, 10, 13–15].

Accurate and robust myokine analyses are essential to avoid critical errors in detection and interpretation that would hinder clinical myokine progress [8]. Numerous technical constraints must, therefore, be considered when measuring and quantifying myokines in a biological sample, from pre-analytical sample collection and handling to the specific choice of downstream analytical approach(es). As such, this chapter aims to provide a technical road map of robust myokine identification and physiological experimentation by overviewing myokines and their known biological roles; outlining and technically critiquing existing approaches for myokine analysis; emphasizing the utility of contemporary informatic methodologies; and appraising wider technical and experimental considerations for designing and conducting myokine experiments.

2 Myokines: Characterization and Biological Function(s)

2.1 Defining Myokines

At the turn of the twenty-first century, Bente Klarlund Pedersen and colleagues demonstrated that working skeletal muscles produce and systemically release the now well-known humoral factor ("cytokine") interleukin (IL) 6 (IL-6) [16]. The term "myokine" was subsequently coined to describe cytokines (a broad class of small signaling molecules typically with masses <30 kDa) produced and released by skeletal muscle, which exert effects in other parts of the body [17]. Nevertheless, other myokine molecule types have since been described, and more recent myokine classifications typically extend to include both cytokines and other non-cytokine proteins [5, 8, 18]. Myokines were also proposed as myofiber-derived factors [19], but, with muscle being a complex heterogeneous tissue [20], the term is frequently used to describe factors produced and released by any muscle subcompartment/organelle, which together encompass the wider muscle "secretome" [5]. Moreover, though myokines have become synonymous with factors produced and released in response to contraction, they can also be regulated by other (patho)physiological stimuli and/or constitutively secreted [9–11].

Contrary to earlier descriptions of myokines as endocrine factors [17], recent evidence suggests that myokines may primarily act as local agents to protect and improve the functionality of muscle by governing intramuscular cross talk between myofibers, immune cells, fibroblasts, vasculature, and bone [5]. Numerous molecules defined as myokines can also be secreted by tissues other than muscle, such that muscle exclusivity cannot be presumed from systemic myokine identification [21]. Nonetheless, given muscles' substantial contribution to total body mass, skeletal muscle is likely a primary, if not the only, source of many secreted factors observed in circulation [10], particularly in response to large-scale muscular work such as whole-body exercise. Lastly, growing evidence indicates that muscle secretory factors include not only protein or peptide molecules but also other regulatory molecule classes such as metabolites/lipids ("myometabokines") and noncoding microRNAs ("myomiRs") [5, 9, 22], a detailed discussion of which is considered beyond the scope of this chapter. All things considered, this chapter employs a broad definition of the term "myokine" to include any secreted protein or peptide that is capable of being produced in skeletal muscle, regardless of stimulus and whether they act in an autocrine, paracrine, or endocrine fashion [5].

2.2 Biological Functions of Myokines

Myokines have been implicated in diverse biological processes, which have been extensively reviewed elsewhere [5, 8, 18, 21, 23–25]. A dominant feature of myokines described thus far is direct intramuscular regulation, influencing skeletal muscle physiology via diverse mechanism(s) [13, 18, 25]. For instance, myostatin (or growth differentiation factor 8; GDF-8), one of the earliest identified myokines [26], governs myogenic and anabolic signaling-mediated control of muscle mass [27]. Numerous other myokines also putatively act in an autocrine/paracrine fashion to regulate myogenesis and/or muscle trophic-related signaling (e.g., IL-4, IL-6, IL-7, IL-15, decorin, insulin-like growth factor 1 (IGF-1), musclin, and leukemia inhibitory factor (LIF)). Myokines also exert important regulation over muscle metabolism (e.g., IL-6, IL-13, IL-15, brain-derived neurotrophic factor (BDNF), and angiopoietin-like 4 (ANGPTL4)) and mitochondrial biogenesis (e.g., musclin, myonectin, and fibroblast growth factor 21 (FGF-21)). Remodeling of the vasculature (e.g., IL-8, vascular endothelial growth factor (VEGF), and cysteine-rich angiogenic inducer 61 (CYR61)) and extracellular matrix (e.g., secreted protein acidic and cysteine rich (SPARC) and follistatin-like 1 (FSTL1)) is also influenced by myokines, as are immune/inflammatory responses (e.g., IL-10, c-c motif chemokine ligand 2 (CCL2), and fractalkine) [5, 8, 12, 13, 18, 23, 25].

The complexity of myokine function is further illustrated by their capacity to communicate with a variety of distal tissues. For example, the dominant metabolic effects of muscle-derived IL-6 (at least after exercise) are suggested to extend to adipose tissue and the liver [28, 29]; IGF-1 is implicated in bone formation [21]; and IL-15 might influence skin aging [8]. However, since the majority of myokines might work primarily in an autocrine/paracrine manner, future research might focus on the muscle-intrinsic effects of novel myokines. For example, muscle-derived BDNF acts locally to mediate muscle fat oxidation, satellite cell differentiation, and muscle regeneration but is unlikely an endocrine factor [8, 23, 25], with secretion from the brain as the major source of circulating BDNF [30, 31]. Nonetheless, despite much progress, biological function is said to have been established for only a small portion (~5%) of identified myokines [8]. Contributing to this incomplete understanding are technical difficulties when conducting myokine experiments and analyses. This chapter will, therefore, discuss key methodological approaches and associated technical considerations that will facilitate the identification of novel myokines. To assist subsequent exploration of myokine roles in regulating muscle and whole-body physiology, we further highlight important aspects of analytical methods to consider when to design and conduct myokine-centric experiments.

3 Primary Approaches Toward Myokine Analyses

This section outlines primary approaches for performing myokine analyses and critical method-specific considerations. While not exhaustive, methods covered focus on those most commonly employed to measure myokines in vivo (i.e., skeletal muscle tissue and blood samples) and in vitro. Further, most of these techniques are not myokine-specific analytical tools but rather are routine wet lab techniques used more generally in molecular research. Thus, while we do highlight method-dependent considerations of importance to myokine detection/quantification, a number of relevant methodological aspects are generalized, myokine-independent technical constraints and will not be discussed in great detail.

3.1 Primary Techniques for Targeted Myokine Analysis

By far the most common form of myokine detection/quantification is hypothesis-driven analysis of one or a finite number of target myokines. Well-established methods for performing targeted myokine analysis center upon polymerase chain reaction (PCR), immunoassays, and immunohistochemistry (IHC).

3.1.1 Real-Time PCR (qPCR)

The qPCR method is used to measure mRNA abundance in a biological sample, permitting myokine detection/quantification at the transcriptional level. Basic principles of this method are reviewed in [32], and a wide range of commercial kits are available from multiple manufacturers for performing both single and multi-molecule qPCR analysis of target myokines and associated functional molecular classes, e.g., the TaqMan™ Array Human Cytokine Network (Cat No. 4418769, Thermo Fisher Scientific), which permits the simultaneous measurement of mRNA abundance for

28 cytokine molecules. qPCR subsequently offers a fast, robust, cost-effective, and relatively high-throughput method for hypothesis-driven myokine measurement, at the same time providing a high degree of sensitivity while remaining sample-efficient [33, 34].

A high degree of sensitivity is particularly advantageous, with many myokines having a low abundance, especially in the healthy basal state [35]. However, this low abundance of many myokines renders some general aspects of robust PCR even more relevant when examining myokine changes. For example, while it is theoretically possible to measure mRNA abundance in absolute terms using this method, greater accuracy is attained when measuring relative differences between conditions [36] (e.g., baseline versus postexercise). Additionally, to further ensure accurate results, mRNA abundance should be normalized to a reference gene that is unperturbed across conditions, to control for differences in starting RNA concentrations and reaction efficiency [36]. However, care should be taken to ensure that chosen housekeeping genes are appropriately stable across conditions, for example with muscle contraction, where commonly used reference genes such as glyceraldehyde-3-phosphate dehydrogenase (*GAPDH*) can be robustly increased by acute exercise [37]. Such reference genes should, therefore, be used with caution due to the potential for masking biologically meaningful expression changes in low-abundance myokines. Overall, PCR offers a gold standard in sensitive quantitative assessment of myokine transcript changes. When appropriately executed and combined with key experimental considerations (*see* Subheading 4), PCR can robustly determine myokine responsiveness to external stimuli and validate gene-level expression of novel candidate myokines identified through untargeted techniques (*see* Subheading 3.2). Accordingly, making sure that the primer aligns to sequences used to map myokines from transcriptomics is an important consideration. Moreover, in developing a multi-myokine PCR assay, one should ensure that all primer-probe sets are compatible with one another in accord with the exact set of myokine targets chosen for analysis [36].

3.1.2 Immunoassays

Immunoassays permit myokine detection/quantification at the protein level and allow measurement of a limited number of target proteins. Immunoassays regularly used for myokine research are the enzyme-linked immunosorbent assay (ELISA), Western blotting, and antibody arrays. Each method associates with certain unique methodological constraints that should be considered when applied to myokine-centered research, briefly outlined below. However, all immunoassay approaches converge on a common principle: quantifying antigen (herein myokine) content via the magnitude of antigen binding with a specific corresponding antibody [38]. Ensuring excellent antibody specificity is thus the single most important technical item across immunoplatforms for robust myokine analysis, failure of which is a frequent source of detection error. A stark reminder of the need for robust antibodies is the recent controversy surrounding the biological relevance of irisin upon its discovery as a myokine, with measures of serum irisin later identified as an artifact of poor ELISA specificity [39]. Thus, appropriate scientific rigor in antibody design and, crucially, proper antibody validation prior to experimentation are essential to avoid significant wasted myokine-related research efforts.

ELISA

First described half a century ago [40], the ELISA technique remains one of the most utilized methods for targeted myokine analysis, particularly in biological fluid samples (blood and cell culture supernatant). Well-validated, commercially available ELISA kits exist for many (candidate) myokines. For example, Invitrogen™ currently offers individual ELISA kits for a broad menu of over 700 cytokine, chemokine, and growth factor targets (thermofisher.com). ELISAs are time-efficient per assay, highly sensitive, and quantitative [41, 42]; however, serial sample dilutions are often required to account for the often narrow dynamic range of many ELISAs, potentially exaggerating differences between samples falling inside versus outside the dynamic range [43]. This factor might be particularly influential when analyzing myokines, since frequently low basal levels may, in some instances, increase up to 1000-fold upon demand [44]. Standard ELISAs also only permit measurement of one analyte at a time, and so, large sample volumes, time commitment, and financial cost would be required to analyze multiple myokines using this approach in isolation [42].

Western Blot

Western blotting (immunoblotting) is a cornerstone of modern physiology research. While less sensitive than ELISAs, Western blotting can afford greater target specificity by virtue of electrophoretic separation of proteins within a sample by their molecular weight and can thus be applied to delineate inactive and active protein forms, as well as different splice variants [33]. Both methodological features are advantageous for myokine analysis, since more than two-thirds of genes encoding secreted proteins have at least one splice variant with different cellular location (s) [14, 45]. Myokine splice variants thus possibly possess divergent biological roles, and, indeed, the biological potency of certain target myokines (e.g., IL-1β) can be distinguishable by molecular weight [33]. Other general technical considerations associated with Western blotting equally apply to myokine studies (comprehensively reviewed by Bass and colleagues [46]). However, specific to myokine analysis, the relative validity of using a non-secreted versus secreted reference/loading control protein in a given context is worth considering (e.g., the validity of using a non-secreted protein as a reference/loading control for Western blotting of biological fluids) [47]. As an alternative that circumvents the need to identify (non)secreted reference proteins, Coomassie staining represents a reliable approach for standardizing total protein load [46].

A feature of many myokines is low molecular weight [19, 48]; thus, more specific considerations for Western blot can include appropriate gel concentration (20% recommended for proteins <20 kDa) and running buffer (2-(N-morpholino)ethanesulfonic acid (MES) suggested for proteins <36 kDa) choices during electrophoresis and membrane pore size (smaller advised for proteins <20 kDa) selection during protein transfer [46]. Unlike ELISAs, Western blotting permits measurement of multiple myokines in a single analysis via membrane strip and reprobing (up to three times being recommended) and/or membrane dissection by molecular weight (assuming appropriate prior antibody validation) [46]. However, as well as ensuring antibody specificity for each analyzed target, extreme care should be taken when choosing myokine target combinations given how close myokines can be across low molecular weight ranges [49].

Antibody Arrays

Antibody arrays are an extension of traditional ELISA methods, which allow simultaneous measurement of multiple myokines in the same sample [42]. These arrays typically exist in one of two forms: planar or bead-based. In planar arrays, individual antibodies are immobilized in microarray format and probed with the sample of interest. Conversely, bead arrays utilize color-coded and size-coded microbeads to permit protein-specific binding and identification and are generally most commonly employed [33]. Exploiting microbeads suspended in a sample versus fixed to a platform offers superior binding kinetics, limits of detection, and broader dynamic detection ranges than traditional ELISAs [42, 50], minimizing serial sample dilutions. However, broad/varying dynamic ranges can be problematic when assaying low- and high-abundance targets in tandem [43]. Sensitivity/comparability with conventional ELISAs may also vary across individual myokines and/or biological samples (e.g., tissue culture supernatant versus plasma samples) [43, 51]. When customizing myokine combinations on a single array, it is further critical to consider non-reactivity of individual targets with all other antibodies, and a reliable uniplex assay is not necessarily suitable for addition to a functioning multiplex assay [42]. After mitigating these factors, antibody arrays can be customized for desired myokine targets, with several arrays focused on specific myokines or functional classes of myokine also commercially available (e.g., the MILLIPLEX® MAP Human Myokine Magnetic Bead Panel containing 15 established myokines (Cat No. HMYOMAG-56K, Merck Millipore) or equivalent cytokine/

chemokine panels that contain upward of 40 targets (Merck Millipore)). As such, antibody arrays can offer substantially higher throughput, accurate myokine analysis versus traditional ELISAs and Western blotting while being cost-, time-, and sample-efficient [33].

3.1.3 Immunohistochemistry (IHC)

IHC is categorized as an immunoassay technique since protein detection is achieved using antigen-antibody binding and is thus subject to many of the above technical caveats. However, IHC also determines target protein location in intact muscle tissue sections [52], which is important given the tight link that exists between protein spatial localization and function [53]. Most IHC technical considerations are generic to biological studies [54], and with appropriate primary/secondary antibody combinations [36], colocalization of multiple target myokines can be assessed in a single tissue section. Like Western blotting, selecting antibodies in accord with biological potency of interest is particularly relevant to myokine analysis, since some (candidate) myokine targets can be synthesized in a specific form (e.g., IL-1β, which can be synthesized in inactive form) [36], with possible functional consequences.

When employed across multiple physiological time points, IHC further complements myokine studies with temporospatial information to infer functionality beyond expression changes alone (e.g., subcellular site of action and colocalization-based inferences of multi-myokine functional similarities/differences). Classical protein secretion involves sequential movement from the cytosol to secretory vesicles via the rough endoplasmic reticulum lumen and Golgi cisternae [55, 56] and is thus presumably the main route of secretion for myokines that contain a signaling sequence. Conversely, myokines (e.g., fibroblast growth factor 2; FGF-2) can lack a signaling sequence but be secreted through alternate means such as via small exosome vesicles [8, 45, 57, 58]. By assessing temporal colocalization of myokines with specific subcellular markers (e.g., CD63—an exosomal marker protein), the temporospatial capacity of IHC could thus also serve as a useful tool to help identify myokine mechanisms of secretion [59–61]. A further such advantage [36] is the ability to assess myokine production by cell type. For example, contraction-induced muscle production of IL-6 occurs preferentially in type 2 versus type 1 myofibers in vivo [62]. Combined, given the capacity for dynamic, location- and cell-specific myokine function, significant new functional information can be gained by including IHC analyses in myokine research.

3.2 Approaches Toward Untargeted Myokine Discovery

Rapid technological advancements have paved the way for a shift toward a new era of modern molecular analyses, termed the "omics" revolution [63, 64]. Omics-based tools make it feasible to simultaneously characterize the molecular state of a sampled cell/tissue at an extremely high resolution. This feature facilitates "untargeted" molecular study and, in turn, provides a powerful platform to generate new hypotheses and identify novel molecular signatures and regulatory control mechanisms underpinning phenotypic change [63, 65–67]. Accordingly, omics applications have emerged as prime tools toward discovering new myokine candidates for subsequent hypothesis-driven work [45]. Main omics tracts for unbiased myokine analyses are transcriptomics and proteomics: the global study of mRNA and protein levels in a sample, respectively [66]. Transcriptomic data are typically generated by means of microarray or RNA-sequencing (RNA-seq). While microarrays are cost-effective and can enable the expression of tens of thousands of genes to be quantified simultaneously, RNA-seq affords significantly higher resolution and lower limit of detection, allowing quantification of previously annotated genes plus novel features and, in turn, theoretically permitting measurement of all mRNA species in a cell/tissue [68–72]. Proteomic data are normally attained using mass spectrometry, either via a top-down or bottom-up strategy. Top-down methods directly analyze intact proteins, whereas bottom-up approaches analyze proteins on the peptide level; the latter are perhaps preferable for studying myokines, whereby low abundance can be problematic due to greater sensitivity, larger dynamic range of detection independent of protein size, isoelectric point or hydrophobicity, and a greater degree of automation [22, 24]. The relative advantages/disadvantages of transcriptomics and proteomics for untargeted myokine discovery are further discussed in Subheading 5.

An extensive description on conventional transcriptomic/proteomic data analysis is beyond the scope of this work, for which the reader is directed to several excellent online resources [73–78]. Transcriptomic/proteomic experiments typically yield a list of genes/proteins that are differentially regulated across two or more sets of samples and/or central, highly connected ("hub") molecules in the context of a large-scale molecular interaction network. Thereafter, prediction of myokine candidates can be achieved by unbiasedly screening differentially regulated/hub molecule lists for possible secreted factors—a process termed "secretome" analysis. A number of predictive bioinformatic tools then exist that can facilitate untargeted myokine discovery. A notable mention is SignalP, an online bioinformatic tool that uses sequence alignments and machine learning-based algorithms to predict secreted proteins [22, 79]. Curated knowledge bases can also serve as useful for identifying secreted factors. These include the Gene Ontology (GO), the world's largest source of information on gene function that is organized into logical classes [80]. For myokine identification, putative secreted molecules can be filtered under extracellular-related cellular component classes (e.g., "extracellular region" and "extracellular space") [15]. Additional knowledge bases include the Universal Protein Resource (UniProt) database [81], which contains reviewed and computationally predicted secreted molecule lists (under UniProt keyword "Secreted").

In SignalP, secreted proteins are predicted based on the presence of an N-terminal cleavable signal sequence, and thus, such an approach is effective for elucidating myokine candidates secreted via classical means. However, as already noted, many secreted factors including myokines do not carry a signal sequence and are, therefore, likely secreted through nonclassical means, such as via exosomes [8, 45, 57, 58]. In which case, specific bioinformatic tools for sequence-based prediction of nonclassically secreted proteins are further available [82], such as the SecretomeP web server [83], as well as other tools more aimed at specifically predicting protein secretion by exosomes (e.g., ExoPred [84]). Many proteins secreted into the endoplasmic reticulum are also sorted to various intracellular compartments (mitochondria, lysosomes, etc.) or retained in the endoplasmic reticulum/Golgi cisternae, rather than being secreted from the cell [85]. Further filtering by subcellular localization may thus also be informative, which can be aided by several predictive tools including DeepLoc [86] and TargetP [87]. False-positive prediction of classical or unconventional secretion can further arise from proteins containing transmembrane domain [82, 88]. Accordingly, updated iterations of SignalP include algorithms that aim to better discriminate signal peptides from transmembrane regions with higher precision (e.g., [89]). Additional stand-alone tools that directly predict transmembrane helices in protein sequences can also be integrated to eliminate transmembrane molecules from predicted secretory factor lists, such as the transmembrane hidden Markov model (TMHMM) web server [90]. A recommended [82] and common [15, 35, 91, 92] approach is, therefore, to integrate multiple bioinformatic tools when screening for putative secretory factors. Further screening of myokine candidates for preferential skeletal muscle expression may help add an extra layer of confidence in muscle as the predominant origin of production/release [15]. Experimentally tractable sets of novel candidate myokines, responsive to particular (patho)-physiological stimuli, can henceforth be discovered for subsequent hypothesis-driven work and validation as definitive myokines.

With omics approaches now well at the forefront of modern molecular discovery, the current chapter focuses on providing extended methodological detail for performing omics-driven myokine prediction, found in the next section. Notably, we outline a handful of protein sequence-based prediction tools that can be incorporated into untargeted myokine discovery pipelines. It is important to highlight, however, that the selection of tools given is by no means exhaustive, with those presented chosen to give the reader a flavor of both utilizing and integrating sequence-based prediction tools of differing purposes toward a workflow for untargeted myokine prediction. Nevertheless, even tools that hold a similar purpose (e.g., identifying proteins with a signal sequence) can vary in the underlying algorithm and may therefore generate marginally different results [14, 85]. As such, we also suggest a viable alternative/complementary tool that could help contribute to "standardized" myokine predictive screening (at least when analyzing human samples).

4 Methodology for Omics-Driven Myokine Prediction

4.1 Protein Sequence-Based Prediction of Candidate Myokine Targets

4.1.1 Obtaining Protein Sequences for Input into Prediction Tools

Given a (typically long) list of molecules derived via omics analyses (differentially expressed genes/proteins, network "hub" genes/proteins, etc.) of in vivo (muscle tissue and/or blood) or in vitro (muscle cell line and/or supernatant) samples, a required first step toward filtering for potential myokines is obtaining associated protein sequences for input into sequence-based prediction tools. Various knowledge bases exist that provide such detail across a variety of different organisms. A precise choice could simply be a matter of user preference and/or driven by other pertinent factors, such as methodological specificities. For example, if genome build and annotation files for RNA-seq analysis were used from a particular database, obtaining associated protein sequences directly from the same source may provide greater consistency and/or optimize database coverage (exhaustive discussion on all pertinent factors is considered beyond the scope of this work). With the predominant focus of this section being on the screening application per se, we therefore restrict focus on describing how to obtain associated protein sequences for gene/protein lists using two popular databases: Ensembl [93] and UniProt [81]. It is worthwhile noting that both tools provide associated sequences at the isoform level. If omics data have been summarized on a per gene/protein basis (rather than on an individual isoform basis), the user may thus wish to either focus on a specific isoform of interest or analyze all isoform-specific sequences per gene/protein molecule and then define a molecule as a myokine candidate if at least one variant pertains to the required criteria. The latter is perhaps more straightforward from a screening viewpoint, though the possibility of myokine splice variants having divergent biological roles (or even regulation and/or localization) may remain a potential caveat.

Ensembl

The BioMart tool within Ensembl (available through the Ensembl homepage: https://useast.ensembl.org/index.html) can be used to obtain protein sequences for multiple targets of interest in batch. Initially, one is required to choose a database option from the "Dataset" field (e.g., "Ensembl Genes" → "Human genes"). The "Filter" field can then be used to restrict the query to only those targets of interest (e.g., "GENE:" → "Input external references ID list [Max 500 advised]" → "Gene Stable ID(s) [e.g. ENSG00000000003]" then paste or upload list of corresponding ID's), and the "Attributes" field is used to specify inclusion of associated peptide sequences in the output (select "Sequences" then "SEQUENCES:" → "Peptide"). Peptide sequences per transcript per gene can consequently be downloaded in FASTA format ("Results" → "Export all results to" → "File" → "FASTA" → "Go"), a common text-based file format for representing protein sequences. Such a tool can accept several different types of gene and protein identifier as input.

UniProt

UniProt supports batch protein sequence retrieval via its "Retrieve/ID mapping" function (accessible via the UniProt homepage: https://www.uniprot.org), with pasted or uploaded lists able to be submitted upon "From" (target list ID type, e.g., "UniProt Gene name," etc.), "To" (specific UniProt database to map to, e.g., "UniProtKB"), and "Organism" (e.g., "Homo sapiens (Human)") option selections. Consequent results can be further filtered ("Filter by" parameter) to include manually annotated and reviewed (Swiss-Prot) and/or automatically annotated but unreviewed (TrEMBL) database records with the "Columns" option enabling for protein sequences to be appended to the results table ("Columns" → check "Sequences" under the "Sequences" subfield of "Add more columns"). UniProt-defined canonical only or canonical plus other isoform protein sequences for inputted targets can then be readily downloaded in FASTA file format ("Download" → "Download all" → "File format" either "FASTA (canonical)" or "FASTA (canonical and isoform)" → choose "Compressed" or "Uncompressed" file format → "Go"). Like Ensembl, such a tool supports a variety of different gene and protein ID inputs.

4.1.2 Signal Peptide Prediction via SignalP

As noted, classically secreted proteins carry a signal peptide with SignalP, a popular sequence scanning tool for establishing the presence and cleavage site location of possible signal peptides in proteins. A recent iteration of SignalP at time of writing is SignalP 5.0 (accessible at http://www.cbs.dtu.dk/services/SignalP/) [94], which can distinguish between "standard" secretory signal peptides (Sec/SPI), "lipoprotein" signal peptides (Sec/SPII), and "Tat" signal peptides (Tat/SPI). Desired protein sequences can be either pasted or uploaded in FASTA format with further options to define "Organism group" ("Eukarya" (default), "Gram-positive" (bacteria), "Gram-negative" (bacteria), or "Archaea"—n.b.: discrimination between the given three types of signal peptide is only possible for bacteria and archaea) and "Output format" (i.e., whether ("Long format") or not ("Short format") to include figures illustrating predicted cleavage site location) prior to submission (n.b.:

input protein sequences should be no less than 10 and no longer than 10,000 amino acids, with the maximum number of input sequences being 5000). A summary text file that includes the type of prediction (Sec/SPI, Sec/SPII, Tat/SPI, or OTHER if not predicted to contain a signal sequence) and location of cleavage site (if any) for all uploaded protein sequences can then be readily downloaded ("Downloads" → "Prediction summary"). Additional developer guidance for running SignalP 5.0 can be found at http://www.cbs.dtu.dk/services/SignalP/instructions.php.

4.1.3 Prediction of Nonclassically Secreted Proteins Using SecretomeP

To further probe for nonclassically secreted proteins (i.e., secreted factors that do not carry a signal sequence) by means of sequence-based prediction, SecretomeP represents a popular tool [83], the most recent iteration at time of writing being SecretomeP 2.0 (http://www.cbs.dtu.dk/services/SecretomeP/). Again, protein sequences can be either pasted or uploaded for submission in FASTA format (restricted to 100 sequences and 200,000 amino acids per submission, with each sequence being no less than 40 and no more than 4000 amino acids) and with further option to define organism prior to submission (specific options being "Gram-negative bacteria," "Gram-positive bacteria," and "Mammalian"). The provided results per sequence include an "NN-score" and "Warning." As described in the results output, nonclassically secreted proteins should have an NN-score ("SecP score") above a threshold of 0.5 for bacterial sequences and 0.6 for mammalian sequences and not flag a warning of being predicted to contain a signal peptide. Result outputs can be copy/pasted and stored in text file format for subsequent reference. Additional developer guidance for using SecretomeP 2.0 can be found at http://www.cbs.dtu.dk/services/SecretomeP/instructions.php.

4.1.4 Prediction of Protein Subcellular Localization Using DeepLoc

As mentioned, not all proteins secreted into the endoplasmic reticulum necessarily end up being secreted from the cell. DeepLoc [86] consequently presents as a protein sequence scanning tool for predicting subcellular localization, integration of which can thus add extra confidence that a protein may indeed be secreted. The most recent version at time of writing is DeepLoc 1.0 (available at http://www.cbs.dtu.dk/services/DeepLoc/), which can discriminate between ten different localizations: cell membrane, chloroplast, cytoplasm, endoplasmic reticulum, extracellular, Golgi apparatus, lysosome/vacuole, mitochondrion, nucleus, and peroxisome. Pasted or uploaded FASTA format protein sequences (not less than 10 or more than 6000 amino acids in length) can be submitted for scanning with an additional option ("Protein encoding") to define preference for either accurate ("Profiles (accurate, 50 sequences maximum)") or fast ("BLOSUM62 (fast, 500 sequences maximum)") processing. Generated results include an overall prediction of subcellular localization as well as whether a given protein is soluble or membrane with resultant data for all inputted sequences available to download as a single .csv file ("Downloads" → "CSV summary"). From a secretome analysis standpoint, extracellular localization presents as a logical filtering choice to include. Further guidance provided by the developers of DeepLoc 1.0 can be found at http://www.cbs.dtu.dk/services/DeepLoc/instructions.php.

4.1.5 Prediction of Transmembrane Helices in Proteins Using TMHMM

As elucidated, transmembrane proteins can exaggerate false-positive secretion hits, and additional confidence may thus be gained by incorporating stand-alone sequence-based transmembrane region prediction tools. A prominent option is TMHMM [90]; the most recent version (v2.0) of which can currently be located at http://www.cbs.dtu.dk/services/TMHMM/. As above, protein sequences can be either pasted or uploaded in FASTA format with restrictions being a maximum of 10,000 sequences and 4,000,000 amino acids per submission plus each sequence no more than 8000 amino acids. Prior to submission, there is also an option to define "Output format" ("Extensive, with graphics," "Extensive, no graphics," or "One line per protein"), as well as "Other options" (viz. whether to use old model—i.e., TMHMM v1). The "One line per protein" output includes per input protein, the length of protein sequence, expected number of amino acids in transmembrane helices, expected number of amino acids in transmembrane helices in the first 60 amino acids, number of predicted transmembrane helices, and associated predicted topology. Per input protein, extensive results further include a statistic on the total probability that the N-terminal is on the cytoplasmic side of the membrane, a warning if a predicted transmembrane helix in the N-terminal could in fact be a signal peptide (occurs when the expected number of amino acids in transmembrane helices in the first 60 amino acids of a given protein exceeds 10) and a list (if any) of locations of predicted membrane helices and intervening loop regions. In which case, if the whole sequence is labeled as inside or outside, the concomitant prediction is that the given protein contains no transmembrane helices (and thus recommended not to interpret as a prediction of location). If the expected number of amino acids in transmembrane helices exceeds 18, then it is said likely to be a transmembrane protein (or have a signal peptide). The "Extensive, with graphics" option further provides plots showing the posterior probabilities of inside/outside/transmembrane helix. Result outputs of the "One line per protein" option are highly suitable for being copy/pasted and stored in text file format for subsequent reference. Full guidance provided by the developer can be found at http://www.cbs.dtu.dk/services/TMHMM/TMHMM2.0b.guide.php.

4.1.6 Prediction of Exosome-Secreted Proteins Using ExoPred

Many tools for predicting nonclassically secreted proteins, including SecretomeP, incorporate algorithms that were not trained toward the specific intent of predicting protein secretion by exosomes and thus may not perform optimally in doing so [84]. Moreover, some myokine candidates that neither contain a signal sequence nor display extracellular localization may still yet be secreted by means of exosome transportation [45]. In which case, a recent complementary sequence-based tool for predicting myokines unconventionally secreted via exosomes is ExoPred [84], currently available at http://imath.med.ucm.es/exopred/. Such a tool too accepts either pasted or uploaded protein sequences in FASTA format with additional options to include subcellular location as defined by UniProt and/or predicted by Psort (another protein subcellular localization prediction tool [95]) in the results. Clicking run will consequently generate the results table, wherein the "ExoPred" column details whether ("Y"; yes) or not ("N"; no) input proteins can be secreted by exosomes (n.b.: as part of its processing, "ExoPred" also detects proteins with signal sequences (via SignalP) and/or transmembrane regions (via TMHMM), of which are consequently assigned a "NA" (not available) prediction in the "ExoPred" column). Result outputs can be copy/pasted and stored in text file format for subsequent reference. Further developer guidance for using ExoPred can be found at http://imath.med.ucm.es/exopred/instructions.html#UG.

Given the abovementioned tools, a logical workflow for their integration within omics-driven myokine prediction is outlined in Fig. 1 (though input could quite easily be associated protein sequences of any viable gene/protein list of interest). As noted, however, a variety of other pertinent sequence-based tools exist with aims adjunct to the specific tools described above, and distinct tools with a similar purpose may not necessarily provide identical results. For example, SignalP 4.0, Phobius, and SPOCTOPUS signal peptide prediction tools have shown to yield a differing number of hits to one another upon whole-proteome analysis [85]. Thus, while the workflow per se presented in Fig. 1 may be viable for candidate myokine screening, a more "standardized" approach could be to integrate several different sequence-based tools and even knowledge base annotations (e.g., GO cellular component terms for protein localization filtering) at each stage. In which case, a viable option that may help contribute to standardized candidate myokine screens (at least in humans) is the Human Protein Atlas-defined human secretome [14, 85].

4.2 Putative Myokine Screening via the Human Protein Atlas

4.2.1 Human Protein Atlas-Defined Human Secretome

The Human Protein Atlas-defined human secretome is a readily available dataset containing 1708 genes conservatively predicted to have at least one secreted protein variant as based on whole-proteome analysis of all Ensembl transcripts using a compendium of bioinformatic tools, including a vast array of different sequence-based prediction approaches. Notably, this list comprises genes

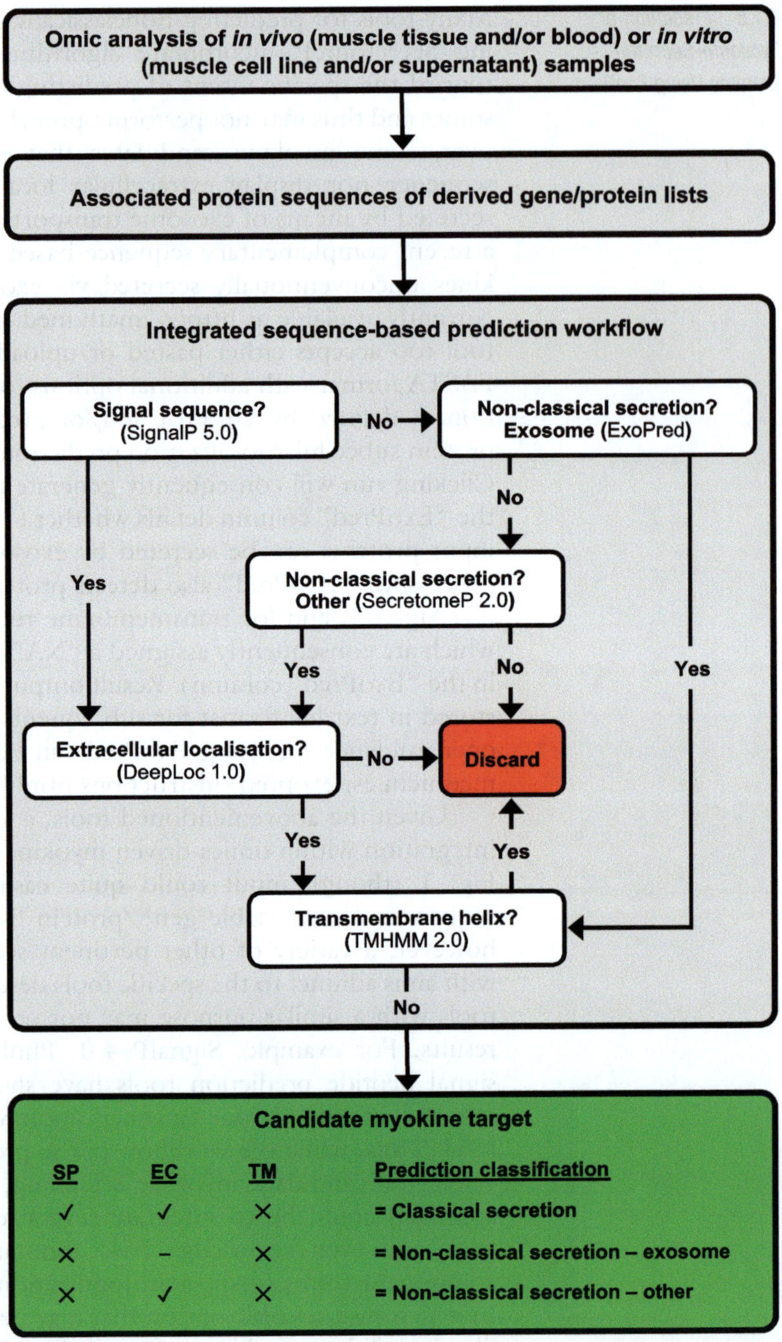

Fig. 1 Omics-driven myokine prediction workflow. Central flowchart highlights example workflow for screening candidate myokines by means of protein sequence-based prediction tools. As noted in-text, since ExoPred itself omits proteins with transmembrane regions, positive hits obtained using this tool should naturally pass the subsequent TMHMM transmembrane helix screening stage. Abbreviations: EC extracellular, SP signal peptide, TM transmembrane

having at least one protein variant: (i) predicted to contain a signal peptide based on a majority decision method of three sequence-based tools (SignalP 4.0, Phobius, and SPOCTOPUS) with no predicted transmembrane region based on majority decision method of seven sequence-based tools (MEMSAT3, MEMSAT-SVM, Phobius, SCAMPI, SPOCTOPUS, THUMBUP, and TMHMM) or otherwise annotated under the UniProt keyword "Secreted" and (ii) not annotated as intracellular and/or membrane-bound (e.g., endoplasmic reticulum or Golgi residing, mitochondrial, lysosomal, membrane-associated, etc.). Such a tool therefore substantially minimizes the need for the user to directly employ multiple prediction tools and may thus present as one possible "to hand" source for helping standardize large-scale myokine predictions across independent studies. It is worthwhile noting however that, since the Human Protein Atlas-derived secretome largely stems from predictions made on signal peptide presence, unconventionally secreted proteins may not be fully encompassed, and it may therefore be most suitable as a standard for screening of classically secreted myokines. In which case, an updated iteration or stand-alone process that incorporates multi-method prediction of unconventionally secreted proteins (e.g., SecretomeP, SRTpred, SecretP, SPRED, OutCyte, ExoPred, etc. [84, 96]) may be desired as an alternative/complementary standard.

The complete list of molecules that comprise the Human Protein Atlas-defined human secretome can be found at https://www.proteinatlas.org/humanproteome/tissue/secretome, by clicking on "Predicted secreted proteins" within Table 2. The resultant data table can then be downloaded as a .tsv file in its entirety (i.e., including columns for all pertinent Human Protein Atlas database annotations; click ".TSV" option) or with custom data columns (click "Custom TSV/JSON" then "Download TSV"; default data columns included this way, along with Ensembl Gene ID's, are "Gene," "Gene description," and "Evidence").

4.2.2 Using the Human Protein Atlas to Establish Skeletal Muscle Specificity

The Human Protein Atlas can also be used to further screen myokine candidates for preferential skeletal muscle tissue expression, [15] which, as noted, may help to provide an extra layer of confidence in muscle as the predominant origin of production/release. The Human Protein Atlas has a defined skeletal muscle-specific proteome locatable at https://www.proteinatlas.org/humanproteome/tissue/skeletal+muscle, which comprises a total of 907 associated protein-encoding genes that show elevated expression in skeletal muscle compared to other tissues. The full list can be readily downloaded by first clicking on "907" in the lower right-hand corner of Table 1 from the given website link, then following the same procedure as described above for the Human Protein Atlas-defined human secretome.

5 General Myokine Experimental Considerations

5.1 *In Vivo and In Vitro Sampling*

While subject to similar experimental considerations and constraints as any life sciences experiment aimed at identifying and exploring the function of novel molecules, certain factors are noteworthy in the context of myokine-centric studies. An obvious example is whether to analyze blood versus muscle samples, with muscle being the de facto gold standard for ascertaining "true" myokines. Practical limitations clearly factor heavily into sampling choices (i.e., the technical capacity to obtain muscle biopsies) and presumably explain the literature predominance of myokine quantification in blood [22]. Systemic appearance is clearly necessary for establishing endocrine myokines. Nonetheless, given the apparent auto/paracrine predominance of myokines [5], focusing solely on blood-based myokine changes is prone to omitting a functionally important class of myokines and prevents temporal associations between muscle-blood myokine appearance, a key facet to consider when it comes to delineating molecules that reside in blood versus those secreted from cells [45]. Thus, an optimal approach to disentangling myokine function is measuring both muscle and blood, though arteriovenous (a-v) differences across muscle beds offer a strong albeit technically challenging [97] surrogate, particularly with respect to elucidating possible endocrine relevance of myokines. Indeed, a-v difference studies identified muscle as the predominant source of circulating IL-6 [16] but not BDNF [30] or FGF-21 [98], upon contraction. Measuring a-v differences in protein-containing extracellular vesicles can further inform on mechanism(s) of myokine secretion [61].

Numerous extraneous variables such as level of food intake (i.e., fasted versus fed state) and time of day (i.e., diurnal oscillations) can also affect myokine production/release [43, 99–103]. Sampling must therefore be timed accordingly and consistently to obtain reliable myokine readouts in accord with the (patho)physiological context of interest. If collecting muscle biopsies, serial samples should utilize incision sites separated at least 2.5 cm apart because the biopsy procedure may impact muscle myokine levels [104–107]. Similarly, for multiple blood withdrawals, local indwelling cannula placement is advantageous but versus venipuncture may provoke local production of certain myokines [43]. Sample storage is another consideration because several myokines (e.g., IL-4, IL-5, IL-6, IL-8, IL-10, IL-13, and IL-15) can degrade in blood samples stored at −80 °C for durations >1–2 years [108]. Finally, all myokine detection/quantification methods described herein can theoretically be applied to muscle tissue and blood samples. However, since blood concentrations of secreted factors are typically very low (i.e., often pico or even femto levels) [5, 22], the greater sensitivity of ELISA/antibody arrays may be well suited for detecting/quantifying myokines in fluid samples.

Although establishing the human applicability of myokine function remains a key end goal, complementary use of a wide variety of model systems will clearly expedite myokine discoveries. Though beyond the scope here to discuss in detail the wealth of models available, transgenic animals with tissue (muscle)-specific manipulation of putative myokine levels will serve as an excellent experimental approach to establishing "true" myokines. Convincing transgenic mouse models can, however, be challenging, as with paradoxical observations of greater postexercise circulating IL-6 levels in muscle-specific IL-6 knockout mice [5, 109]. Nonetheless, if successful, such approaches will also provide a model for functional assessment of myokine gain/loss of function [45] (NB-specific analytical tools, such as MILLIPLEX® bead arrays, provide scope for species-specific myokine analyses). Moreover, because an inherent limitation of in vivo myokine measurement is unambiguously establishing muscle-specific myokines, in vitro models might be beneficial. For example, a combined examination of muscle cell cultures and associated supernatant provides a substantially simpler alternative to in vivo approaches for assessing myocyte-specific myokine production and secretion [23]. Frequently used muscle cell lines include the commercially available mouse C2C12 and rat L6 cell lines, as well as primary human skeletal muscle cells [24, 110]. However, key differences in transcriptomic and proteomic profiles across/between rodent/human cell lines and intact muscle [45, 111] might limit the forward translation of some newly identified myokines [23, 35].

5.2 Quantification of Myokine Gene Versus Protein Changes

A relevant factor influencing the choice of analytical approach is the relative suitability of assessing myokine protein versus mRNA levels. Owing to the presence of high-abundance proteins in blood (e.g., albumin) and muscle (e.g., contractile proteins), associated proteomes cover a vast dynamical range, in turn hindering the ability to detect low-abundance myokines using current proteomic tools [45]. Proteomic studies of skeletal muscle or plasma/serum are, therefore, generally constrained by limited proteome coverage that favors high-abundance proteins [112–114]. In contrast, established microarray and RNA-seq transcriptomic approaches offer substantially greater detection coverage, ranging from tens of thousands up to all mRNA species [115]. This, coupled with the fact that several myokines may never reach the circulation [22], perhaps implicates skeletal muscle transcriptomics-based secretome analysis as an effective initial strategy toward novel myokine discovery [35]. Deep proteomic strategies are nevertheless emerging for greater proteome coverage of muscle/ blood, though quantity of detected entities and/or capacity to detect low-abundance proteins may still be suboptimal [116–119].

Additionally, although mRNA and protein abundance often display strong agreement [114], the two can correlate poorly [45], including myokines [120]. Associations between mRNA and protein levels of secretory factors are particularly key to consider, since secretion may in some cases depend on posttranscriptional and/or posttranslational regulation (e.g., influence of specific secretory pathways, binding to neutralizing molecules, activating proteolytic cleavage, storage prior to release) [22, 33, 120]. Certainly, myokines can be regulated posttranscriptionally via translational (e.g., IL-4, IL-10, and SPARC) or posttranslational (e.g., IL-1 and IL-18) mechanisms [36, 121]. Thus, knowledge of myokine regulation within the context of the central dogma has clear implications on the relative merit of (targeted) gene- versus protein-level measurement. Toward this end, protein depletion strategies might help myokine detection [122, 123], and employing alternative in vitro models can improve proteome coverage due to attenuated dilution by high-abundance contractile proteins [45].

6 Summary: Road Map for Myokine Discovery and Analyses

In summary, there exist a number of analytical approaches that can facilitate the discovery and quantification of myokines in a variety of experimental designs. Each methodology offers distinct merits and limitations; the choice of which will be further driven by specific research goals (e.g., identifying new myokine targets versus deeper mechanistic study of a defined myokine target) and/or current state of biological knowledge (e.g., regulation of secretion pre- versus posttranscriptionally). While many defined approaches present as complementary (e.g., IHC-based localization analysis of myokines displaying altered expression via Western blot, etc.), no single method in isolation establishes myokine detection/function. Thus, resources permitting, multiple avenues should be combined [124], an analytical "road map" of which is outlined in Fig. 2. Establishing the full complexity of the muscle secretome promises exciting physiological and clinical advances, and this chapter should provide a foundation for informed experimental decisions to optimize myokine discovery and analyses.

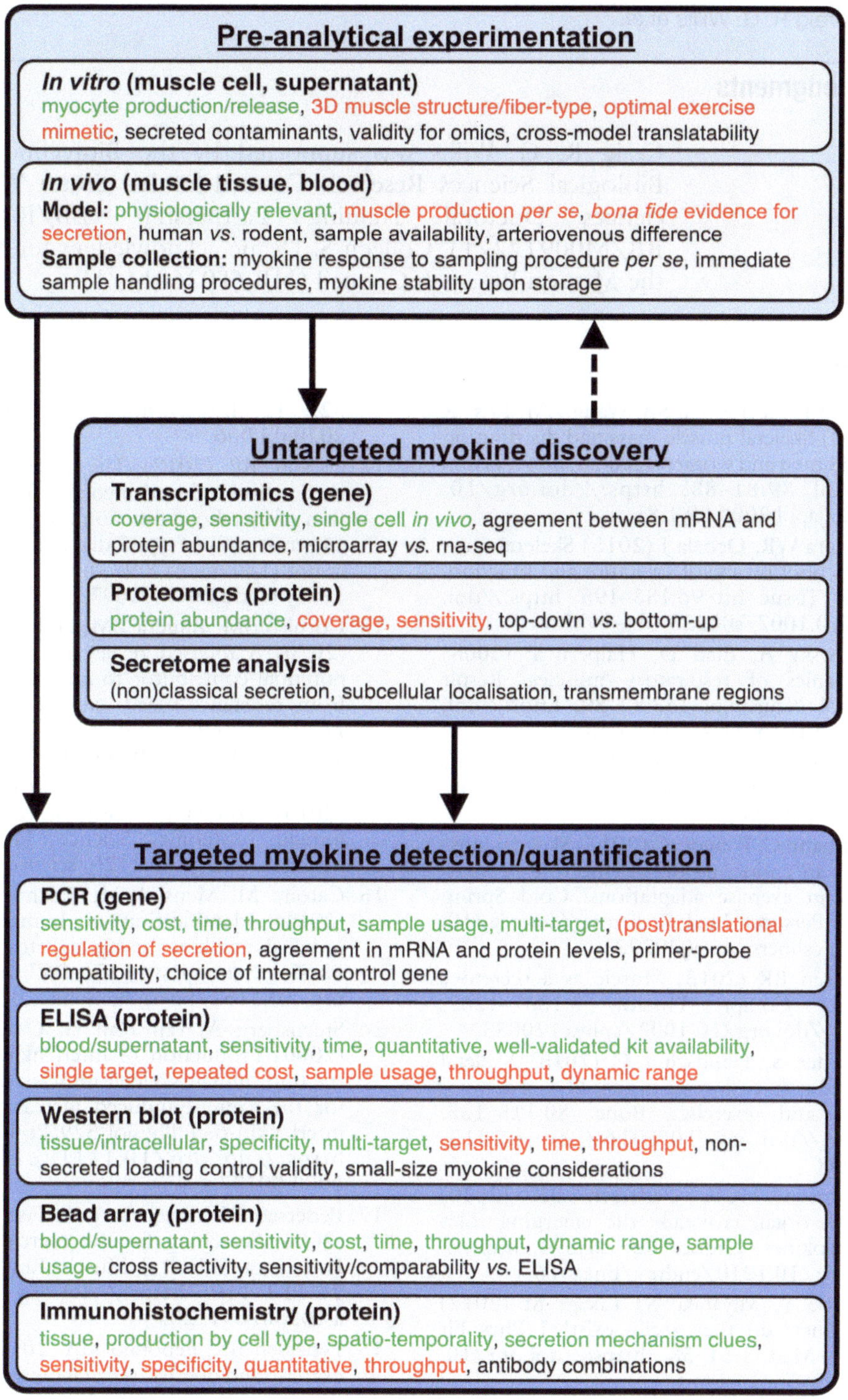

Fig. 2 Summary road map for myokine discovery/quantification. Green and red text indicate a positive/recommendable or limiting aspect, respectively (including relative to other techniques/approaches outlined). Black text indicates any other pertinent technical point worth considering in each case, as outlined at various points throughout the chapter. Filled arrows demonstrate typical analytical workflow. For example, from sample collection toward targeted detection directly or indirectly via untargeted myokine candidate identification. Dashed arrow highlights possible route for multimodel incorporation during untargeted myokine analysis phase. For example, screening for new myokine candidates in vitro and feeding results back toward a hypothesis-driven (targeted) experiment in vivo and vice versa

Acknowledgments

Craig R. G. Willis was supported by the Biotechnology and Biological Sciences Research Council-funded South West Biosciences Doctoral Training Partnership (BB/J014400/1; BB/M009122/1). Colleen S. Deane acknowledges funding from the Medical Research Council (MR/T026014/1).

References

1. Janssen I, Heymsfield SB, Wang ZM, Ross R (2000) Skeletal muscle mass and distribution in 468 men and women aged 18–88 yr. J Appl Physiol 89:81–88. https://doi.org/10.1152/jappl.2000.89.1.81
2. Frontera WR, Ochala J (2015) Skeletal muscle: A brief review of structure and function. Calcif Tissue Int 96:183–195. https://doi.org/10.1007/s00223-014-9915-y
3. Ratnovsky A, Elad D, Halpern P (2008) Mechanics of respiratory muscles. Respir Physiol Neurobiol 163:82–89. https://doi.org/10.1016/j.resp.2008.04.019
4. MacIntosh BR, Gardiner PF, McComas AJ (2006) Skeletal muscle: form and function. Human Kinetics
5. Hoffmann C, Weigert C (2017) Skeletal muscle as an endocrine organ: the role of myokines in exercise adaptations. Cold Spring Harb Perspect Med 7. https://doi.org/10.1101/cshperspect.a029793
6. Pedersen BK (2013) Muscle as a secretory organ. Compr Physiol 3:1337–1362. https://doi.org/10.1002/cphy.c120033
7. Schnyder S, Handschin C (2015) Skeletal muscle as an endocrine organ: PGC-1α, myokines and exercise. Bone 80:115–125. https://doi.org/10.1016/j.bone.2015.02.008
8. Severinsen MCK, Pedersen BK (2020) Muscle-organ crosstalk: the emerging roles of myokines. Endocr Rev 41:594. https://doi.org/10.1210/endrev/bnaa016
9. Manabe Y, Miyatake S, Takagi M (2012) Myokines: do they really exist? J Phys Fit Sport Med 1:51–58. https://doi.org/10.7600/jpfsm.1.51
10. Piccirillo R (2019) Exercise-induced myokines with therapeutic potential for muscle wasting. Front Physiol 10:287. https://doi.org/10.3389/fphys.2019.00287
11. Suriano F, Van Hul M, Cani PD (2020) Gut microbiota and regulation of myokine-adipokine function. Curr Opin Pharmacol 52:9–17. https://doi.org/10.1016/j.coph.2020.03.006
12. Carson BP (2017) The potential role of contraction-induced myokines in the regulation of metabolic function for the prevention and treatment of type 2 diabetes. Front Endocrinol (Lausanne) 8:97. https://doi.org/10.3389/fendo.2017.00097
13. Cornish SM, Bugera EM, Duhamel TA et al (2020) A focused review of myokines as a potential contributor to muscle hypertrophy from resistance-based exercise. Eur J Appl Physiol 120:941–959. https://doi.org/10.1007/s00421-020-04337-1
14. Uhlén M, Fagerberg L, Hallström BM et al (2015) Proteomics. Tissue-based map of the human proteome. Science 347:1260419. https://doi.org/10.1126/science.1260419
15. Catoire M, Mensink M, Kalkhoven E et al (2014) Identification of human exercise-induced myokines using secretome analysis. Physiol Genomics 46:256–267. https://doi.org/10.1152/physiolgenomics.00174.2013
16. Steensberg A, van Hall G, Osada T et al (2000) Production of interleukin-6 in contracting human skeletal muscles can account for the exercise-induced increase in plasma interleukin-6. J Physiol 529(Pt 1):237–242. https://doi.org/10.1111/j.1469-7793.2000.00237.x
17. Pedersen BK, Steensberg A, Fischer C et al (2003) Searching for the exercise factor: is IL-6 a candidate? J Muscle Res Cell Motil 24:113–119. https://doi.org/10.1023/a:1026070911202
18. Pedersen BK, Febbraio MA (2012) Muscles, exercise and obesity: skeletal muscle as a secretory organ. Nat Rev Endocrinol 8:457–465. https://doi.org/10.1038/nrendo.2012.49
19. Pedersen BK, Akerström TCA, Nielsen AR, Fischer CP (2007) Role of myokines in exercise and metabolism. J Appl Physiol 103:1093–1098. https://doi.org/10.1152/japplphysiol.00080.2007

20. Rubenstein AB, Smith GR, Raue U et al (2020) Single-cell transcriptional profiles in human skeletal muscle. Sci Rep 10:229. https://doi.org/10.1038/s41598-019-57110-6
21. Hamrick MW (2012) The skeletal muscle secretome: an emerging player in muscle-bone crosstalk. Bonekey Rep 1:60. https://doi.org/10.1038/bonekey.2012.60
22. Weigert C, Lehmann R, Hartwig S, Lehr S (2014) The secretome of the working human skeletal muscle—a promising opportunity to combat the metabolic disaster? Proteomics Clin Appl 8:5–18. https://doi.org/10.1002/prca.201300094
23. Catoire M, Kersten S (2015) The search for exercise factors in humans. FASEB J 29: 1615–1628. https://doi.org/10.1096/fj.14-263699
24. Florin A, Lambert C, Sanchez C et al (2020) The secretome of skeletal muscle cells: a systematic review. Osteoarthr Cartil Open 2: 100019. https://doi.org/10.1016/j.ocarto.2019.100019
25. Lee JH, Jun H-S (2019) Role of myokines in regulating skeletal muscle mass and function. Front Physiol 10:42. https://doi.org/10.3389/fphys.2019.00042
26. McPherron AC, Lawler AM, Lee SJ (1997) Regulation of skeletal muscle mass in mice by a new TGF-beta superfamily member. Nature 387:83–90. https://doi.org/10.1038/387083a0
27. Rodriguez J, Vernus B, Chelh I et al (2014) Myostatin and the skeletal muscle atrophy and hypertrophy signaling pathways. Cell Mol Life Sci 71:4361–4371. https://doi.org/10.1007/s00018-014-1689-x
28. Pedersen BK, Fischer CP (2007) Beneficial health effects of exercise—the role of IL-6 as a myokine. Trends Pharmacol Sci 28:152–156. https://doi.org/10.1016/j.tips.2007.02.002
29. Pedersen BK (2011) Muscles and their myokines. J Exp Biol 214:337–346. https://doi.org/10.1242/jeb.048074
30. Rasmussen P, Brassard P, Adser H et al (2009) Evidence for a release of brain-derived neurotrophic factor from the brain during exercise. Exp Physiol 94:1062–1069. https://doi.org/10.1113/expphysiol.2009.048512
31. Matthews VB, Aström M-B, Chan MHS et al (2009) Brain-derived neurotrophic factor is produced by skeletal muscle cells in response to contraction and enhances fat oxidation via activation of AMP-activated protein kinase. Diabetologia 52:1409–1418. https://doi.org/10.1007/s00125-009-1364-1
32. Nolan T, Hands RE, Bustin SA (2006) Quantification of mRNA using real-time RT-PCR. Nat Protoc 1:1559–1582. https://doi.org/10.1038/nprot.2006.236
33. Kupcova Skalnikova H, Cizkova J, Cervenka J, Vodicka P (2017) Advances in proteomic techniques for cytokine analysis: focus on melanoma research. Int J Mol Sci 18. https://doi.org/10.3390/ijms18122697
34. Giulietti A, Overbergh L, Valckx D et al (2001) An overview of real-time quantitative PCR: applications to quantify cytokine gene expression. Methods 25:386–401. https://doi.org/10.1006/meth.2001.1261
35. Pourteymour S, Eckardt K, Holen T et al (2017) Global mRNA sequencing of human skeletal muscle: search for novel exercise-regulated myokines. Mol Metab 6:352–365. https://doi.org/10.1016/j.molmet.2017.01.007
36. Amsen D, de Visser KE, Town T (2009) Approaches to determine expression of inflammatory cytokines. Methods Mol Biol 511:107–142. https://doi.org/10.1007/978-1-59745-447-6_5
37. Mahoney DJ, Carey K, Fu M-H et al (2004) Real-time RT-PCR analysis of housekeeping genes in human skeletal muscle following acute exercise. Physiol Genomics 18:226–231. https://doi.org/10.1152/physiolgenomics.00067.2004
38. Darwish IA (2006) Immunoassay methods and their applications in pharmaceutical analysis: basic methodology and recent advances. Int J Biomed Sci 2:217–235
39. Albrecht E, Norheim F, Thiede B et al (2015) Irisin—a myth rather than an exercise-inducible myokine. Sci Rep 5:8889. https://doi.org/10.1038/srep08889
40. Engvall E, Perlmann P (1972) Enzyme-linked immunosorbent assay, Elisa. 3. Quantitation of specific antibodies by enzyme-labeled anti-immunoglobulin in antigen-coated tubes. J Immunol 109:129–135
41. ThermoFisher Scientific (2018) A comparison of protein detection and quantification techniques. https://assets.thermofisher.com/TFS-Assets/BID/Application-Notes/comparison-protein-detection-quantification-techniques-application-note.pdf
42. Leng SX, McElhaney JE, Walston JD et al (2008) ELISA and multiplex technologies

for cytokine measurement in inflammation and aging research. J Gerontol A Biol Sci Med Sci 63:879–884. https://doi.org/10.1093/gerona/63.8.879

43. Zhou X, Fragala MS, McElhaney JE, Kuchel GA (2010) Conceptual and methodological issues relevant to cytokine and inflammatory marker measurements in clinical research. Curr Opin Clin Nutr Metab Care 13:541–547. https://doi.org/10.1097/MCO.0b013e32833cf3bc

44. Stenken JA, Poschenrieder AJ (2015) Bioanalytical chemistry of cytokines—a review. Anal Chim Acta 853:95–115. https://doi.org/10.1016/j.aca.2014.10.009

45. Whitham M, Febbraio MA (2016) The ever-expanding myokinome: discovery challenges and therapeutic implications. Nat Rev Drug Discov 15:719–729. https://doi.org/10.1038/nrd.2016.153

46. Bass JJ, Wilkinson DJ, Rankin D et al (2017) An overview of technical considerations for Western blotting applications to physiological research. Scand J Med Sci Sports 27:4–25. https://doi.org/10.1111/sms.12702

47. Collins MA, An J, Peller D, Bowser R (2015) Total protein is an effective loading control for cerebrospinal fluid western blots. J Neurosci Methods 251:72–82. https://doi.org/10.1016/j.jneumeth.2015.05.011

48. Kwon JH, Moon KM, Min K-W (2020) Exercise-induced myokines can explain the importance of physical activity in the elderly: an overview. Healthc (Basel, Switzerland) 8. https://doi.org/10.3390/healthcare8040378

49. Akdis M, Aab A, Altunbulakli C et al (2016) Interleukins (from IL-1 to IL-38), interferons, transforming growth factor β, and TNF-α: Receptors, functions, and roles in diseases. J Allergy Clin Immunol 138:984–1010. https://doi.org/10.1016/j.jaci.2016.06.033

50. Schipper HS, de Jager W, van Dijk MEA et al (2010) A multiplex immunoassay for human adipokine profiling. Clin Chem 56:1320–1328. https://doi.org/10.1373/clinchem.2010.146118

51. Elshal MF, McCoy JP (2006) Multiplex bead array assays: performance evaluation and comparison of sensitivity to ELISA. Methods 38: 317–323. https://doi.org/10.1016/j.ymeth.2005.11.010

52. Kim S-W, Roh J, Park C-S (2016) Immunohistochemistry for pathologists: protocols, pitfalls, and tips. J Pathol Transl Med 50: 411–418. https://doi.org/10.4132/jptm.2016.08.08

53. Digre A, Lindskog C (2021) The Human Protein Atlas-Spatial localization of the human proteome in health and disease. Protein Sci 30:218–233. https://doi.org/10.1002/pro.3987

54. Balaji S, Li H, Steen E, Keswani SG (2019) Considerations for immunohistochemistry. In: Kennedy G, Gosain A, Kibbe M, LeMaire SA (eds) Success in academic surgery: basic science. Springer International Publishing, Cham, pp 105–144

55. Lodish H, Berk A, Zipursky SL et al (2000) Section 17.3, Overview of the secretory pathway. In: Freeman WH (ed) Molecular cell biology, 4th edn, New York

56. Barlowe CK, Miller EA (2013) Secretory protein biogenesis and traffic in the early secretory pathway. Genetics 193:383–410. https://doi.org/10.1534/genetics.112.142810

57. Kim J, Gee HY, Lee MG (2018) Unconventional protein secretion—new insights into the pathogenesis and therapeutic targets of human diseases. J Cell Sci 131. https://doi.org/10.1242/jcs.213686

58. Prudovsky I, Tarantini F, Landriscina M et al (2008) Secretion without Golgi. J Cell Biochem 103:1327–1343. https://doi.org/10.1002/jcb.21513

59. Trovato E, Di Felice V, Barone R (2019) Extracellular vesicles: delivery vehicles of myokines. Front Physiol 10:522. https://doi.org/10.3389/fphys.2019.00522

60. Safdar A, Saleem A, Tarnopolsky MA (2016) The potential of endurance exercise-derived exosomes to treat metabolic diseases. Nat Rev Endocrinol 12:504–517. https://doi.org/10.1038/nrendo.2016.76

61. Whitham M, Parker BL, Friedrichsen M et al (2018) Extracellular vesicles provide a means for tissue crosstalk during exercise. Cell Metab 27:237–251.e4. https://doi.org/10.1016/j.cmet.2017.12.001

62. Hiscock N, Chan MHS, Bisucci T et al (2004) Skeletal myocytes are a source of interleukin-6 mRNA expression and protein release during contraction: evidence of fiber type specificity. FASEB J 18:992–994. https://doi.org/10.1096/fj.03-1259fje

63. Kedaigle A, Fraenkel E (2018) Turning omics data into therapeutic insights. Curr Opin Pharmacol 42:95–101. https://doi.org/10.1016/j.coph.2018.08.006

64. Manzoni C, Kia DA, Vandrovcova J et al (2018) Genome, transcriptome and proteome: the rise of omics data and their integration in biomedical sciences. Brief Bioinform 19:286–302. https://doi.org/10.1093/bib/bbw114
65. Neufer PD, Bamman MM, Muoio DM et al (2015) Understanding the cellular and molecular mechanisms of physical activity-induced health benefits. Cell Metab 22:4–11. https://doi.org/10.1016/j.cmet.2015.05.011
66. Horgan RP, Kenny LC (2011) 'Omic' technologies: genomics, transcriptomics, proteomics and metabolomics. Obstet Gynaecol 13:189–195. https://doi.org/10.1576/toag.13.3.189.27672
67. Davidsen PK, Turan N, Egginton S, Falciani F (2016) Multilevel functional genomics data integration as a tool for understanding physiology: a network biology perspective. J Appl Physiol 120:297–309. https://doi.org/10.1152/japplphysiol.01110.2014
68. Zhao S, Fung-Leung W-P, Bittner A et al (2014) Comparison of RNA-Seq and microarray in transcriptome profiling of activated T cells. PLoS One 9:e78644. https://doi.org/10.1371/journal.pone.0078644
69. Mortazavi A, Williams BA, McCue K et al (2008) Mapping and quantifying mammalian transcriptomes by RNA-Seq. Nat Methods 5:621–628. https://doi.org/10.1038/nmeth.1226
70. Mantione KJ, Kream RM, Kuzelova H et al (2014) Comparing bioinformatic gene expression profiling methods: microarray and RNA-Seq. Med Sci Monit Basic Res 20:138–142. https://doi.org/10.12659/MSMBR.892101
71. van der Kloet FM, Buurmans J, Jonker MJ et al (2020) Increased comparability between RNA-Seq and microarray data by utilization of gene sets. PLoS Comput Biol 16:e1008295. https://doi.org/10.1371/journal.pcbi.1008295
72. Bolón-Canedo V, Sánchez-Maroño N, Alonso-Betanzos A et al (2014) A review of microarray datasets and applied feature selection methods. Inf Sci (Ny) 282:111–135. https://doi.org/10.1016/j.ins.2014.05.042
73. Slonim DK, Yanai I (2009) Getting started in gene expression microarray analysis. PLoS Comput Biol 5:e1000543. https://doi.org/10.1371/journal.pcbi.1000543
74. Conesa A, Madrigal P, Tarazona S et al (2016) A survey of best practices for RNA-seq data analysis. Genome Biol 17:13. https://doi.org/10.1186/s13059-016-0881-8
75. Schmidt A, Forne I, Imhof A (2014) Bioinformatic analysis of proteomics data. BMC Syst Biol 8 Suppl 2:S3. https://doi.org/10.1186/1752-0509-8-S2-S3
76. Chen C, Hou J, Tanner JJ, Cheng J (2020) Bioinformatics methods for mass spectrometry-based proteomics data analysis. Int J Mol Sci 21. https://doi.org/10.3390/ijms21082873
77. Pei G, Chen L, Zhang W (2017) WGCNA application to proteomic and metabolomic data analysis. Methods Enzymol 585:135–158. https://doi.org/10.1016/bs.mie.2016.09.016
78. Langfelder P, Horvath S (2008) WGCNA: an R package for weighted correlation network analysis. BMC Bioinformatics 9:559. https://doi.org/10.1186/1471-2105-9-559
79. Nielsen H, Engelbrecht J, Brunak S, von Heijne G (1997) Identification of prokaryotic and eukaryotic signal peptides and prediction of their cleavage sites. Protein Eng 10:1–6. https://doi.org/10.1093/protein/10.1.1
80. Ashburner M, Ball CA, Blake JA et al (2000) Gene ontology: tool for the unification of biology. The Gene Ontology Consortium. Nat Genet 25:25–29. https://doi.org/10.1038/75556
81. UniProt Consortium (2019) UniProt: a worldwide hub of protein knowledge. Nucleic Acids Res 47:D506–D515. https://doi.org/10.1093/nar/gky1049
82. Caccia D, Dugo M, Callari M, Bongarzone I (2013) Bioinformatics tools for secretome analysis. Biochim Biophys Acta 1834:2442–2453. https://doi.org/10.1016/j.bbapap.2013.01.039
83. Bendtsen JD, Kiemer L, Fausbøll A, Brunak S (2005) Non-classical protein secretion in bacteria. BMC Microbiol 5:58. https://doi.org/10.1186/1471-2180-5-58
84. Ras-Carmona A, Gomez-Perosanz M, Reche PA (2021) Prediction of unconventional protein secretion by exosomes. BMC Bioinformatics 22:333. https://doi.org/10.1186/s12859-021-04219-z
85. Uhlén M, Karlsson MJ, Hober A et al (2019) The human secretome. Sci Signal 12. https://doi.org/10.1126/scisignal.aaz0274
86. Almagro Armenteros JJ, Sønderby CK, Sønderby SK et al (2017) DeepLoc: prediction of protein subcellular localization using deep learning. Bioinformatics 33:3387–3395.

https://doi.org/10.1093/bioinformatics/btx431

87. Almagro Armenteros JJ, Salvatore M, Emanuelsson O et al (2019) Detecting sequence signals in targeting peptides using deep learning. Life Sci Alliance 2. https://doi.org/10.26508/lsa.201900429
88. Zhao L, Poschmann G, Waldera-Lupa D et al (2019) OutCyte: a novel tool for predicting unconventional protein secretion. Sci Rep 9: 19448. https://doi.org/10.1038/s41598-019-55351-z
89. Petersen TN, Brunak S, von Heijne G, Nielsen H (2011) SignalP 4.0: discriminating signal peptides from transmembrane regions. Nat Methods 8:785–786. https://doi.org/10.1038/nmeth.1701
90. Krogh A, Larsson B, von Heijne G, Sonnhammer EL (2001) Predicting transmembrane protein topology with a hidden Markov model: application to complete genomes. J Mol Biol 305:567–580. https://doi.org/10.1006/jmbi.2000.4315
91. Yoon JH, Yea K, Kim J et al (2009) Comparative proteomic analysis of the insulin-induced L6 myotube secretome. Proteomics 9:51–60. https://doi.org/10.1002/pmic.200800187
92. Deshmukh AS, Cox J, Jensen LJ et al (2015) Secretome analysis of lipid-induced insulin resistance in skeletal muscle cells by a combined experimental and bioinformatics workflow. J Proteome Res 14:4885–4895. https://doi.org/10.1021/acs.jproteome.5b00720
93. Howe KL, Achuthan P, Allen J et al (2021) Ensembl 2021. Nucleic Acids Res 49:D884–D891. https://doi.org/10.1093/nar/gkaa942
94. Almagro Armenteros JJ, Tsirigos KD, Sønderby CK et al (2019) SignalP 5.0 improves signal peptide predictions using deep neural networks. Nat Biotechnol 37:420–423. https://doi.org/10.1038/s41587-019-0036-z
95. Horton P, Park K-J, Obayashi T et al (2007) WoLF PSORT: protein localization predictor. Nucleic Acids Res 35:W585–W587. https://doi.org/10.1093/nar/gkm259
96. Nielsen H, Petsalaki EI, Zhao L, Stühler K (2019) Predicting eukaryotic protein secretion without signals. Biochim Biophys Acta Proteins Proteomics 1867:140174. https://doi.org/10.1016/j.bbapap.2018.11.011
97. Rudrappa SS, Wilkinson DJ, Greenhaff PL et al (2016) Human skeletal muscle disuse atrophy: effects on muscle protein synthesis, breakdown, and insulin resistance—A qualitative review. Front Physiol 7:361. https://doi.org/10.3389/fphys.2016.00361
98. Hansen JS, Clemmesen JO, Secher NH et al (2015) Glucagon-to-insulin ratio is pivotal for splanchnic regulation of FGF-21 in humans. Mol Metab 4:551–560. https://doi.org/10.1016/j.molmet.2015.06.001
99. Altara R, Manca M, Hermans KCM et al (2015) Diurnal rhythms of serum and plasma cytokine profiles in healthy elderly individuals assessed using membrane based multiplexed immunoassay. J Transl Med 13:129. https://doi.org/10.1186/s12967-015-0477-1
100. Duglan D, Lamia KA (2019) Clocking in, working out: Circadian regulation of exercise physiology. Trends Endocrinol Metab 30: 347–356. https://doi.org/10.1016/j.tem.2019.04.003
101. Mooren FC, Krueger K, Ringseis R et al (2020) Combined effects of moderate exercise and short-term fasting on markers of immune function in healthy human subjects. Am J Physiol Regul Integr Comp Physiol 318:R1103–R1115. https://doi.org/10.1152/ajpregu.00341.2019
102. Petersen PS, Lei X, Seldin MM et al (2014) Dynamic and extensive metabolic state-dependent regulation of cytokine expression and circulating levels. Am J Physiol Regul Integr Comp Physiol 307:R1458–R1470. https://doi.org/10.1152/ajpregu.00335.2014
103. Jia W-H, Wang N-Q, Yin L et al (2019) Effect of skeletal muscle phenotype and gender on fasting-induced myokine expression in mice. Biochem Biophys Res Commun 514:407–414. https://doi.org/10.1016/j.bbrc.2019.04.155
104. Malm C, Nyberg P, Engstrom M et al (2000) Immunological changes in human skeletal muscle and blood after eccentric exercise and multiple biopsies. J Physiol 529(Pt 1): 243–262. https://doi.org/10.1111/j.1469-7793.2000.00243.x
105. Aronson D, Wojtaszewski JF, Thorell A et al (1998) Extracellular-regulated protein kinase cascades are activated in response to injury in human skeletal muscle. Am J Phys 275: C555–C561. https://doi.org/10.1152/ajpcell.1998.275.2.C555
106. Murton AJ, Billeter R, Stephens FB et al (2014) Transient transcriptional events in human skeletal muscle at the outset of concentric resistance exercise training. J Appl Physiol 116:113–125. https://doi.org/10.1152/japplphysiol.00426.2013

107. Guerra B, Gómez-Cabrera MC, Ponce-González JG et al (2011) Repeated muscle biopsies through a single skin incision do not elicit muscle signaling, but IL-6 mRNA and STAT3 phosphorylation increase in injured muscle. J Appl Physiol 110:1708–1715. https://doi.org/10.1152/japplphysiol.00091.2011
108. de Jager W, Bourcier K, Rijkers GT et al (2009) Prerequisites for cytokine measurements in clinical trials with multiplex immunoassays. BMC Immunol 10:52. https://doi.org/10.1186/1471-2172-10-52
109. Gudiksen A, Schwartz CL, Bertholdt L et al (2016) Lack of skeletal muscle IL-6 affects pyruvate dehydrogenase activity at rest and during prolonged exercise. PLoS One 11:e0156460. https://doi.org/10.1371/journal.pone.0156460
110. Ost M, Coleman V, Kasch J, Klaus S (2016) Regulation of myokine expression: role of exercise and cellular stress. Free Radic Biol Med 98:78–89. https://doi.org/10.1016/j.freeradbiomed.2016.02.018
111. Abdelmoez AM, Sardón Puig L, Smith JAB et al (2020) Comparative profiling of skeletal muscle models reveals heterogeneity of transcriptome and metabolism. Am J Physiol Cell Physiol 318:C615–C626. https://doi.org/10.1152/ajpcell.00540.2019
112. Geyer PE, Holdt LM, Teupser D, Mann M (2017) Revisiting biomarker discovery by plasma proteomics. Mol Syst Biol 13:942. https://doi.org/10.15252/msb.20156297
113. Ignjatovic V, Geyer PE, Palaniappan KK et al (2019) Mass spectrometry-based plasma proteomics: considerations from sample collection to achieving translational data. J Proteome Res 18:4085–4097. https://doi.org/10.1021/acs.jproteome.9b00503
114. Joanisse S, Lim C, McKendry J et al (2020) Recent advances in understanding resistance exercise training-induced skeletal muscle hypertrophy in humans. F1000Research 9. https://doi.org/10.12688/f1000research.21588.1
115. Vikman P, Fadista J, Oskolkov N (2014) RNA sequencing: current and prospective uses in metabolic research. J Mol Endocrinol 53:R93–R101. https://doi.org/10.1530/JME-14-0170
116. Deshmukh AS, Steenberg DE, Hostrup M et al (2021) Deep muscle-proteomic analysis of freeze-dried human muscle biopsies reveals fiber type-specific adaptations to exercise training. Nat Commun 12:304. https://doi.org/10.1038/s41467-020-20556-8
117. Deshmukh AS, Murgia M, Nagaraj N et al (2015) Deep proteomics of mouse skeletal muscle enables quantitation of protein isoforms, metabolic pathways, and transcription factors. Mol Cell Proteomics 14:841–853. https://doi.org/10.1074/mcp.M114.044222
118. Wang D, Eraslan B, Wieland T et al (2019) A deep proteome and transcriptome abundance atlas of 29 healthy human tissues. Mol Syst Biol 15:e8503. https://doi.org/10.15252/msb.20188503
119. Dey KK, Wang H, Niu M et al (2019) Deep undepleted human serum proteome profiling toward biomarker discovery for Alzheimer's disease. Clin Proteomics 16:16. https://doi.org/10.1186/s12014-019-9237-1
120. Little HC, Tan SY, Cali FM et al (2018) Multiplex quantification identifies novel exercise-regulated myokines/cytokines in plasma and in glycolytic and oxidative skeletal muscle. Mol Cell Proteomics 17:1546–1563. https://doi.org/10.1074/mcp.RA118.000794
121. Aoi W, Naito Y, Takagi T et al (2013) A novel myokine, secreted protein acidic and rich in cysteine (SPARC), suppresses colon tumorigenesis via regular exercise. Gut 62:882–889. https://doi.org/10.1136/gutjnl-2011-300776
122. Gianazza E, Miller I, Palazzolo L et al (2016) With or without you—Proteomics with or without major plasma/serum proteins. J Proteome 140:62–80. https://doi.org/10.1016/j.jprot.2016.04.002
123. Yakimchuk K (2018) Low abundance proteins. Mater Methods 8. https://doi.org/10.13070/mm.en.8.2658
124. Raschke S, Eckardt K, Bjørklund Holven K et al (2013) Identification and validation of novel contraction-regulated myokines released from primary human skeletal muscle cells. PLoS One 8:e62008. https://doi.org/10.1371/journal.pone.0062008

Chapter 13

Skeletal Muscle Satellite Cell Physiology and Function: Complimentary In Vitro and In Vivo Models and Methods

Mark Viggars, Andy Nolan, Adam Sharples, and Claire Stewart

Abstract

Skeletal muscle is a highly adaptable tissue, displaying features of hypertrophy (growth) and atrophy (loss), as a result of both internal and external stimuli. The mechanisms underpinning skeletal muscle adaptation have featured heavily in research for almost two centuries. Since the discovery of the skeletal muscle satellite/stem cell in the early 1960s, the function of these cells has been extensively studied, both in vivo and in vitro. For the purposes of this chapter and to facilitate such research, we detail methods relating to satellite cell study both in vitro (from biopsy harvest to cell culture) and in vivo (from biopsy harvest to immunohistochemistry). Methods relating to exercise adaptation in terms of stretch and load, as well as electrical stimulation, are also detailed in an in vitro cell model. Investigating the behavior of the versatile skeletal muscle satellite cell, in diverse (in vitro and in vivo) models, is fundamental to furthering our understanding of skeletal muscle adaptation.

Key words Muscle biopsy, Satellite cell, Myoblast, Electrical stimulation, Stretch and load, Exercise, Immunohistochemistry

1 Introduction

Muscle comprises the largest mass of tissue in the human body, with two subtypes in existence: smooth and striated. The former is present in the walls of the gastrointestinal system, the genitourinary urinary tract, and blood vessels and is under involuntary control. Cardiac and skeletal muscles are striated, with the former being under involuntary and the latter under voluntary control. As its name suggests, cardiac muscle is found only in the heart. Although the arrangement of its contractile proteins is equivalent to that of skeletal muscle, the arrangement of their fibers differs. Approximately 650 different skeletal muscles (which is the focus of this chapter) are present in the human body, and it comprises approximately 40–50% of the healthy young adult human mass. Skeletal muscle is essential for growth, posture, locomotion, and storage/provision of energy. Muscle grows in mass during childhood and

Philip J. Atherton and Daniel J. Wilkinson (eds.), *Neuromuscular Assessments of Form and Function*, Neuromethods, vol. 204, https://doi.org/10.1007/978-1-0716-3315-1_13,

puberty and following, for example, resistance training interventions. In contrast, aging, disease, and disuse result in losses of muscle mass and strength. Finally, an intact skeletal muscle mass is essential for survival, with suggestions that losses in total mass of 40% would result in death.

Given the important metabolic, physiological, and functional roles that skeletal muscle plays, its adaptability is key to health and quality of life. As a result, this chapter will provide a brief overview of skeletal muscle and its myogenic (originating in muscle) stem cell population called the "satellite" cells, including the methods required for their study. The focus will be on (1) in vitro cellular methodology to harvest cells from human biopsy samples and (2) the in vivo histological methods for assessing the cells in situ. The chapter starts with a focus on skeletal muscle development, the discovery of the satellite cell, and its role in muscle adaptation, before moving to materials and methods for the culture and histochemical study of the satellite cell.

1.1 Skeletal Muscle Development

Skeletal muscle development, both during embryogenesis and postnatally, is a highly orchestrated and precisely coordinated process. Although some discussion abounds regarding the precise source of myogenic cells of the trunk and limb, the general consensus is that they are derived from the somites (temporary mesoderm-derived segments), located in pairs on either side of the neural tube in developing vertebrate embryos [1]. More specifically, progenitor cells believed to originate from the hypaxial domain of the somites reportedly migrate into the limb buds, where they proliferate and express muscle cell-specific determination factors, under the influence of signals from neighboring tissues [2, 3]. In addition to limb muscles, the cells of the hypaxial somites are also believed to give rise to the abdominal muscles and the diaphragm, whereas most muscles of the head originate from the cranial mesoderm [4]. The focus of this chapter will be on muscle cells derived from the hypaxial mesoderm.

Briefly, primary muscle fibers form at approximately 11–14 embryonic days in murine limbs. This is followed at embryonic Days 14–16 by the formation of secondary fibers in parallel and prior to commencement of innervation [5–8]. A similar but slower pattern is reported in human embryonic development. Myogenic progenitor cells derived from pluripotent embryonic stem cells migrate to limb buds and differentiate to myoblasts at approximately 6 weeks of gestation. Following several rounds of proliferation, at 7–9 weeks of gestation, these cells fuse to form multinucleated primary myotubes attached via tendons to bones. With time, additional myoblasts fuse with the primary myotubes to form secondary myotubes, which initially share a common basement membrane. At this stage, the first histologically recognizable muscle is well defined, with synthesis of the contractile proteins, actin, and myosin occurring and striations becoming visible. From 11 weeks, the fibers grow in length and in cross-sectional area, and it is believed that by 6 months of gestation in the human embryo, the number of fibers is believed to be set, with any further growth occurring as a consequence of increased cross-sectional area rather than increased fiber numbers [9].

1.2 Key Milestones in Skeletal Muscle Research

The study of myotube formation, growth, repair, and loss is not new. Highly novel work in Germany from the 1830s to 1860s provided extensive knowledge of muscle adaptation, long before any knowledge of adult muscle stem cells. Briefly, as early as 1839, the concept of muscle cell fusion was proposed [10], and from the mid-1850s to mid-1860s, experimental models of muscle degeneration and regeneration were reported [11–13]. This knowledge continued to expand, with scientists providing the first evidence of in vitro myotube formation before the end of the second decade of the twentieth century [14]. The concept that myonuclei are incapable of replication was first reported in the late 1950s [15], suggesting that skeletal muscle fibers are terminally differentiated and are also incapable of replication (for a comprehensive historical perspective, *see* Scharner and Zammit [16]).

1.3 The Satellite Cell

Both historically and currently, it is well documented that adult skeletal muscle displays significant adaptability to environmental cues, including hypertrophy, atrophy, and repair. Therefore, the question arises, if the fibers are incapable of replication (otherwise known as hyperplasia), how is this adaptation achieved? In 1961, using electron microscopy to examine the tibialis anticus of the frog, Alexander Mauro reported the presence of single cells beneath the basal lamina of mature muscle fibers, which he designated satellite cells [17], due to their peripheral location. Mauro went on to speculate that these cells were quiescent myoblasts (muscle cells) that had not fused and were present to provide a resident cell pool to facilitate growth and repair, following damage. Although not frequently quoted, in the same month of the same year, while examining frog muscle spindles, Bernard Katz [18] also reported the presence of satellite cells on the periphery of the muscle, which he speculated were associated with hypertrophy of muscle spindles.

Indeed, it has since been substantiated that these resident muscle cells, which remain as single cells beneath the basal lamina of mature muscle fibers, are indeed responsible for the growth, differentiation, and repair potential of adult skeletal muscle. In the last 60 years, since the initial reports of satellite cells, extensive progress has been made relating to the behavior and capacity of resident, postnatal, muscle cells. Commercially available cell lines (most notably the L6s, Sol8s, C2, and C2C12 muscle cells) and primary muscle cell cultures derived from muscle biopsies have been extensively studied, and data relating to mechanisms underpinning adaptation have been published.

1.4 Example Satellite Cell Models

Importantly, recent findings suggest that cell models such as those mentioned above respond differently to similar stimuli in terms of their transcriptome and metabolome, and as such, data need to be considered carefully before translational conclusions are drawn [19]. Nonetheless, several questions remain unanswered relating to the role that satellite cells play in muscle adaptation, warranting further investigation. For example, controversy exists as to the numbers and relevance of different muscle cell populations; the mechanisms regulating quiescence and activation; the impact of age and disease on numbers and functional capacity; and cross talk between myonuclei and satellite cells and the precise roles that they play in vivo.

Muscle cells retain a memory of the donor environment from which they were derived such as from cachectic, physically active, sarcopenic, or obese individuals (reviewed in [20, 21]). Consequently, the study of the cells themselves (from a single biopsy sample) provides appropriate temporal understanding of external and internal influencers of their behavior. Such cellular studies enable extensive time courses to be investigated once cells have been isolated where equivalent time course studies (taking multiple biopsies) could be problematic to perform in humans in vivo. Cells therefore provide relevant in vitro models and biopsies an actual snapshot in time. Therefore, if the cells are paired with biopsies from which they were derived, an opportunity to monitor in vivo and in vitro adaptation with exercise and nutrition, age, disease, disuse, and injury is provided. Finally, stimulation of the human muscle cells with e.g. growth factors, cytokines, nutraceuticals, or serum following in vivo interventions; damaging protocols; loading or electrical stimulation, all provide opportunities for models of hypertrophy, atrophy and injury thereby providing knowledge to better inform in vivo questions. The purpose of this chapter, therefore, is to provide details of methodologies relating to the study of muscle-derived cells as well as histological analyses of the satellite cells in the biopsies from which they are derived.

In terms of deriving muscle cells in culture, there are two main schools of thought in terms of isolation techniques. In the first, the satellite cells are pulled out from other cells in the biopsy slurry using targeted antibodies to muscle-specific cell surface markers and magnetic cell sorting. This provides a purer population of muscle cells to study but is limited by the fact that the cellular milieu is no longer intact. The second method omits this purification step, retaining all cells from the muscle niche and thereby enabling studies that assess cross talk between myoblasts and the associated fibroblasts, which would lay down relevant extracellular matrices that would otherwise be missing. However, when performing whole cell lysate mRNA analyses or Western immunoblotting, it is not possible to ascribe any changes to a specific cell population but rather to the whole cellular environment (unless advanced techniques such as single-cell RNA sequencing are utilized). For the purposes of this chapter, the latter method of human muscle cell isolation, with the mixed population, is described. What is key to cell isolation, regardless of which methodology is deemed relevant to the questions being asked, is to be consistent. Be aware of interpretation and always determine the proportion of myogenic to non-myogenic cells, regardless of the isolation method of choice.

Once cells have been isolated and expanded for experimentation, then any cell-related questions can be addressed, including investigation of replication, metabolism, migration, fusion, hypertrophy, apoptosis, senescence, and atrophy. Once ethical approval has been gained to harvest biopsies and isolate the cells, so long as the research is covered by the ethics application and complies with the Human Tissue Act (in the UK), then all and any of these outcome measures are feasible at a molecular, biochemical, and cellular level.

Extensive literature is available relating to methods to address these questions. What is less widely available is information relating to the application of stretch and load or electrical stimulation, as interventions to mimic exercise stimuli and to manipulate muscle cell behavior. These two methodologies will therefore be presented here, in relation to the cell culture techniques.

1.5 Stretch and Load

Seminal in vivo work first demonstrated that mechanically loading (stretching) the plantaris and soleus muscles via removal of the gastrocnemius tendon (termed synergistic ablation, SA) in rats increased muscle mass by 20% and 40%, respectively, within just 5 days post SA [22]. Subsequently, mechanical loading of muscle has been used to study the cellular and molecular mechanisms of muscle adaptation. Using both undifferentiated and differentiated muscle cells (myotubes) in vitro, researchers have undertaken mechanical loading/stretch experiments using different loading regimes, such as moderate cyclic stretch for longer duration (experiments attempting to mimic an "endurance"-type exercise stimulus) or more extensive stretching for intermittent durations (to mimic a more "resistance"-type exercise stimulus).

The first mechanical loading stimulus of cultured cells in monolayer (single layer of cells in a two-dimensional/2D flat-bottomed culture dish) was undertaken in 12-day-old embryonic chick myotubes cultured on an elastic silicone membrane that was stretched multiaxially (different directions) for 18 h. This resulted in increased amino acid uptake, total/myofibrillar protein synthesis, and myotube size [23, 24]. Furthermore, uniaxial (same direction) cyclic stretch of C2C12 cells in 2D monolayer cultures increased phosphorylation of the well-characterized protein translation initiator, p70S6K [25], a kinase; the activity of which correlates with muscle hypertrophy, following chronic resistance exercise [26].

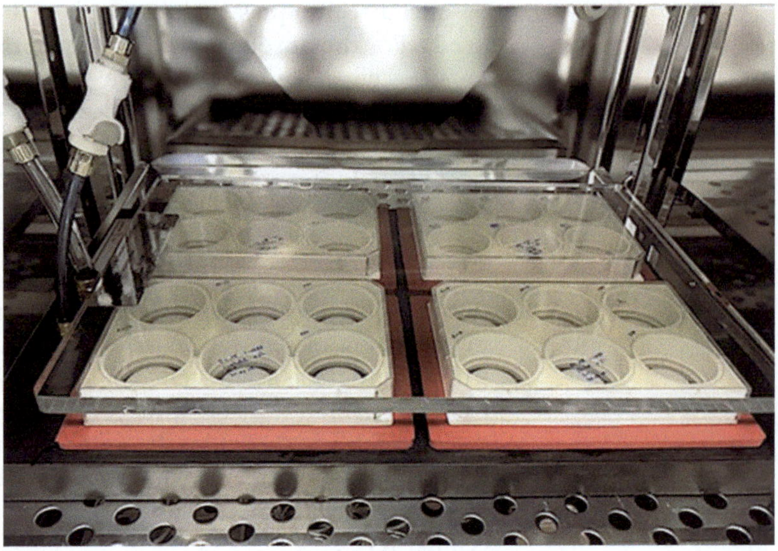

Fig. 1 Vacuum pressured, flexible-bottomed culture plates housed in a cell culture incubator (Flexcell® FX-5000™ tension system, Dunn Labortechnik, Germany)

The loading of 2D monolayer (as well as three-dimensional (3D) bioengineered muscle cultures—detailed below) and the ability for these models to mimic exercise stimuli in vivo are extensively reviewed here [27]. With respect to methods, the most characterized system to apply both uniaxial and multiaxial load to monolayer muscle cells, cultured on flexible-bottomed culture plates, is the commercially available Flexcell® FX-5000™ tension system (Dunn Labortechnik, Germany). The loading is achieved via regulated vacuum pressure, which deforms the flexible silicon membrane on which cells are cultured (Fig. 1).

Three-dimensional muscle cultures in vitro (tissue-engineered or bioengineered muscle), which differentiate muscle-derived cells in a 3D matrix (such as collagen or fibrin), have also been used to investigate the effect of loading on skeletal muscle adaptation. A number of "bioreactors" are used to enable mechanical loading of 3D bioengineered muscle. For example, muscle cells are grown on 3D elastic membranes and stretched multiaxially via expansion of a surrounding frame [23]. More recent bioreactors apply uniaxial stretch to engineered muscle using computerized step motors [28, 29], thereby allowing precise manipulation of the degree of stretch and the timing and frequency of the load/stretch regime with a similar system also available that allows higher throughput of engineered muscle constructs in six-well culture dishes [30]. Finally, the most up-to-date bioreactors allow for even greater throughput and a highly controlled and stable culture environment. For example, using the TC-3 tension bioreactor system (EBERS Medical

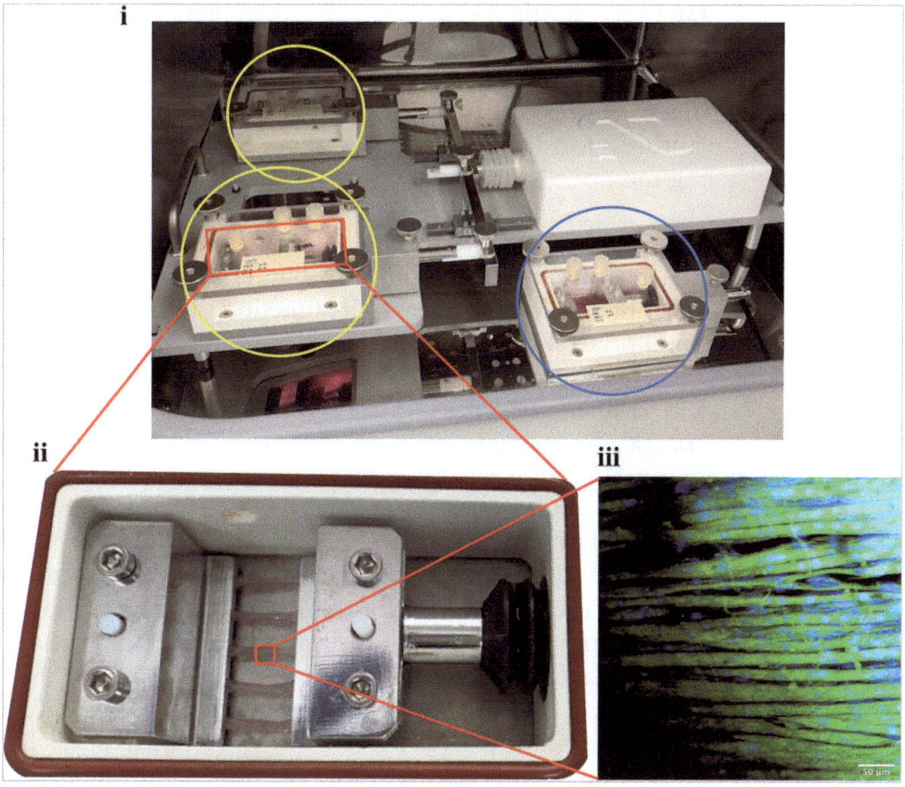

Fig. 2 Diagrammatic representation of mechanical loading of bioengineered skeletal muscle. (**a**) The TC-3 bioreactor system (EBERS Medical Technology, Spain) used to mechanically load 3D bioengineered skeletal muscle and consists of three chambers that can house approximately five bioengineered muscles per chamber (total approx. 15 muscles in one experiment). In the image, the "loaded" muscles are highlighted with yellow circles, and the "non-loaded" controls are highlighted in the blue circle (*see* Subheading 3.1.5 for details). (**b**) 5× fibrin bioengineered skeletal muscle constructs clamped within a single bioreactor chamber. (**c**) Microscopic image taken of a single bioengineered muscle construct immunostained for f-actin (phalloidin-FITC, green) and myonuclei (DAPI, blue) and imaged using confocal microscopy. (The figure is taken from Turner et al. [31], as an open access article where permissions are not required, provided the work is properly cited. *The Journal of Cellular Physiology* (John Wiley and Sons). Figure Biii (immuno-image) is originally taken from Seaborne et al. [32] with permission (Copyright-2019) from the authors. *The Journal of Physiology* (Copyright-2019 The Physiological Society))

Technology, Spain), researchers have uniaxially stretched up to 15 bioengineered muscles (Fig. 2 [31]) at one time in tightly controlled and highly specific loading regimes. This includes the ability to manipulate degree of stretch, frequency, timing, repetitions, and sets consistently across all muscle cultures in specialized cell chambers containing the bioengineered muscles and relevant medias (as well as perfusion if required) within a sterile cell culture incubator (Fig. 2 [31]). Both the loading of monolayer and 3D cultures have allowed important insight into the effects of exercise stimuli on cells at the molecular level, with the advantage of combining additional manipulation of the cellular environment (steroids, nutrients, drugs, gene therapies, and pharmacological inhibitors/activators). From an exercise perspective, the major limitation of these in vitro cultures is the inability to perform "training." This is because there are issues with maintaining sterile cultures for long periods of time to enable repeated loading regimes.

1.6 Electrical Stimulation

In addition to the important roles that loading/unloading play in skeletal muscle hypertrophy/atrophy, electrical stimulation of muscle contraction is also key to its function. The delivery of electrical impulses to skeletal muscle is paramount for muscular development and muscle maturation. Early in vivo work demonstrated that type I and II fibers, upon being cut and cross aromatized, displayed the contractile characteristics of the nerve supplier, indicating the importance of electrical innervation in governing muscle phenotype [33]. In attempts to understand the impact of motor neuronal signals on in vitro muscle cell behavior, the electrical stimulation of monolayers and 3D tissue constructs has been used as a method for studying the molecular mechanisms that underpin muscular contraction and providing a model of muscle damage without the use of pharmacological agents.

Indeed, the application of chronic stimulation may be implemented to mimic exercise and study the associated mechanisms of adaptation in a biologically relevant model. Although it could be argued that electrical stimulation of monolayer cultures culminates in non-uniaxial force production (as a consequence of nonaligned myotubes) and is hence not representative of contraction in vivo. However, electrical stimulation of myoblast monolayers has been reported to elicit similar signaling responses to those evoked in skeletal muscle by exercise. These adaptations include increased AMP-activated protein kinase (AMPK), enhanced glucose metabolism, and fatty acid oxidation, in response to continuous low-frequency stimulation (reviewed in [27]). Furthermore, high-frequency stimulation of myotubes culminates in increased protein synthesis (for review *see* [27]). Therefore, since electrically induced contractile activity comprises a key component of skeletal muscle function, it warrants further investigation in cell culture models.

The first reported use of electrical stimulation in vitro involved the stimulation of primary cells isolated from the breast of 12- to 13-day-old chick embryos cultured for 4–5 days until myotube formation occurred before undergoing a 34 h stimulation protocol (0.6 s train of 10- to 20-ms biphasic pulses, every 4 s) [34]. This resulted in an almost 40% increase in total myosin accumulation alongside a 21% increase in contractile protein synthesis. Furthermore, electrical stimulation patterns of skeletal muscle isolated from newborn rat hindlimbs (mixed fiber type) designed to produce a more "endurance"-like response (250 ms burst, 15, 40, or 100 Hz repeated every 1 or 4 s) enhanced the abundance of slow myosin heavy chains, whereas "resistance"-like protocols (250 ms burst, 15, 40, or 100 Hz repeated every 100 s) enhanced abundance of faster myosin heavy chains [35]. Furthermore, chronic low-frequency stimulation of murine C2C12 skeletal myoblasts for 24 h (2 ms burst, 1 Hz, and 40 V) resulted in the activation of the AMPK cascade, which drives the mitochondrial adaptations to endurance exercise, highlighting skeletal muscle‘s plasticity in response to electrochemical stimuli [36]. The stimulation of 2D monolayers and 3D tissue constructs and how these can be used to study the mechanisms of skeletal muscle contraction are detailed in these reviews [27, 37].

The most common method for measuring the skeletal muscle responses to programmed electrical stimulation is via the use of IonOptix C-Pace EM® (IonOptix, Dublin, Ireland) and its predecessors, which distributes an electrical current through the cell culture medium via a series of carbon electrodes, which protrude into the well of the cell culture plate, generating uniformly distributed electrical current across the cell population. The amplifiers can produce a voltage output of ±40 V, a frequency of 0.01–99 Hz, and a pulse duration of 0.4 to 24 ms, which allow for the study of endurance-like low-frequency stimulation and resistance training-like high-frequency stimulation protocols. Non-commercial, in-house stimulators are also typically used due to their cost-effectiveness and consist of a conductive electrode present within the dish (Fig. 3).

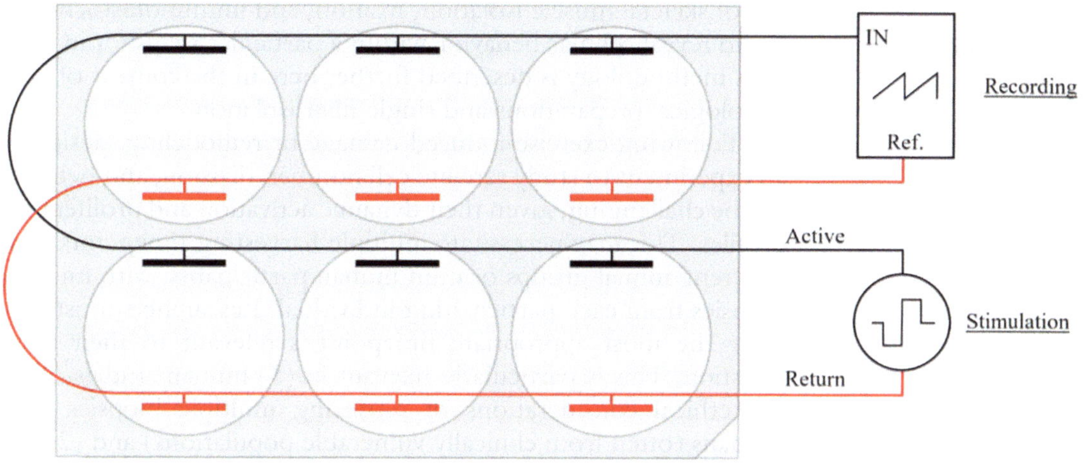

Fig. 3 Diagram of an example in-house programmed electrical stimulator with an output running from the stimulator through to the oscilloscope and returning via the input creating a uniform electrical stimulation in each well

1.7 Introduction to the Study of Satellite Cells In Vivo

To overcome the nuances of removing satellite cells and their dividing progeny into culture, it is possible to study satellite cells directly from their host in the in vivo environment. Intravital imaging allows for the direct observation of a multitude of organs and their resident cells within living animals, including the liver, brain, heart, and skeletal muscle. Intravital imaging of skeletal muscle is predominately performed on the easily accessible dorsiflexors at the front of the lower hindlimb performed on anesthetized animals, with optical access made to the muscle through an opening within the skin [38–41]. The combination of higher resolution microscopy and engineering of fluorescent reporter gene constructs (Green fluorescent protein (GFP), mCherry), as well as luminescent reporters (luciferin-luciferase assays), to a protein of interest is advancing how scientists study cells in their in vivo environment. This technology enables the study of cell dynamics and motility, cell–cell interactions, and cell distribution and localization within skeletal muscle, as well as cell death, division, and differentiation. In addition to being able to observe cells/proteins of interest during an acute period, repeated and longitudinal imaging is also possible without repeated surgery or the closing and opening of sutures given the development of intravital imaging windows that allow repeated optical access to the same cells or nuclei [39, 42]. While novel and with great potential, this methodology is not yet widely available across laboratories for the study of skeletal muscle, and while it is important for readers to be aware of its development, it will not be described further.

1.8 Immunohistochemistry Considerations

More commonly and more widely accessible to researchers is the use of skeletal muscle isolation, fixation, and immunohistochemistry to reveal cellular behavior within a particular snapshot in time. This methodology is described further here in the context of both histological preparations and single fiber isolation.

Following exercise-induced damage or remodeling, designing an experiment to study satellite cell number, division, and behavior can be challenging, given their dynamic activation and proliferation profiles. This may necessitate multiple harvesting timepoints from different animal groups or even human participants, with multiple biopsies from each participating individual. Researchers must consider the most appropriate timepoint/s relevant to their study question. This is particularly relevant in (1) human studies, given the ethical considerations of retrieving multiple biopsies from humans (often from clinically vulnerable populations) and (2) animal models, where all work should also consider the 3R's framework for the replacement, reduction, and refinement of using animal models in biomedical research. Strict care should be taken under the supervision of the Animals (Scientific Procedures) Act 1986 and approved by the British Home Office if undertaken within the UK or by the European Guidelines for the care and use of laboratory animals if conducted within Europe.

1.9 Histological Analyses

The most common method used to study satellite cells in their in vivo environment is via fixation of muscle tissue and preparation of ultrathin slices, which can be chemically stained or immunofluorescently labeled to reveal particular proteins of interest, such as the paired box transcription factor family member, Pax7, present in muscle satellite cells. This method can be completed along with the immunohistochemical identification of specific myosin heavy chain isoforms (I, IIA, IIX, and IIB) to determine the number of satellite cells associated with different fiber types. This provides essential information relating to the cells within their natural environment compared with cells isolated for culture, where details of the biopsy origin may be lost.

Tissue is fixed in formalin (usually 10% formaldehyde in water), subsequently embedded in paraffin wax, and stored at room temperature until further analyses. This method while quickly penetrating and fixing tissues, therefore preserving morphology, has several limitations. Along with the associated health risks of paraformaldehyde use, formalin fixation also cross-links all proteins within the muscle, which can subsequently affect immunoreactivity with target-specific antibodies, especially when using monoclonal antibodies that recognize only one epitope of an antigen of interest. While antigen retrieval is possible through heating or enzymatic procedures, it can be timely and is a difficult process to optimize.

Alternatively, human biopsies can be aligned in parallel; embedded in optimal cutting temperature (OCT) compounds such as Tissue-Tek (Sakura Finetek Europe, Zoeterwoude, the Netherlands) on a cork disk and snap frozen, with agitation (to prevent ice crystal formation) in isopentane (2-methylbutane); and precooled in liquid nitrogen before storage at −80 °C for future use. This can also be performed on whole muscles, isolated from small animals following euthanasia, usually by cutting out the mid-belly of the muscle and completely embedding it in OCT/Tissue-Tek while placed on a cork. Muscle samples can be laid out perpendicular to the cork disk for transverse sectioning or in parallel for longitudinal sectioning. Freezing muscle tissue in this manner allows tissue to be preserved quickly without damaging morphology while also preventing the actions of proteases that may degrade proteins of interest.

1.10 Identifying Satellite Cells and Myonuclei in Histological Muscle Samples

Investigating the role of satellite cells and their fusion to existing myofibers during hypertrophic events has received considerable attention, with a debate existing as to whether satellite cell accretion is obligatory for hypertrophy across the life span [43]. Difficulties can arise between the identification of mature myonuclei, cells resident within the sarcolemma of the muscle fiber, and the muscle satellite cell that rests between the sarcolemma and basal lamina. The identification is also made more difficult due to the close proximity of interstitial cells, including vascular/endothelial, stromal, and inflammatory cells. The identification of myonuclei has traditionally been completed by careful observation of nuclei location under light microscopy, identifying as a myonuclei only if it lays beneath the sarcolemma, usually demarcated by labeling with laminin or dystrophin. This method is, however, time-consuming and brings in observer bias, calling for the need of a myonuclear-specific marker to study the number and location of myonuclei when investigating muscle plasticity following, for example, exercise, disuse, and in muscle fusion-defective myopathy research. Current proposed "specific" markers include pericentriolar material 1, (PCM1) [44] and transmembrane protein 38A (TMEM38A) [45]. However, it is unclear how markers such as PCM1, which is ubiquitously expressed, may affect myonuclear counting due to its localization in non-myogenic nuclei, which reside in close proximity to the muscle fiber membrane and the myonuclei themselves.

2 Materials

2.1 Adult Human Satellite Cell Isolation

Following ethical approval and informed consent, all procedures relating to human muscle biopsies must be performed according to aseptic practice and by individuals with relevant training. Prior to the biopsy being taken, approval also needs to be provided by the participants' physician to confirm that their current health status means that it is safe to do so and that the individual may indeed participate. Inclusion and exclusion criteria should be identified and adhered to, relevant (to the institution where the work is taking place) risk assessments should also be prepared and updated, as well as procedures and practices put in place to deal with adverse events, should they arise.

2.1.1 Preparation and Biopsy Harvest

When preparing for the biopsy, it is important to make sure that all work surfaces and the biopsy area are cleaned and sterile, that any packaged items are only touched on the outside, and that sterile protective surgical gloves and lab coats or aprons are worn. Additional protective equipment should be used in accordance with local and government guidelines. Equipment and consumables are required as follows: sterile wound cleaning and dressing packs containing swabs, disposable sheets and a tray, Hydrex Surgical Scrub (e.g., Ecolab©, UK), and disposable hospital razors. Anesthetic is 0.5% bupivacaine hydrochloride (local anesthetics can induce myotoxic effects, believed to be a result of increased intracellular calcium release and simultaneous inhibition of calcium reuptake) prescribed by a designated physician/medical doctor (allergies to be checked prior to administration). Fine, sterile, intramuscular injection needles and single-use, sterile 5 mL syringes are required, as well as single-use, disposable sterile scalpels, a sterile biopsy needle, or biopsy forceps, e.g., Monopty®, disposable core biopsy instrument (e.g., CR Bard, Crawley, UK), or a Weil-Blakesley conchotome (e.g., Gebrueder Zepf Medizintechnik, Duerbheim, DEU), respectively; if the latter is employed, this will require sterilization between each use—local hospital sterilization services can be used to ensure appropriate cleanliness between samplings. Following biopsy removal, the wound should be closed using 6 × 75 mm Steri-Strips and covered using 7.5 cm × 5 cm sterile wound dressing plasters with transparent waterproof film.

2.1.2 Biopsy Transfer

Muscle is placed in a 2 mL Eppendorf tube containing approximately 1.5 mL cold transfer medium stored on ice Ham's F-10 medium, 1 mM L-glutamine, 0.1% heat-inactivated fetal bovine serum (hiFBS), 0.1% heat-inactivated new-born calf serum (hiNBCS), 100 U/mL penicillin, 100 μg/mL streptomycin, and 2.5 μg/mL amphotericin. Note, hiFBS and hiNBCS are purchased from South America (EU approved) and New Zealand. It is important to maintain a uniform "origin of source" serum because of potential differences in the systemic hormonal environments (hiFBS from US vs. EU origin), which will impact cellular behavior. When possible, also try to batch match serum.

2.1.3 Human Primary Cell Isolation

To undertake this work, a scientific Herasafe Class II biological safety cabinet or equivalent is required and a cell culture humidified CO_2 incubator. Also required are sterile disposable scalpels (autoclaved at 122 °C for 30 min), fine mincing scissors, disposable sterile petri dishes (100 mm diameter), 15 mL and 50 mL flat-bottomed Falcon tubes or sterile 100 mL specimen containers, sterile (autoclaved at 122 °C for 30 min) 0.01 M phosphate-buffered saline (1× PBS) stored and utilized at room temperature, 0.5% Trypsin-0.2% EDTA at 37 °C, heat-inactivated horse serum (hiHS, to neutralize the trypsin) at 37 °C, autoclaved (122 °C for 30 min), magnetic stirring bars (6 × 35 mm), and heated magnetic stirring platform (set to 37 °C) to enable the tissue digestion. Also 2 mg/mL porcine skin gelatin (Type A; Sigma-Aldrich, G2500; Note: Do not use Type B gelatin) dissolved in distilled water, autoclaved at 122 °C for 30 min prior to use, and stored at room temperature; growth medium: Ham's F-10 medium, 1 mM L-glutamine, 10% hiFBS, 10% hiNBCS, 2 mM L-glutamine, 100 U/mL penicillin, 100 μg/mL streptomycin, 2.5 μg/mL amphotericin B, stored at 4 °C and pre-warmed to 37 °C prior to use, and sterile, vented T25 Falcon, Nunc, or Corning flasks are all needed to culture the isolated cells.

2.1.4 Cell Expansion and Storage

Sterile, vented T75 Falcon, Nunc, or Corning flasks; 2 mg/mL porcine skin gelatin (Type A; Sigma-Aldrich, G2500) dissolved in distilled water, autoclaved at 122 °C for 30 min prior to use, and stored at room temperature; 0.01 M phosphate-buffered saline (1×

PBS) stored and utilized at room temperature; 0.5% Trypsin-0.2% EDTA at 37 °C; Growth medium: Ham's F-10 medium, 1 mM L-glutamine, 10% hiFBS, 10% hiNBCS, 2 mM L-glutamine, 10 U/mL penicillin, 100 μg/mL streptomycin, 2.5 μg/mL amphotericin B stored at 4 °C and pre-warmed to 37 °C prior to use; cell culture grade dimethyl sulfoxide (DMSO); cryopreservation freezer container (e.g., "Mr. frosty"), −80 °C freezer; and liquid nitrogen storage facility with relevant risk assessments and handling protocols in place are required to expand and freeze cell cultures for experimentation and storage, respectively.

2.1.5 Cell Loading

Should the impact of mechanical loading wish to be studied, then one possible method utilizes 2D monolayer cultures and the Flexcell® FX-5000™ tension system (Dunn Labortechnik, Germany), introduced in Subheading 1.5. This system enables the application of predetermined equibiaxial, uniaxial, and gradient mechanical strain (cyclic or static) to cells, which are cultured on specialized fibronectin-coated (10 μg/mL in PBS, Sigma-Aldrich, UK) flexible-bottomed culture dishes (BioFlex®, Dunn Labortechnik, Germany). In 3D cultures, the TC-3 tension bioreactor system (EBERS Medical Technology, Spain—introduced above in Subheading 1.5) permits uniaxially stretching of up to approximately 15 bioengineered muscles at one time.

To achieve loading of 3D-engineered cultures, sylgard-coated (Sylgard™ 184 Elastomer Kit, Dow Corning, USA) 35 mm culture dishes are used (Easy-Grip, BD Falcon®, VWR, UK). Then, 2 × 6 mm silk suture threads (Ethicon Mersilk, 2.0, Portugal) are pinned 12 mm apart using 0.15 mm Minutien pins (Entomoravia, Czech Republic). The wells are cleaned with 70% ethanol and air-dried under UV for 1 h in a Class II hood. "Basal" Growth Medium (GM) is high glucose (4.5 g/L) Dulbecco's Modified Eagle's Medium, including 4 mM L-Glutamine (LG, Sigma-Aldrich), 10% heat-inactivated foetal bovine serum (hiFBS; SLS), and 10% heat-inactivated newborn calf serum (hiNBCS; Fisher Scientific, Denmark), supplemented with an additional 2 mM LG (Lonza), 100 U/mL penicillin, and 100 μg/mL streptomycin (PS; Lonza). Subsequently, 500 μL Growth Media containing 50 μL/mL thrombin (T4648, Sigma-Aldrich, UK), 8 μL/mL aprotinin (A3428, Sigma-Aldrich, UK), and 0.5 mg/mL 6-Aminocaproic acid (6AA) are added to the culture dishes and agitated to ensure coverage of each well. A total of 200 microliters of fibrinogen (20 mg/mL; F8630, Sigma-Aldrich, UK) is added dropwise to the culture well and left to incubate at room temperature for 10 min followed by 1 h at 37 °C to facilitate polymerization. Human-derived muscle cells are seeded onto the fibrin gels at 200,000 cells/mL in 2 mL GM containing 0.5 mg/mL 6AA and 50 μM of both L-ascorbic acid (LAA; A4403, Sigma-Aldrich, UK) and L-proline (LP; P8865, Sigma-Aldrich, UK). Cells are grown to ~90% confluency (changing the GM is every 48 h if required). Once confluent, GM is removed and differentiation media added (DM; same composition as GM with the exception of increased 6AA to 1 mg/mL and reduced serum to 2% using heat-inactivated horse serum, hiHS). Following 48 h, DM is switched to maintenance medium (MM; same components as GM with reduced 3.5% hiFBS and 3.5% hiNBCS serum and increased 1 mg/mL 6AA) for the remainder of the experiment.

2.1.6 Electrical Stimulation

Should the impact of electrical stimulation wish to be studied, then one possible method utilizes 2D monolayer cultures and the IonOptix C-Pace EM® (IonOptix, Dublin, Ireland). This system enables fully programmable, multistep sequence protocols, which are useful for cell-based exercise simulation. In-house built stimulators remain a cost-effective, viable alternative to commercially available kits, allowing for greater flexibility in the generation of personalized electrostimulation protocols. There is currently no commercially based system available to electrically stimulate 3D cultures. However, several groups have developed custom-made systems or modified commercially available systems, outlined in Subheading 1.5, that would enable electrical stimulation of 3D bioengineered cultures.

2.2 Histological Preparation

2.2.1 Histological Immunofluorescence

For the purposes of this chapter, we provide guidance for the immunohistochemical labeling of myonuclei and Pax7 positivity in satellite cells on transverse cross sections after they have already been sliced and placed on glass slides. For collection of sliced muscle cross sections, it is recommended to use Thermo Scientific™ SuperFrost Plus™ Adhesion slides (Fisher Scientific, Pittsburgh, PA) to improve section adhesion during antibody addition, washing, and any fixation/antigen retrieval steps. Antibodies should be diluted as recommended in Subheading 3 in freshly made immunobuffer (IB) or IImmunobuffer with Triton X (best kept for ~1 month maximum).

Immunobuffer (IB): PBS (10 mM phosphate pH 7.4, 150 mM NaCl), 50 mM glycine (Merk 1.02401_1000), 0.25% Bovine Serum Albumin (BSA), 0.03% saponin (Sigma 100 g S-7900), and 0.05% sodium azide. IB + Triton (IB + T): PBS (10 mM phosphate pH 7.4, 150 mM NaCl), 50 mM glycine (Merk 1.02401_1000), 0.25% BSA, 0.03% saponin (Sigma 100 g S-7900), and 0.05% sodium azide. For 500 mL of IB, add 0.5 mL of Triton X. Note, as Triton X is viscous, snipping 10 mm off the end of a 1 mL pipette tip with scissors can ease pipetting and help to ensure the appropriate volume is added [46].

Immunofluorescence incubations (Table 1) are best performed in damp, lightproof boxes (such as the StainTray 30 Place with Black Lid (Newcomer Supply, WI, USA)) to protect sections from

Table 1
Tested antibodies used on skeletal muscle cross sections from humans, rats, and mice

Primary Antibody	Secondary Antibody	Tested on: Human	Rat	Mouse
Pax7 (DSHB Supernatant-Kawakami, A.) 1:100	MIgG1, kappa light chain	✓	✓	✓*
Myogenin (Myog) (F5d) (DSHB Supernatant-Wright, W.E.) 1:100	MIgG1, kappa light chain	✓	✓	
MyoD (554130, BD Pharmingen, San Diego, CA) 1:100	MIgG1, kappa light chain	✓	✓	
PCM1 HPA023370 Sigma Aldrich (Merck) 1:1000	Goat anti-rabbit IgG H&L	✓	✓	✓
MANDYS8 (8H11) (DSHB Supernatant-Morris, G.E) 1:100	Goat anti-mouse IgG H&L	✓	✓	✓
Dystrophin (PA5-32388. Thermofisher Scientific) 1:500	Goat anti-rabbit IgG H + L	✓	✓	✓
Collagen IV AB769 (Merck Millipore, Darmstadt, Germany) 1:500	Donkey anti-goat IgG H&L	✓		
BA-D5 (Type I MHC) (DSHB Supernatant. Schiaffino, S.) 1:100	MIgG2b	✓	✓	✓*
SC-71(Type IIA MHC) (DSHB Supernatant. Schiaffino, S.) 1:100	MIgG1	✓	✓	✓*
6H1 (Type IIX MHC) (DSHB Supernatant. Lucas, C.) 1:100	MIgM	✓	✓	✓*
BF-F3- MIgM (DSHB Supernatant. Schiaffino, S.) 1:100	MIgM		✓	✓*

drying and to prevent fluorescent conjugates from bleaching. We recommend using an ImmEdge Hydrophobic Barrier PAP Pen (Vector Laboratories, CA, USA) for circling of tissue sections to decrease the antibody volume required. Nuclear counterstaining can be performed using 4’,6-diamidino-2-phenylindole (DAPI solution) (#62248 Thermo Fisher Scientific, USA).

Please note that batches of antibodies will have different working ranges for dilutions and should be tested individually. For antibodies against Pax7, Myog, and MyoD, it is advised that the tissue cross section is fixed with 4% freshly made formaldehyde for 5 min followed by epitope retrieval using a sodium citrate and polysorbate 20 (Tween) (10 mmol L^{-1}, pH 6.5, 0.05% Tween 20) at 90–95 °C for 30 min (*Requires appropriate mouse on mouse blocking kit).

2.2.2 Single Extracted Muscle Fibers Immunofluorescence

Zamboni fixative (Newcomer Supply, WI, USA) should be kept at 4 °C or made fresh on day of use with paraformaldehyde when fixing fresh biopsies or rodent muscles. During extraction and mounting of single fibers, it is recommended to use the finest ended pincettes/tweezers and that fibers are handled at their ends to avoid tearing or snapping.

3 Methods (*See* Subheading 2.1 for Materials Specific to Satellite Cell Isolation and Stimulation)

3.1 Adult Human Satellite Cell Isolation

All procedures must be conducted using aseptic techniques, by fully trained and appropriately insured (for the biopsy harvest) individuals. All human samples that are stored and/or disposed of must be handled in accordance with relevant government guidelines and safety legislation (e.g., in the UK, the Human Tissue Act (2004) and Control of Substances Hazardous to Health (2002) regulations, respectively). Health screening of participants for blood-borne viral infections is advised and inclusion/exclusion/safety criteria implemented as relevant. Researchers working with primary human cells in the lab should be vaccinated against hepatitis C. For a summary of the key protocols described below, *see* Fig. 4.

3.1.1 Preparation and Biopsy Harvest/Transfer

1. Following cleaning of the skin with an alcohol wipe and working in an aseptic manner, prepare to harvest the muscle biopsy using either a bioptic gun (this will yield ~20–30 mg sample from one pass) or a Weil-Blakesley conchotome (this will yield ~100–300 mg sample from one pass). For detailed materials of the biopsy harvest, *see* Subheading 2.1.1 above and the following reference [47].
2. Following biopsy removal, it should then be transferred on ice (*see* Subheading 2.1.2, Fig. 4) to a Class II cell culture hood for primary cell isolation.

3.1.2 Human Primary Cell Isolation

1. Before undertaking any work in the Class II hood, it is important to ensure it is clutter-free (only have items in the hood you plan to use), cleaned with 70% ethanol, and ideally UV-irradiated (if the hood has that capacity) for 20 min prior to use. Any items taken into the hood for cell isolation must be sterile, and the outer packaging must be sprayed down with 70% ethanol before being placed in the hood. The person isolating the cells must wear a lab coat and sterile gloves, which are sprayed with 70% ethanol every time before placing hands in the hood. Avoid using needles (where they are used, do not resheath prior to disposal in a sharps bin) and minimize the use of glass where possible.

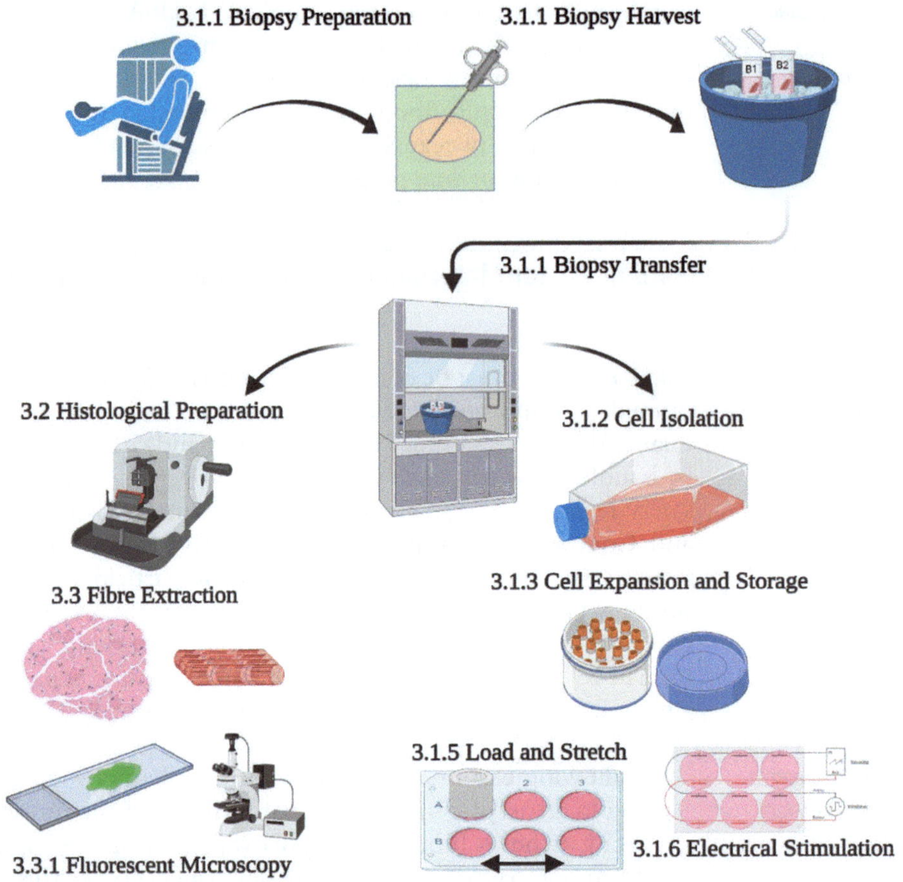

Fig. 4 Schematic of key methods detailed below. Created using biorender.com

2. With the hood set up and cleaned in advance of cell isolation, transfer the biopsy into an irradiated sterile petri dish, and remove any transfer medium. Wash the biopsy 3 × in 5 mL PBS at room temperature, or pre-warm to 37 °C. Do not use cold PBS as this will interfere with the enzyme activity of the trypsin. Use a hand pipette gun rather than a vacuum aspirator to remove the PBS between washes, to avoid the small biopsy from accidentally being aspirated to waste.
3. During the last PBS wash, remove any visible connective and/or adipose tissue using fine scissors (pre-autoclaved) or a sterile scalpel—take care with this step.
4. Remove the last PBS wash; add 4.5 mL trypsin-EDTA (pre--warmed to 37 °C) to the muscle biopsy, and finely mince it, using either two sterile scalpels or fine scissors for at least 1 min.
5. Transfer the trypsin-EDTA biopsy suspension to a flat-bottomed, sterile Falcon tube or specimen pot containing a sterile magnetic stirrer, and place it on a heated magnetic

stirring platform (with gentle mixing) at 37 °C for 10 min. If a heated stirrer is not available, place the mixing platform into an oven or incubator set at 37 °C, and mix the samples for 10 min.

6. While the sample is mixing, add 1 mL hiHS (pre-warmed to 37 °C) to a sterile 15 mL Falcon tube. After 10 min of stirring, transfer the trypsin-EDTA supernatant (using a pipette gun and 5 mL Stripette) to the Falcon tube containing the hiHS in order to neutralize the trypsin. Store at 37 °C in the CO_2 incubator. It is important that the remaining, undigested muscle tissue is not transferred as this will be put through the trypsin digestion process again.
7. Repeat **steps 4** (from adding the 4.5 mL trypsin-EDTA (pre--warmed to 37 °C) to the muscle biopsy and fine mincing) to 6 two more times (three times in total). Add the trypsin-EDTA cell suspension (after mixing) to the 15 mL Falcon tube in the incubator, which already contains the hiHS and the first batch of supernatant (produced in **step 6**, Round 1). At the end of three rounds of digestion and mixing, the Falcon tube in the incubator will contain ~14.5 mL fluid (1 mL hiHS plus 3 × 4.5 mL trypsin-EDTA cell slurries). If the sample digests very easily, it may be that only two rounds of digestion (not three) are required, resulting in 10 mL total fluid.
8. Centrifuge the Falcon tube containing the cell supernatant at 340 × *g* for 5 min (ensure that the sample is balanced in the centrifuge) at 4 °C. While the slurry is being centrifuged, any remaining muscle tissue, post-trypsin digestion, should be transferred to a pre-gelatinized T25 vented flask containing 7.5 mL growth medium (Fig. 4) and cultured in the incubator, without disturbing, for 5–7 days. This additional step will enable any cells, not successfully isolated, to migrate out of the residual muscle explant. If the digestion was successful, this additional step will yield few to no cells, and the explanted tissue can be disposed of according to local guidelines and legislation. If cells are liberated from the residual explant, they can be cultured and treated in the same way as those harvested from the cell slurry (*see* **steps 11–13** below).
9. A cell pellet will be visible at the bottom of the tube post-centrifugation. Carefully remove the cell supernatant (without disturbing the pellet), transfer it to a pre-gelatinized T75 vented flask containing 15 mL pre-warmed growth medium, and incubate, without disturbing, for 5–7 days. This additional step rarely yields any cells but is a backup precaution, in case the cell pellet and residual cells are lost, should the supernatant simply be aspirated to waste.
10. Resuspend the cell pellet in 7.5 mL pre-warmed growth medium and transfer it to a pre-gelatinized T25 vented flask.

Incubate, without disturbing or removing from the incubator, for 5–7 days. This flask will contain the majority of the isolated muscle cells.

11. Following this initial incubation period, top up the growth medium and return the plates to the incubator for a further 5–7 days. After 10 to 14 days, cells will be visible. At this stage, carefully remove the supernatant, wash 1× with warm PBS, and add fresh growth medium to the dishes (7.5 mL to a T25 and 15 mL to a T75). Monitor cell growth every 48 h; if cell density is low, top up the medium once per week, changing only weekly. If the growth is more rapid, change the medium (without washing) every 2–3 days.
12. Once cells reach approximately 80% confluency, they can be split, following PBS washing and trypsinization (*see* Subheading 3.1.3, **steps 1–3** below), to a single pre-gelatinized, vented T75 flask (or 3× pre-gelatinized, vented T25 flasks) to expand numbers for future experimentation. Ideally 1× T25 will liberate at least 500,000 cells. Before passaging further, freeze ¼ to ½ of the cells from the T25 as a backup. The culture of primary cells takes longer to establish large numbers than C2C12 cells; therefore, a risk of contamination and loss of cells prevails, the longer the process progresses. Having these cells stored in liquid nitrogen (*see* Subheading 3.1.3 below) at Passage 1 means that the biopsy is still a viable source of experimentation.
13. Repeat **step 12** (split one to three once confluency has been achieved) until sufficient cells are available for experimentation and additional cells have been frozen for longer term use at each passage stage. Other than for P1, any remaining frozen cells should be frozen at 5×10^5 to 1×10^6 cells/mL. Care must be taken when passaging cells, as a balance is needed between sufficient cells for the experiments, and over-passaging. Human skeletal muscle cells will become senescent if over-passaged, and the rate at which this occurs correlates well with the age of the donor. If only working with samples from young individuals, it is likely that cells to Passage 20 will not be senescent; however, if working with cells from older participants, it is encouraged to undertake the experiments between Passages 3 and 5. If comparing cells from younger and older participants, all experiments should be performed between Passages 3 and 5, unless senescence forms part of the question being addressed.

3.1.3 Cell Trypsinization for Expansion and Storage

1. Having attained approximately 80% confluency, cells are ready for (1) expansion, (2) plate for experimentation, or (3) cryopreservation (Fig. 4). As these are adherent cultures, cells need to be detached from the plates. There are different methods available; here, cell dissociation using trypsin is described.

2. Following pre-warming (37 °C) of PBS and trypsin-EDTA, cells are removed from the incubator, growth medium is removed by aspiration, and the cells are washed 2–3× with 10 mL PBS/T75 (5 mL/T25) dish. PBS should be gently swirled on the plate before aspiration between washes. Washes should be efficient (cells do not want to cool down too much during this process), and cells should not be left post-aspiration in the absence of any medium. Following washing, 1 mL pre-warmed trypsin-EDTA/T75 (0.5 mL/T25) is added to the cells, swirled to cover the bottom of the plate, and incubated at 37 °C for 5 min. To ensure effective detachment, gently agitate the cells and observe them down an inverted light microscope—cells should be rounded and floating in the trypsin-EDTA solution. If needed, return to the incubator for 1 min intervals to ensure effective dissociation.
3. Following detachment, add 4 mL growth medium to the dish. This will arrest the enzyme activity. If more than one flask/sample exists (e.g., cells have been expanded to more than 1× T75/biopsy participant), pool the cell suspensions (5 mL/T75 dish) within a sterile Falcon tube (15 mL or 50 mL, depending on the number of flasks pooled). Pipette the cells up and down using a 5 mL Stripette to ensure a homogenous suspension exists prior to counting. Once mixed, count straight away—if the cells are left before counting, they will settle to the bottom of the tubes, and an underestimate of cell numbers will be achieved.
4. Transfer an aliquot (e.g., 20–50 μL) of the cells into a microcentrifuge tube and add the same volume of filtered 1× trypan blue. Mix using a 20 μL pipette and count using a hemocytometer. When calculating the cell yield/mL, remember to consider the twofold dilution with the trypan blue. If cells are to be split, plate them at the appropriate cell number onto pre-gelatinized experimental dishes in fresh pre-warmed growth medium. If the cells are to be cryopreserved using the calculated cells/mL, add a relevant amount of fresh, pre-warmed growth medium to the suspension in the Falcon tube to yield a final cell count of 1×10^6 cells/mL, e.g., if 15 mL from three T75 dishes in the Falcon tube has a cell count of 2×10^6 cells/mL, then an additional 15 mL growth medium needs to be added to the Falcon tube to reduce the cell number to 1×10^6 cells/mL.
5. Prior to freezing, 10% (final volume) DMSO (added dropwise with swirling) needs to be added to the cell suspension to prevent formation of ice crystals during freezing and to therefore preserve cell viability. Do not be tempted to add the DMSO quickly as it will be toxic to the cells it contacts in high concentration. To calculate the volume of DMSO needed,

the following equation is helpful for a 30 mL cell suspension (*see* comment in **step 4** above) and a final concentration of DMSO at 10%:

x mL DMSO/(30 mL cells + x mL DMSO) = 0.1 (or 10%)

$x = 0.1 \times (30 + x \text{ mL DMSO})$

$x = 3 + 0.1x$

$0.9x = 3$

x = 3/0.9 = 3.33 mL DMSO added dropwise into 30 mL cell suspension

6. Once resuspended, transfer 1 mL cell solution per labeled (using a permanent marker or label printing system that withstands freezing in liquid nitrogen) cryovial—include researcher name, cell type/biopsy id, passage number, cell count/mL, and date. Transfer the labeled cryovials to a cryopreservation container (i.e., "Mr Frosty" Fig. 4) filled with isopropanol to ensure a gradual (~1 °C/min) decrease in temperature, and place the container into a −80 °C freezer for 24 h prior to storing cells in liquid nitrogen. Cells could be stored at −80 °C for short-term storage; however, for longer term preservation, storage in liquid nitrogen is advised.

3.1.4 Resuscitation of Cryopreserved Cells

1. Pre-warm growth medium to 37 °C, and pre-gelatinize one T25 vented Falcon flask per vial of cells. Ensure appropriate labeling of each flask to match the cryovial label and the participant ID.
2. Remove the cryovial(s) from the liquid nitrogen using appropriate Personal Protective Equipment (PPE) and in line with local regulations. Thaw the cells rapidly at 37 °C, and as soon as they are defrosted, transfer them to the T25 flask containing 5 mL pre-warmed growth medium. Gently move the flask (on the flat surface of the hood) up and down and left to right to ensure even distribution of cells across the culture flask.
3. Transfer the flasks to the incubator (37 °C) and allow them to grow to ~80% confluency; once confluency is attained, wash, trypsinize, count, and split as needed (e.g., to expand or for experimentation).

3.1.5 Cell Loading/Stretching (Fig. 4)

For Mechanical Loading of Cells in Monolayer

1. Myoblasts are grown to confluence in high (20%)-serum media, washed 3× with PBS, and then differentiated in low-serum (2%) media for 7 to 10 days on fibronectin-coated flexible-bottomed culture plates.
2. After confirming adequate myotube formation, resulting myotubes can be subjected to equibiaxial or uniaxial tension using the Flexcell® FX-5000™ tension system with high-frequency intermittent loading regimes or low-frequency continuous loading regimes to mimic resistance or endurance loading

patterns, respectively. Any outputs from loading conditions should be compared with non-loaded control myotubes also grown Flexcell® base plates with the vacuum entry sealed to any loading (*see* Fig. 1 in Subheading 1.5 above).

3. Example mechanical stretch regimes previously summarized in [47] are:
   - *Cyclic stretch* (one cycle, three sets × five reps, each set interspersed with ~30 s rest). Undertake these cycles with 5% strain for 2 days, 10% strain for 2 days, and 15% strain for 4 days in an attempt to model progressive overload that has been demonstrated to evoke myotube hypertrophy [30].
   - *Static stretch*. This is where the construct is stretched to 10% quickly (e.g., 0.4 mm/s resulting in 1.2 mm stretch of a 1.2 cm fibrin muscle after 3 s) and held at 10% stretch for 1 h to provide a model that would equivalent to synergistic ablation [22, 28].
   - *Continuous ramp load* at 0.35 mm/h, which is the same rate at which stretch stimulates in vivo bone elongation during development [48].
   - *10% intermittent stretch* × (one interval, four sets × ten reps at 0.4 mm/s during stretch and return to resting length), 10% stretch is calculated as 10% of resting length of a 3D bioengineered muscle or in monolayer 10% of the resting length of the substrate that the cells are grown on. Each set interspersed with 90 s rest with each interval separated by 3.5 min rest and repeated three to five times in an attempt to replicate a similar number of repetitions performed during a typical in vivo bout of resistance-type exercise that evokes hypertrophy (e.g., four sets × ten reps × three to five exercises per muscle group) [31, 47].

For Bioengineering and Loading of 3D Skeletal Muscle Using Fibrin as the Extracellular Matrix

1. Following growth and transfer to DM, after 48 h in DM, the fibrin muscle constructs are washed twice in PBS, and media is switched to maintenance medium (MM). This media is switched to fresh MM every 48 h, with additional 0.5 mL MM top up on days where media was not changed. The fibrin muscle cell constructs are then left until mature cylindrical-like muscles with uniaxial myotubes are observed (Fig. 5a–d), which occurs approximately after 14 days (sometimes longer in slow growing human cells). At this point, construct width is usually assessed using digital vernier callipers, where the width should be <4 mm at the narrowest point of the fibrin muscle construct. This is because these dimensions have been associated with the most mature myotube phenotypes [49–54]. Prior to loading and in addition to construct width measures, confirmation of myotube maturation can be undertaken

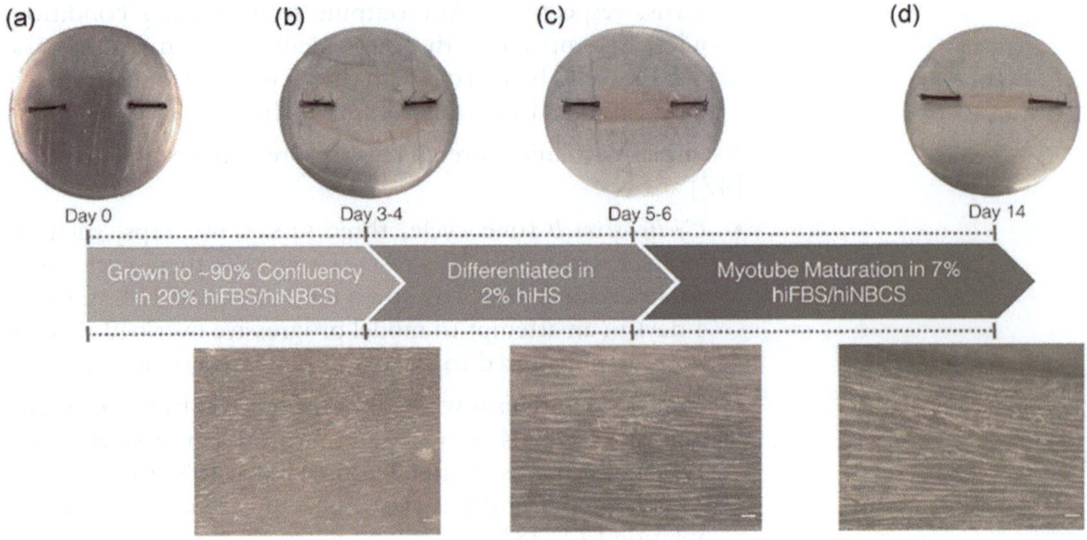

Fig. 5 Schematic representation of methods for bioengineering 3D fibrin skeletal muscle. Macro- and microscopic images of fibrin bioengineered muscle at (**a**) 0 days, (**b**) 3–4 days grown to confluency in media containing 20% heat-inactivated fetal bovine serum (hiFBS)/heat-inactivated newborn calf serum (hiNBCS), (**c**) 5–6 days differentiated in media containing 2% heat-inactivated horse serum (hiHS) for 48 h, and (**d**) myotubes matured up to 14 days in media containing 7% hiFBS/hiNBCS (10× magnification, scale bar = 50 μm). (The figure is taken from Turner et al. [31], as an open access article where permissions are not required, provided the work is properly cited. *The Journal of Cellular Physiology* (John Wiley and Sons))

by histochemical analysis for adult myosin heavy chains: myotube number, diameter, and area measures (by staining with desmin or phalloidin in addition to a nuclear stain), as well as RT-PCR for gene expression of embryonic/neonatal myosin heavy chain expression (which should have reduced) versus adult myosin heavy chain expression (which should have increased). Constructs once mature can often be seen spontaneously twitching under the light microscope. If the lab has access to electrical stimulation for bioengineered muscles, constructs can also be electrically stimulated to measure peak twitch force and tetanic force to confirm positive force-frequency relationships and normal length-tension relationships of skeletal muscle [55, 56].

Mechanical Loading of Bioengineered Skeletal Muscle

1. Once bioengineered muscles have matured, as described above, the TC-3 tension bioreactor system (*see* Fig. 2 in Subheading 1.5 above) can be used to mechanically load bioengineered constructs. Fibrin muscle constructs are then transferred to the bioreactor chambers and submerged in 20 mL MM (*see* Fig. 2 in Subheading 1.5 above). Chambers are then attached to the bioreactor system, which is housed in a humidified incubator (37 °C, 5% CO_2) ready to undergo mechanical loading. To provide relevant non-loaded controls, separate constructs are also kept at resting length within the same

chambers as the constructs to be loaded for the same duration of the experiments performed in loading conditions (*see* Fig. 2 in Subheading 1.5 above). Loaded constructs can then be subject to uniaxial mechanical loading regimes as described in detail above for monolayer experiments, e.g., cyclic stretch, static stretch continuous ramp loading, or intermittent loading.

3.1.6 Electrical Stimulation

1. Cells should be seeded onto pre-gelatinized Falcon, Nunc, or Corning 6-well plates at 80,000 cells per well in growth medium for at least 24 h (until 80% confluency) prior to 3× PBS washes and transferred to low-serum differentiation medium.
2. Following incubation for at least 72 h (confirm myotube formation), the stimulator plate should be placed gently into the cell media, ensuring the electrodes are within the media. Appropriate controls for experiments should consist of electrodes inserted into the cell medium without active current passing through to account for any environmental changes to the medium caused by the electrodes.
3. The stimulator should run through an oscilloscope, which allows viewing of the waveforms of the programmed electrical stimulation alongside providing recordable evidence of the experimental treatment. An example of an "endurance" protocol would be 13 V, 2 ms pulse duration, and 2 Hz continuous for 3 h, which has shown to upregulate glucose oxidative capacity and improve mitochondrial content and function [57]. A "resistance" protocol would be 15 V, 0.4 ms pulse duration with 4 s rest between each contraction, and 100 Hz for 30 min, which has shown to increase phosphorylation of Akt^{Ser473}, $mTOR^{Ser2448}$, and $ERK1/2^{Thr202/Tyr204}$ following stimulation [58].
4. Following each experimental treatment, care must be taken to ensure that the electrodes do not absorb salts, electrolysis by-products, and proteins from the media. Alternatively, if improperly maintained, toxic chemicals can leech from the electrodes into the cell media and cause unspecific effects and/or kill cells. Therefore, electrodes should be soaked in distilled H_2O for 2–3 days with daily water changes to ensure that hazardous by-products are not located on the plate. A measurement of pH can give an indication whether ions or other chemicals are leeching from the electrodes and causing subsequent damage to the cell population. Following prolonged stimulation, the electrodes can accumulate absorbed salts, which over time may cause damage; therefore, a soft toothbrush or sponge can be used to gently scrub the surface of the electrode under rinsing dH_2O to prolong the equipment.

5. Following cleaning, the stimulator needs to be sterilized before coming into contact with the cell population. Autoclaving is not recommended due to equipment sensitivity and the potential for irreparable damage. An alternative is to disinfect with 70% ethanol followed by UV irradiation under a sterile cell culture hood for at least 20 min. Ensure that ethanol has completely air-dried before reusing the stimulator plate.

3.2 Histological Preparation (for Materials, See Subheading 2.2)

1. Once ready to section, histological samples should be moved (on dry ice) to an appropriate cryostat with a freezing chamber set between −20 °C and −24 °C, where the samples should be allowed to equilibrate for 30 min before attempting to cut. Any thawing or freezing/refreezing will significantly affect the quality of the histological sample due to the formation of ice crystal damage. The cork disk and sample can be attached/mounted to the metal chuck with the use of water droplets or OCT.
2. Adjust the thickness setting of the cryostat between 5 and 12 μm depending on application, and trim the block by advancing the sample toward the blade until the surface of the sample and OCT are smooth, presenting the full cross section when producing transverse cross sections.
3. Samples can then be sliced and collected individually or in ribbons onto positively charged, pre-warmed slides, by carefully lowering the glass slide above the sections without contacting the metal roll plate. The muscle samples will be negatively charged, so they should jump/adhere to the glass slide without actual contact.
4. Freeze slides at −80 °C for future use to preserve the antigen integrity on the target proteins of interest.
5. Once appropriately sectioned, remove slides from −80 °C and bring to room temperature before circling tissue sections using a PAP Pen. When the barrier is dry, rehydrate sections in PBS using a Pasteur pipette while preparing fresh immunobuffer (IB) and immunobuffer with 0.1% Triton (IB + T) [46].
6. Remove PBS before adding antibodies of choice, diluted appropriately in fresh IB in accordance with the general guidelines in the materials (Table 1, Subheading 2.1.1) for commonly used antibodies employed in muscle and satellite cell research. Following pipetting of antibodies onto muscle sections, place the samples in a refrigerator overnight at 4 °C in a sealed, slightly moist container/slide box.
7. On the following day, remove slides from the refrigerator, and aspirate the primary antibody solution from the slide before washing sections in IB for 3 × 5 min, aspirating the solution carefully, between each wash. Appropriate secondary

antibodies (Table 1, Subheading 2.1.1) with fluorescent conjugates can be added to IB at predetermined concentrations and pipetted carefully back onto tissue sections. Note: This is best performed in a dark room under a red light to reduce the likelihood of fluorescent conjugates being bleached and reducing the signal intensity when bound to your protein of interest.

8. Incubate with secondary antibodies for 1–2 h, at 4 °C, in the dark, before repeating the IB washing steps, as outlined above (**step** 7). If required, nuclear counterstaining can be performed using DAPI solution, prior to further washing and the application of a mounting medium and coverslip.

9. Following successful staining, antigens of interest will be visible using a fluorescent microscope when equipped with an appropriate selection of secondary fluorophores and appropriate filter/laser sets. Once detected, the signal acquired from the fluorescent marker can be used to identify the location (cytosolic or nuclear), translocation (the movement of a protein between a control and experimental condition), and co-localization (observation of spatial overlapping as an indicative marker that two proteins are interacting) of proteins of interest (Fig. 6). Skeletal muscle researchers often choose to report the number of myonuclei or Pax7+/MyoD+/Myog+ cells per muscle fiber cross section as a measure of their expansion or loss.

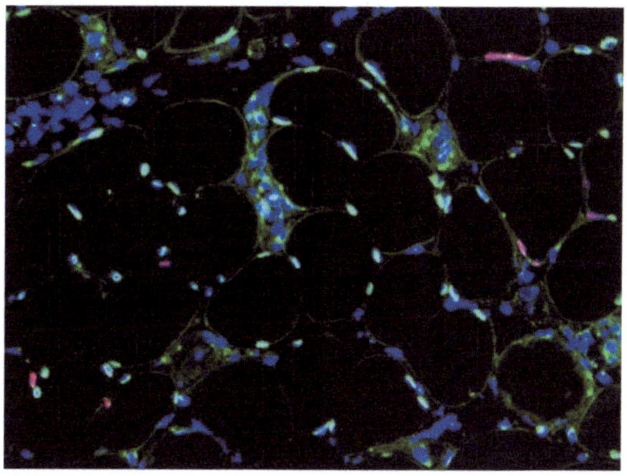

Fig. 6 Transverse section of a skeletal muscle biopsy, taken from a human vastus lateralis muscle, 7 days post-electrical stimulation-induced eccentric damage. The blue channel demarcates the labeling of DNA (nuclei), green demarcating pericentriolar material 1 (PCM1), a nuclear envelope protein present in myonuclei that sit inside the sarcolemma of the muscle fiber and at the centrosome of proliferating macrophages present in the muscle interstitial tissue and necrotic muscle fibers damaged from the eccentric protocol. The red channel labels satellite cells through present between the sarcolemma and basal lamina through the labeling of the Pax7 protein

3.3 Single Fiber Extraction

Whole rodent muscles or long intact bundles of muscle fibers present from a biopsy may also be used to study myonuclear and satellite cell content, usually expressed as the number of nuclei per mm of fiber length. This approach is generally more time-consuming and may not be representative of all muscle fibers within a muscle, which often vary in fiber type and size dependent on locality to the blood supply. For example, it is common to find oxidative portions within the deep regions of muscles, such as in the tibialis anterior and gastrocnemius. However, the capacity to study single fibers longitudinally also allows for better understanding of nuclear localization. The protocol below is advantageous to skinned or enzymatically separated muscle fibers as the muscle fiber connective tissue will remain intact, preserving the attached muscle fiber basement membrane and satellite cells.

1. To prepare single fibers from mice or rats following euthanasia, dissected muscles of interest should be pinned at their resting in vivo length with needles/pins through proximal and distal tendons into a Sylgard gel or silicone-coated 90 mm petri dish, (this usually requires slight tension). For human muscle biopsies, single fiber preparations should only be performed on long, intact bundles of fibers and if appropriate pinned at their resting length. Specimens should be kept moist during this process with the appropriate addition of either PBS or Hanks' Balanced Salt Solution (HBSS).
2. Once pinned, pour on Zamboni fixative if available or 2% formaldehyde in a fume cupboard. Zamboni fixative contains both formaldehyde and picric acid, so its use should be carefully planned with a risk assessment.
3. After 30 min of fixation, transfer the muscle or human fiber bundle into a 5 mL plastic vial (if large rat muscle) or a cryovial if mouse muscle/human biopsy fiber bundle containing fresh Zamboni fixative or 2% formaldehyde. Place vial into a refrigerator at 4 °C for a further 3 h before carefully removing chosen fixative with a Pasteur pipette or by carefully pouring away in line with the preprepared risk assessment.
4. Add an appropriate volume of 50% glycerol in PBS to the container to completely submerge the sample. Leave overnight in the refrigerator before moving to a −20 °C freezer the following day. Please note: Samples must be kept at −20 °C and not −80 °C, as glycerol will freeze if cooled to below −25 °C.

3.3.1 Immunohistochemical Labeling of Single Fibers

1. Once ready, muscles or fiber bundles can be moved to a 90 mm petri dish at room temperature containing IB and gently pulled apart with the use of super-fine surgical grade tweezers/

pincettes under a stereomicroscope. Once single fibers are separated from the bundle, they can be moved to a 24-well plate, containing 1 mL IB (stored at 4 °C, but used at RT) per well. For full, intact rodent muscles, it is advised to remove tendinous areas and thick fascia in order to be able to remove fibers easily. Once the fibers have been removed from the bundle for the desired analyses, samples can be returned to their 50% glycerol container and placed again at −20 °C for future use.

2. Antibodies can be added under guidance of Table 1 (*see* Subheading 2.1.1) to IB + T and pipetted into the wells containing the single fibers before overnight incubation at 4 °C. Researchers often also use fluorescently conjugated phalloidin, a bicyclic peptide produced by the death cap mushroom (*Amanita phalloides*) that labels all actin filaments to reveal muscle sarcomeres. α-Bungarotoxin (bound to a fluorescent conjugate), a neurotoxic protein, is also often used to identify neuromuscular junctions present on skeletal muscle fibers, as the snake venom from the genus *Bungarus* binds to nicotinic acetylcholine receptors (nAChRs).
3. Following overnight incubation, carefully aspirate the IB + T containing the primary antibodies and wash 3 × 5 min on a shaking plate at room temperature. Dilute the appropriate secondary antibodies (Table 1, Subheading 2.1.1) in IB and incubate for 2 h at room temperature in the dark. Ensure that the shaker is gently agitating the fibers but that the single fibers are not being lifted onto the upper part of the plate.
4. Wash 3 × 5 min in IB before changing the final solution to PBS. DAPI counterstaining can be performed as described for the immunohistochemical protocol above.
5. Once washed, single fibers can be individually lifted using super-fine surgical grade tweezers to a mounting medium on a glass slide. Please note: Take care and extra time to ensure that single fibers are aligned, side by side for ease of imaging.
6. Finish the protocol by placing a coverslip carefully down over the single fibers. Note that excess mounting medium will allow the single fibers to float around and cross over rather than being aligned neatly. Proteins/dyes of interest can be observed using a fluorescent microscope and appropriate filter/laser sets (Fig. 7).

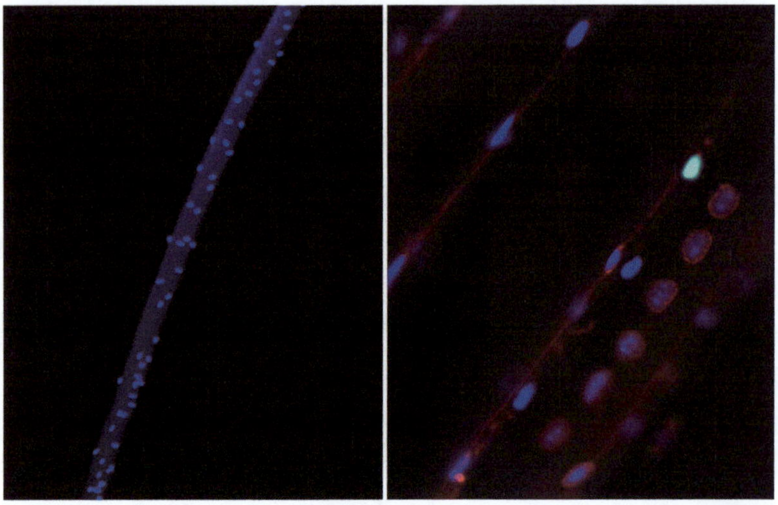

Fig. 7 *Left*: Single fiber extracted and labeled with DAPI to reveal the number of nuclei per mm of fiber length. *Right*: Single muscle fiber extraction from a mouse following synergist ablation-induced mechanical overload. In the middle fiber, myonuclei can be seen at the periphery of the muscle fiber with the labeling of DNA. To the bottom right sits a regenerating fiber, with central myonuclei, peripherally migrating, before relocating to the underneath of the sarcolemma. Their nuclear envelope is labeled with an antibody against PCM1 (red). A Pax7 satellite cell (cyan) sits between the two fibers, capable of dividing to support myogenesis indirectly or by fusing and donating its nuclei to the regenerating muscle fiber to support its growth and development

References

1. Stockdale FE, Nikovits W Jr, Christ B (2000) Molecular and cellular biology of avian somite development. Dev Dyn 219(3):304–321
2. Buckingham M et al (2003) The formation of skeletal muscle: from somite to limb. J Anat 202(1):59–68
3. Hawke TJ, Garry DJ (2001) Myogenic satellite cells: physiology to molecular biology. J Appl Physiol (1985) 91(2):534–551
4. Ono Y et al (2010) Muscle satellite cells are a functionally heterogeneous population in both somite-derived and branchiomeric muscles. Dev Biol 337(1):29–41
5. Cossu G, Biressi S (2005) Satellite cells, myoblasts and other occasional myogenic progenitors: possible origin, phenotypic features and role in muscle regeneration. Semin Cell Dev Biol 16(4–5):623–631
6. Messina G, Cossu G (2009) The origin of embryonic and fetal myoblasts: a role of Pax3 and Pax7. Genes Dev 23(8):902–905
7. Ontell M, Kozeka K (1984) Organogenesis of the mouse extensor digitorum logus muscle: a quantitative study. Am J Anat 171(2):149–161
8. Ontell M, Kozeka K (1984) The organogenesis of murine striated muscle: a cytoarchitectural study. Am J Anat 171(2):133–148
9. Jones DA, Round JM (2007) Muscle development during childhood and adolescence. In: The Young Athlete. Blackwell, pp 18–26
10. Schwann T, Hünseler F (1910) Mikroskopische Untersuchungen über die Ubereinstimmung in der Struktur und dem Wachstume der Tiere und Pflanzen. W. Engelmann
11. Weber C (1863) Ueber die Regeneration quergestreifter Muskelfasern. Vorläufige Mittheilung Centralblatt für die medicinischen Wissenschaften 34:529–531
12. Weber C, Virchow R (1854) Anatomische Untersuchung einer hypertrophischen Zunge nebst Bemerkungen über die Neubildung

quergestreifter Muskelfasern. Virchows Arch Pathol Anat Physiol Klin Med 7(1):115–125

13. Zenker FA (1864) Über die Veränderungen der Willkührlichen Muskeln im Typhus Abdominalis: nebst einem excurs über die Pathologische neubildung quergestreiften Muskelgewebes. FCW Vogel
14. Lewis WH, Lewis MR (1917) Behavior of cross striated muscle in tissue cultures. Am J Anat 22(2):169–194
15. Lash JW, Holtzer H, Swift H (1957) Regeneration of mature skeletal muscle. Anat Rec 128(4):679–697
16. Scharner J, Zammit PS (2011) The muscle satellite cell at 50: the formative years. Skelet Muscle 1(1):28
17. Mauro A (1961) Satellite cell of skeletal muscle fibers. J Biophys Biochem Cytol 9:493–495
18. Katz B (1961) How cells communicate. Sci Am 205:209–220
19. Abdelmoez AM et al (2020) Comparative profiling of skeletal muscle models reveals heterogeneity of transcriptome and metabolism. Am J Physiol Cell Physiol 318(3):C615–C626
20. Sharples AP, Stewart CE, Seaborne RA (2016) Does skeletal muscle have an 'epi'-memory? The role of epigenetics in nutritional programming, metabolic disease, aging and exercise. Aging Cell 15(4):603–616
21. Sharples AP, Seaborne RA, Stewart CE (2018) Chapter 19—Epigenetics of skeletal muscle aging. In: Moskalev A, Vaiserman AM (eds) Epigenetics of aging and longevity. Academic Press, Boston, pp 389–416
22. Goldberg AL (1967) Work-induced growth of skeletal muscle in normal and hypophysectomized rats. Am J Phys 213(5):1193–1198
23. Vandenburgh H, Kaufman S (1979) In vitro model for stretch-induced hypertrophy of skeletal muscle. Science 203(4377):265–268
24. Vandenburgh H, Kaufman S (1980) Protein degradation in embryonic skeletal muscle. Effect of medium, cell type, inhibitors, and passive stretch. J Biol Chem 255(12): 5826–5833
25. Baar K et al (2000) Autocrine phosphorylation of p70(S6k) in response to acute stretch in myotubes. Mol Cell Biol Res Commun 4(2): 76–80
26. Baar K, Esser K (1999) Phosphorylation of p70 (S6k) correlates with increased skeletal muscle mass following resistance exercise. Am J Phys 276(1):C120–C127
27. Kasper AM et al (2018) Mimicking exercise in three-dimensional bioengineered skeletal muscle to investigate cellular and molecular mechanisms of physiological adaptation. J Cell Physiol 233(3):1985–1998
28. Player DJ et al (2014) Acute mechanical overload increases IGF-I and MMP-9 mRNA in 3D tissue-engineered skeletal muscle. Biotechnol Lett 36(5):1113–1124
29. Eastwood M et al (1998) Effect of precise mechanical loading on fibroblast populated collagen lattices: morphological changes. Cell Motil Cytoskeleton 40(1):13–21
30. Powell CA et al (2002) Mechanical stimulation improves tissue-engineered human skeletal muscle. Am J Physiol Cell Physiol 283(5): C1557–C1565
31. Turner DC et al (2021) Mechanical loading of bioengineered skeletal muscle in vitro recapitulates gene expression signatures of resistance exercise in vivo. J Cell Physiol 236:6534
32. Seaborne RA et al (2019) UBR5 is a novel E3 ubiquitin ligase involved in skeletal muscle hypertrophy and recovery from atrophy. J Physiol 597(14):3727–3749
33. Salmons S, Sreter FA (1976) Significance of impulse activity in the transformation of skeletal muscle type. Nature 263(5572):30–34
34. Brevet A et al (1976) Myosin synthesis increased by electrical stimulation of skeletal muscle cell cultures. Science 193(4258): 1152–1154
35. Naumann K, Pette D (1994) Effects of chronic stimulation with different impulse patterns on the expression of myosin isoforms in rat myotube cultures. Differentiation 55(3):203–211
36. Nedachi T, Fujita H, Kanzaki M (2008) Contractile C2C12 myotube model for studying exercise-inducible responses in skeletal muscle. Am J Physiol Endocrinol Metab 295(5): E1191–E1204
37. Nikolic N et al (2017) Electrical pulse stimulation of cultured skeletal muscle cells as a model for in vitro exercise—Possibilities and limitations. Acta Physiol (Oxf) 220(3):310–331
38. Bruusgaard JC et al (2003) Number and spatial distribution of nuclei in the muscle fibres of normal mice studied in vivo. J Physiol 551 (Pt 2):467–478
39. Bruusgaard JC, Gundersen K (2008) In vivo time-lapse microscopy reveals no loss of murine myonuclei during weeks of muscle atrophy. J Clin Invest 118(4):1450–1457
40. Konagaya Y et al (2020) Intravital imaging reveals cell cycle-dependent myogenic cell migration during muscle regeneration. Cell Cycle 19(22):3167–3181
41. Mercier L et al (2016) In vivo imaging of skeletal muscle in mice highlights muscle defects in

a model of myotubular myopathy. Intravital 5(1):e1168553
42. Jacquemin G et al (2021) Longitudinal high-resolution imaging through a flexible intravital imaging window. Sci Adv 7(25):eabg7663
43. Murach KA et al (2018) Starring or supporting role? Satellite cells and skeletal muscle fiber size regulation. Physiology (Bethesda) 33(1):26–38
44. Winje IM et al (2018) Specific labelling of myonuclei by an antibody against pericentriolar material 1 on skeletal muscle tissue sections. Acta Physiol (Oxf) 223(4):e13034
45. Cutler AA, Corbett AH, Pavlath GK (2017) Biochemical isolation of myonuclei from mouse skeletal muscle tissue. Bio Protoc 7(24):e2654
46. Mackey AL, Kjaer M (2017) The breaking and making of healthy adult human skeletal muscle in vivo. Skelet Muscle 7(1):24
47. Turner DC et al (2019) Exercising bioengineered skeletal muscle in vitro: biopsy to bioreactor. Methods Mol Biol 1889:55–79
48. Vandenburgh HH, Karlisch P (1989) Longitudinal growth of skeletal myotubes in vitro in a new horizontal mechanical cell stimulator. In Vitro Cell Dev Biol 25(7):607–616
49. Khodabukus A, Baar K (2009) Regulating fibrinolysis to engineer skeletal muscle from the C2C12 cell line. Tissue Eng Part C Methods 15(3):501–511
50. Khodabukus A, Baar K (2012) Defined electrical stimulation emphasizing excitability for the development and testing of engineered skeletal muscle. Tissue Eng Part C Methods 18(5): 349–357
51. Khodabukus A et al (2015) Role of contraction duration in inducing fast-to-slow contractile and metabolic protein and functional changes in engineered muscle. J Cell Physiol 230(10): 2489–2497
52. Khodabukus A, Baar K (2015) Glucose concentration and streptomycin alter in vitro muscle function and metabolism. J Cell Physiol 230(6):1226–1234
53. Khodabukus A, Baar K (2015) Contractile and metabolic properties of engineered skeletal muscle derived from slow and fast phenotype mouse muscle. J Cell Physiol 230(8): 1750–1757
54. Khodabukus A, Baar K (2015) Streptomycin decreases the functional shift to a slow phenotype induced by electrical stimulation in engineered muscle. Tissue Eng Part A 21(5–6): 1003–1012
55. Dennis RG, Kosnik PE 2nd (2000) Excitability and isometric contractile properties of mammalian skeletal muscle constructs engineered in vitro. In Vitro Cell Dev Biol Anim 36(5): 327–335
56. Huang YC et al (2005) Rapid formation of functional muscle in vitro using fibrin gels. J Appl Physiol (1985) 98(2):706–713
57. Tamura Y et al (2020) Electrically stimulated contractile activity-induced transcriptomic responses and metabolic remodeling in C2C12 myotubes: twitch vs. tetanic contractions. Am J Physiol Cell Physiol 319(6): C1029–C1044
58. Valero-Breton M et al (2020) Acute and chronic effects of high frequency electric pulse stimulation on the Akt/mTOR pathway in human primary myotubes. Front Bioeng Biotechnol 8:565679

Chapter 14

Using the Model Organism *Caenorhabditis elegans* to Explore Neuromuscular Function

Samantha Hughes and Nathaniel Szewczyk

Abstract

The 1 mm long worm *Caenorhabditis elegans* is a free-living nonparasitic nematode found around the world. It offers a simple example of animal biology. The striking level of homology between *C. elegans* and mammals at the genetic, cellular, tissue, and organ level means that the worm offers a unique opportunity to explore various aspects of development. By working with *C. elegans*, it is possible to link molecular, cellular, and tissue level mechanisms with movement and behavior. This chapter aims to provide a brief introduction to *C. elegans* and describes the main approaches that assess the physical performance of *C. elegans* using methods analogous to those used in human performance assessment, such as cardiorespiratory endurance, muscle strength, and body composition. Recent advances have allowed such measurements to be obtained in nematodes, and the experiments described here allow physical performance or fitness to be measured in *C. elegans* as a proxy for humans, which also provides a means by which to investigate how age, diet, drugs, and the environment may affect muscle biology. This chapter provides an introductory guide to those who are not currently in the nematode field but are interested in using *C. elegans* to explore the structure and function of muscle with a view to elucidate human neuromuscular function, which is essential if we are to fully understand human health and disease.

Key words *C. elegans*, Mitochondria, Muscle, Oxygen consumption, Movement, Burrowing, Pluronic gel, Mobility

1 Introduction

Caenorhabditis elegans was introduced to the scientific community by Sydney Brenner in 1965 as a model to study development and neurobiology [1] and today has become the choice preclinical system with which to study aging [2, 3], innate immunity [4], toxicology [5], drug discovery [6], and human disease [7, 8]. Due to the high translational relevance of *C. elegans* and the ability to use it to predict effects in higher animals, the nematodes are regularly used in pharmacological and toxicological research [6, 9–14].

Philip J. Atherton and Daniel J. Wilkinson (eds.), *Neuromuscular Assessments of Form and Function*, Neuromethods, vol. 204, https://doi.org/10.1007/978-1-0716-3315-1_14,

1.1 Caenorhabditis elegans

C. elegans is a free-living nonparasitic nematode found around the world and offers a simple example of animal biology [15]. It is relatively easy and cheap to maintain in large numbers in the laboratory. The 1 mm long self-fertilizing hermaphrodite is maintained in the laboratory on agar plates containing the bacterial food source *Escherichia coli OP50* [1]. The development of *C. elegans* from an embryo to an egg-laying adult takes 3–4 days at 20 °C, during which a single nematode will produce approximately 300 genetically identical progeny; the worm will live for a further 2–3 weeks until death [16]. *C. elegans* has a precise and predetermined anatomy comprising of just 959 cells, but significantly, the nematode does have many tissues and organs found in other higher eukaryotes (Fig. 1a, b). *C. elegans* is biologically well characterized, with remarkable conservation to humans at the genetic and molecular level [9, 17, 18]. Many of the physiological and stress responses observed in nematode worms are the same as in mammals; for example, during growth and development, there is striking conservation of the developmental steps between *C. elegans* and vertebrates, including cell division, differentiation, migration, and cell

A B C

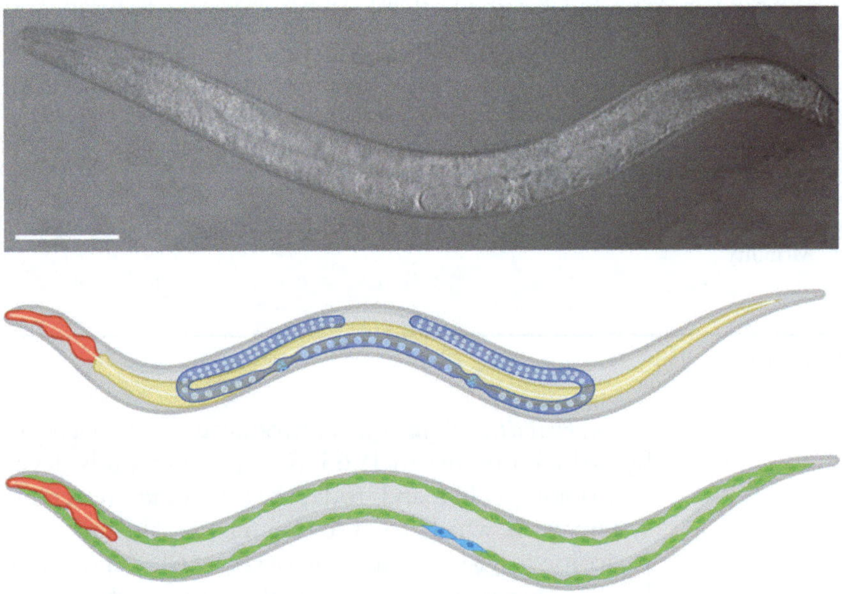

Fig. 1 *Caenorhabditis elegans* and its musculature. (**a**) Brightfield image of *C. elegans*. The transparent cuticle allows easy visualization of the internal structures including the intestine and the germline. Anterior is to the left, ventral to the bottom. Scale bar is 100 μm. (**b**) Schematic diagram of the nematode showing the gonad arms (blue), intestine (yellow), and pharynx (red). (**c**) Schematic showing the major muscle groups in *C. elegans*. The body wall muscles (green) are striated and the functional equivalent of mammalian skeletal muscles. The 20 pharyngeal muscles (red) and vulval muscles (blue) are nonstriated muscles, along with stomato-intestinal, anal, and uterine muscles (not shown). (Images were generated in BioRender and adapted from Fox et al. [93])

death [19]. The genome of *C. elegans* has been fully sequenced [20], which has revealed significant genetic and proteomic homology to humans [21, 22]. Due to the sequencing effort, many loss- and gain-of-function mutants are available through an United States National Institutes of Health (NIH)-supported stock center at the University of Minnesota, which collects, maintains, and distributes strains (http://cbs.umn.edu/cgc/). Further, the investigation of single and combined gene knockdowns is possible in *C. elegans* through RNA interference (RNAi) by directly injecting dsRNA into the gonad, by soaking worms in a solution containing dsRNA, or by feeding dsRNA producing bacteria to *C. elegans* [23–25]. The latter option is the most time- and labor-effective method as well as being the simplest method. Due to its transparency, *C. elegans* is ideal for use with fluorescent reporters to enable the visualization of specific proteins, structure of organelles, and gene expression patterns [26]. Consequently, the use of RNAi, mutants, and reporter strains meaning *C. elegans* is an ideal tool for the in-depth elucidation of molecular pathways that may further the understanding of human health.

A comprehensive description of *C. elegans* anatomy and a complete neuronal diagram of the worm have been available since 1986 [27]. More recently, the connectome of both hermaphrodites and males has been completed [28], making *C. elegans* the only example where the nervous system has been fully mapped at the cellular level. The 302 neurons acting on 143 muscle cells [27] define a catalogue of behaviors, each with a devoted motor circuit and muscle group, e.g., locomotion [29], egg-laying [30], feeding, and defecation [31, 32]. The well-described behaviors of *C. elegans* combined with the detailed neuroanatomy enable readouts of the nervous system at the level of the whole organism so that individual neurons involved in different behaviors, including mechanosensation, chemosensation, and thermosensation, have been identified. Despite being highly complex, such sensorimotor integration is conserved across systems, and *C. elegans* provides a unique opportunity to understand the causal link between neural activity and physical behavior [33].

1.2 C. elegans Musculature

C. elegans is a highly useful organism to study the development of muscle, due to its high similarity of the body wall muscle to the vertebrate muscle. The protein components of the nematode sarcomere have vertebrate counterparts, although there are some slight differences in composition and organization [34–36]. The body wall muscle of *C. elegans* is obliquely striated muscle cells (a total of 95 cells in four quadrants) with highly organized sarcomeric assemblies of actin, myosin, and other accessory proteins, many of which have homologs in mammalian cardiac and skeletal muscles [37]. However, *C. elegans* body wall muscles are not multinucleated, lack muscle satellite cells, and extend out to neurons via muscle arms as well as there being no distinction between slow- and fast-twitch fibers. The regular contraction and relaxation of the muscle cells result in the sinusoidal movement of the worms, which can be observed as crawling on an agar plate or thrashing in a liquid [38]. The conservation of nematode musculature with mammals goes beyond the striated muscle, with nonstriated muscles also present in the nematode (Fig. 1c). Nonstriated muscles in *C. elegans* consist of 20 pharyngeal muscle cells, 2 stomato-intestinal muscles, 1 anal depressor muscle, 1 anal sphincter muscle, 8 vulval muscles, 8 uterine muscles, and 10 contractile gonadal sheath cells. Due to the parallels between *C. elegans* and humans, studies using the nematode to explore organization, metabolism, and function of muscle provide an understanding of the regulation of human muscle. Further, through the use of genetic mutants, it is possible to explore the underlying causes of myopathies in diseases, including heart disease [39].

Approaches that assess the physical performance of *C. elegans* using methods analogous to those used in human performance assessment are essential if we are to fully understand human health and disease. Therefore, this chapter will provide an introductory guide to those interested in using *C. elegans* to explore the structure and function of muscle but are currently working outside the field.

2 Materials

2.1 Reagents

Many sources are available that detail the specific methods and reagents required for general *C. elegans* husbandry [40, 41], and the *C. elegans* community has excellent online platforms for additional information (wormbase.org, wormatlas.org, and wormbook.org). Table 1 lists a summary of the key reagents required for working with *C. elegans*.

2.2 Obtaining Nematodes

C. elegans strains and the bacterial food source are typically obtained from the *Caenorhabditis* Genetics Centre (CGC, USA) and maintained on nematode growth media (NGM), as described in Table 1 [41]. When solid, the NGM can be stored at 4 °C until needed or is seeded directly with *E. coli OP50* (*see* **Note 1**), which is allowed to dry overnight at room temperature before being used [1, 42]. Seeded plates can be stored at room temperature and used until they show signs of drying out, which is typically around 2 weeks.

2.3 Microscopes and Cameras

C. elegans is small in size with an invariant cell lineage, which makes it an ideal organism for imaging, especially as individual cells can be followed using differential interference contrast (DIC) microscopy. As it is transparent, fluorescent reporters can be easily observed in

Table 1
Solutions and buffers for *C. elegans* experiments

Specific reagents and solutions	Supplier	Required for
Platinum wire (0.2 mm)	Agar Scientific	Picking single worms
LB medium	Carl Roth	LB cultures
BactoAgar	Formedium	NGM
BactoPeptone	Organo Technie	NGM
Cholesterol	Sigma-Aldrich	NGM
5% sodium hypochlorite	Honeywell	Bleaching
Sodium Azide	Sigma-Aldrich	Preparing microscope slides
Pluronic® F127 Gel	Sigma-Aldrich	Burrowing assay
Serotonin	Sigma-Aldrich	Pumping assay
Buffers and media	**Ingredients**	**Required volume**
Phosphate buffer pH 6.0	108.3 g/L KH_2PO_4 35.4 g/L K_2HPO_4	25 mL
Cholesterol (5 mg/mL)	Dissolve in (98–99%) ethanol and store at −20 °C	1 mL
NGM	17 g BactoAgar, 3 g NaCl, 2.5 g BactoPeptone to 1 L with water. Sterilize by autoclaving and then add 1 mL 1 M $CaCL_2$, 1 mL cholesterol, 1 mL 1 M $MgSO_4$, and 25 mL phosphate buffer	1 L
M9-buffer	3 g KH_2PO_4, 6 g Na_2HPO_4, and 5 g NaCl to 1 L with water, sterilize by autoclaving, and add 1 mL 1 M $MgSO_4$	1 L
Bleaching solution	3.3 mL 5% sodium hypochlorite, 0.5 mL 5 M KOH, and 10 mL water	14 mL, per single bleaching reaction

Note: All solutions and buffers are made with sterile distilled water and then sterilized by autoclaving

the worm. Indeed, *C. elegans* offers exciting opportunities for whole organism imaging throughout the lifetime of the worm at single-cell resolution [43]. However, the majority of research is able to fully utilize *C. elegans* with basic equipment.

Light microscopy is useful for examining cells and cellular substructures, and consequently, visible light can be used to examine *C. elegans*. Benchtop light microscopes, such as the Lecia S8aP0 binocular with a KL300LED attachment, are ideal as this microscope offers a convenient working distance, high magnification, and large depth of field. To make short movies and simple images from this microscope, a Lecia DMC 2900 camera can be added to the system. To visualize fluorescent markers in a similar setup, worms have to be located using brightfield and then observed for a fluorescent signal, for example, with the Leica TL4000 RC. To observe more detail within the worm and take detailed images, a higher power microscope is required, such as a Zeiss Imager.M2 with an ×63 oil objective or for confocal imaging an inverted Zeiss Axio LSM700 equipped with a 488 nm diode laser and LP490 filter.

3 Methods

3.1 C. elegans Husbandry

To select and transfer worms, a 3 cm piece of platinum wire is inserted into a glass Pasteur pipette and sealed over a Bunsen burner. The other end is flattened as needed to provide a "scoop" allowing a single worm to be picked. Alternatively, "chunking" can be used to transfer a large number of worms. In this case, a sterilized spatula is used to cut out and remove a small square of NGM and transferring it to a newly seeded NGM plate, where it is placed upside down so the worms can crawl off the chunk onto the new plate.

To generate a synchronous population of worms, gravid worms were washed from plates with M9-buffer [1] (Table 1) and pelleted by centrifugation (2500 rpm, 90 s). The M9-buffer is removed and replaced with alkaline hypochlorite solution (bleaching solution, Table 1). The tubes are vigorously shaken for 3 min, and following centrifugation, the bleaching solution is replaced with M9-buffer. After washing away the bleach with M9-buffer a total of three times, released eggs are left to hatch overnight at room temperature in M9-buffer in the absence of food (*see* **Note 2**). The following day, the worms are pelleted by centrifugation and plated out onto *OP50*-seeded NGM as synchronized L1 larvae, which can be used in assays at the desired stage.

3.2 Preparing Worms for Microscopy

When using inverted microscopes, the worms must be mounted onto agarose pads. The pads are made by boiling a solution of 1% agarose in water in the microwave. When cool to touch, a small droplet of melted agarose is placed on a clean microscope slide using a Pasteur pipette, as shown in Fig. 2. Immediately, a second slide is placed over the top of the first at a 90° angle with the use of painter's tape, ensuring that the agar pad is just 0.4 mm thick. After about 20 s, the agar will have solidified. The slides can be separated by sliding them along each other, and the agar will stick to one of the slides.

The slide with the agar pad is turned so that the pad faces upward, and 2 μL M9-buffer is added to the center of the pad.

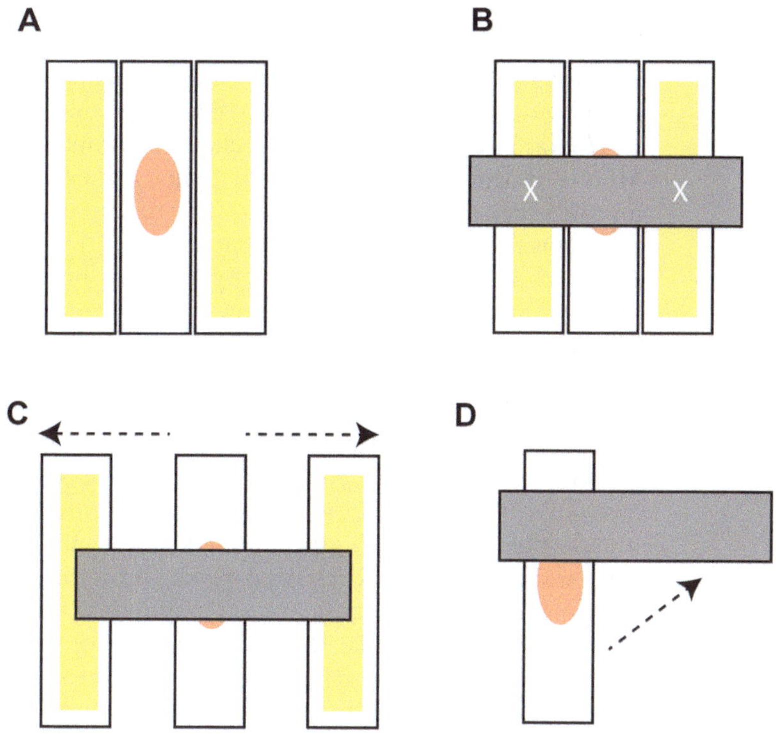

Fig. 2 Preparing slides for microscopy. (**a**) Place three microscopy slides slide by slide on a flat surface. The outer slides contain a small piece of painter's tape (yellow), facing up. A drop of molten 1% agarose (orange) is placed on the center slide using a Pasteur pipette. (**b**) Cover the agarose droplet with a fourth glass slide (grey) at a right angle to the other slides. Press down gently on the sides (indicated by a white "X") so that the agar drop is flattened to a circle of approximately 0.4 mm thick (the thickness of the tape spacers). Avoid getting bubbles in the agar pad. (**c**) Gently slide the two side slides apart in the direction of the arrows. (**d**) Slide apart the microscope slides so that the agar pad is attached to one. The pad should be used directly

Worms are selected and transferred into the droplet of M9-buffer using a worm pick as quickly as possible, taking care not to damage the pad or transfer bacteria. Then, 2 μL 1% sodium azide is added to the M9-buffer before placing a coverslip over the pad. Most of the worms will lie on their side and stop moving in 10–20 s. It is now possible to observe the worms using either an inverted microscope or a confocal. Once finished, the slides should be discarded in an appropriate manner.

3.3 Detailed Assessments of Neuromuscular Function

3.3.1 Sarcomere Morphology

Muscles regulate locomotion, posture, and metabolism, and so the development and maintenance of healthy muscle require the coordination of multiple cellular components and processes. Dysregulation as a consequence of aging and disease can be observed as altered muscle function and structure. Therefore, measuring aspects that are related to muscle deterioration, such as sarcomere disruption and mitochondrial function, can provide an indication of the mechanisms that underpin muscle development.

The body wall muscle of *C. elegans* is obliquely striated muscle with organized sarcomeric assemblies of actin, myosin, and other accessory proteins. The muscle thick filaments in *C. elegans* are composed of myosin, paramyosin, and associated proteins. *C. elegans* has four myosin heavy chain isoforms, with MHCA and MHCB in the body wall muscles. Using animals, which express a myosin green fluorescent protein (GFP) fusion under the control of the myosin heavy chain A promoter (*myo-3*P), such as the strain *RW1596*, it is possible to directly observe how the body wall muscles are organized and how they change over time in *C. elegans* (Fig. 3ai).

Individual worms can be mounted on an agarose pad in sodium azide/M9-buffer solution and the individual muscle fibers observed using fluorescent light (*see* Subheading 3.2). The morphology of the muscle fibers can be classified according to their level of (dys-)organization. Wild-type and young animals have muscle fibers that are highly ordered, linear, and parallel, with a clear and consistent fluorescence signal throughout (Fig. 3aii). As worms age, or as a consequence of disease, the sarcomeres begin to show disruption, in so far that the filaments become nonlinear and have a shortened appearance, termed "intermediate" (Fig. 3aiii), which continues until the fibers become highly shortened and irregular with an inconsistent fluorescent signal across the muscle fibers. At this point, the sarcomeric fibers are defined as "dysregulated" (Fig. 3aiv). It should be noted that occasionally, "wave-like" fibers are observed, but these are where the sodium azide has yet to take effect, and the animals are not fully paralyzed. Higher resolution images can be generated of the muscle structures using polarized light microscopy [44, 45], coherent Raman imaging [46], and structured illumination microscopy [47]; however, as these methods are highly challenging and labor-intensive and require specific microscopes, such descriptions are beyond the scope of this chapter.

3.3.2 Mitochondrial Morphology

Located within the striated muscle cells are the mitochondria, present in organized networks with the appearance of parallel tubes lying on top of (or beneath, depending on dorsal or ventral orientation) the dense bodies of myofilament I-bands [48]. Using a strain of *C. elegans* where mitochondria are visualized with a mitoGFP reporter in the body wall muscles (such as strain *SD1347*), it is possible to classify the structure of the mitochondrial network into three distinct phenotypes: linear, intermediate, and fragmented (Fig. 3b) [49]. It should be noted that using sodium azide as a means to immobilize the worms may induce mitochondrial fragmentation as sodium azide is a mitochondrial inhibitor. Alternative paralyzing agents can be used, such as levamisole or nanobeads, which allow for the mechanical immobilization of worms [50].

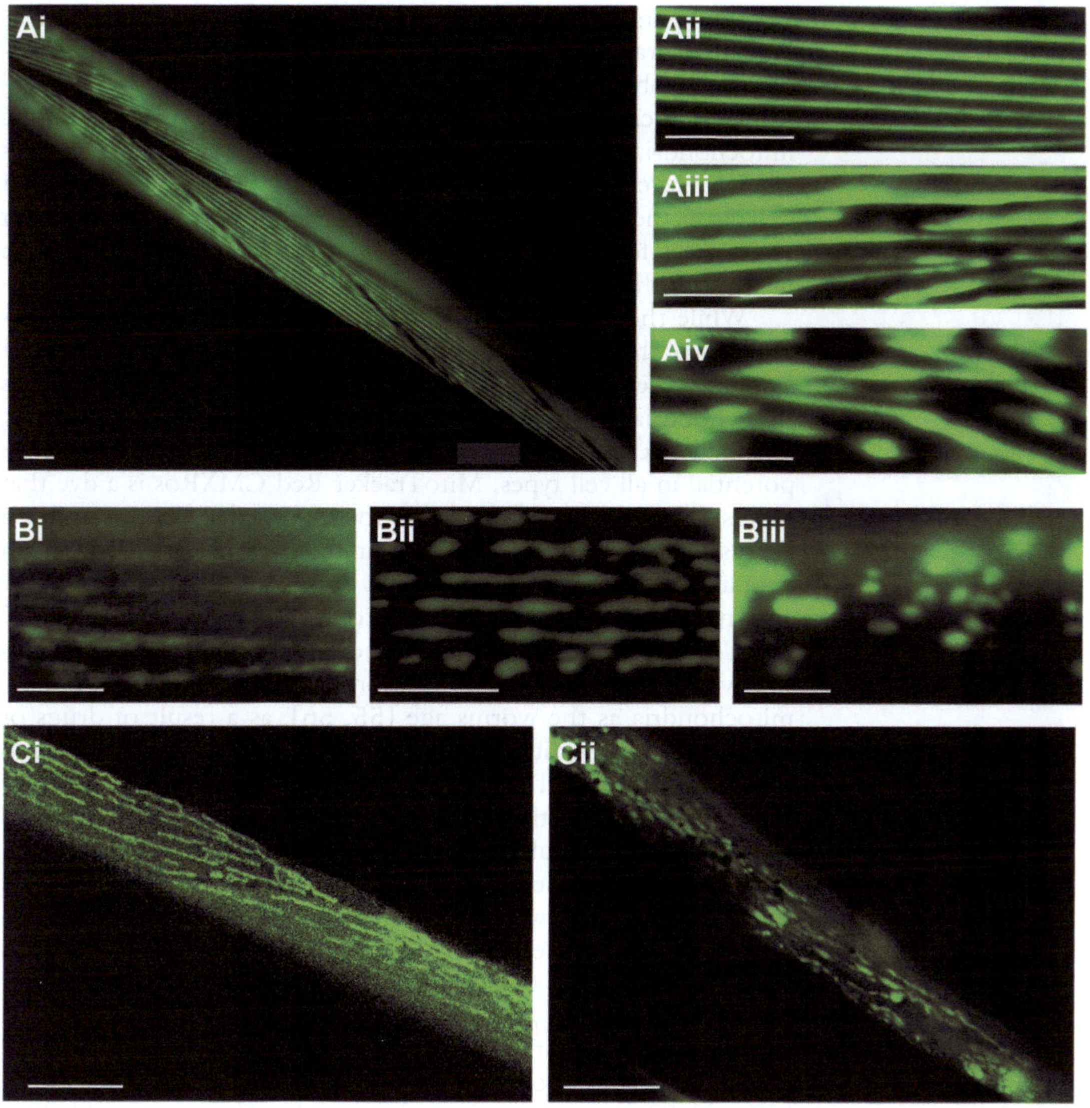

Fig. 3 Representative images of muscle and mitochondria morphology. (**a**) Muscle fibers are visualized using a worm strain, which expressed a *myo-3::gfp* reporter (strain, *RW1596)* using an inverted fluorescence microscope. (i) The muscles can clearly be observed along the length of the worm and can be categorized as (ii) linear, (iii) intermediate, or (iv) dysregulated. Scale bar in all images is 10 μm. (**b**) The strain *SD1346* expresses a mitochondrial GFP reporter driven in the body wall muscles and allows differences in the mitochondrial network to be observed. The network can be classified into (i) linear, (ii) intermediate, or (iii) dysregulated morphology. Scale bar in all images is 10 μm. (**c**) Observations of the mitochondrial network is best done using a confocal microscope. The complete network of the mitochondria can be observed and how it changes over time from (i) a young worm to (ii) an older worm. Scale bar is 500 μm

While an inverted fluorescent microscope provides rudimentary structural details of the mitochondrial network, increased resolution is obtained using confocal microscopy (Fig. 3c). Images from confocal images can be used to infer the volume of the mitochondria as well as the degree of network complexity, as has been achieved with human cells [51]. It should be noted that defects in the mitochondrial network are not necessarily sufficient to account for mobility defects and do not exclude issues with nerves or other tissues.

While the mitoGFP provides a means to determine the morphology of the mitochondria within the body wall muscles, it is not suitable to determine the state of mitochondrial function. To assess mitochondrial function in vivo, cationic dyes can be utilized, enabling the assessment of impaired mitochondrial membrane potential in all cell types. MitoTracker Red CMXRos is a dye that enters but does not exit the mitochondria [52] and has been successfully used to observe the changes in membrane potential in nematodes (*see* **Note 3**), specifically under stress conditions [53], and in a Duchenne's muscular dystrophy (DMD) model [54].

Using the muscle and mitochondrial GFP strains together, it is relatively simple to assess the morphology of the muscles and mitochondria as the worms age [55, 56], as a result of drugs or environmental conditions [57, 58] and in disease models such as that for DMD [54]. The number of worms displaying the morphologies in each condition can be quantified and used as a sound assessment for muscular health. Once the morphology of the mitochondria and sarcomere has been determined, it can be informative to find a correlation with the extent of movement defects. For this reason, such morphological studies are usually undertaken with mobility assays.

3.4 Mobility Assays

The analysis of relationships between environment, genes, and behavior depends upon reliable assays to detect changes in behavior [59, 60]. There are a number of approaches for automated behavioral analysis that have been developed, including, but not limited to, Wormtracker [61, 62], NeMo [63], OptoTracker [64], CoL-BeRT [65], and WormScan, which also enables the quantitative analysis of survival, movement, size, and reproduction [66]. All these systems cost between 300 and 16,000 USD and require various levels of skill to use.

Due to the high cost of automated analysis, many laboratories choose to manually count the number of head-to-tail body bends as a readout of mobility [67]. Single worms are picked onto an unseeded NGM plate, which is at room temperature to remove bacteria from the worm for 1 min. Then, after transferring a single worm to a droplet (~40 μL) of M9-buffer and leaving it to acclimatize for approximately 1 min, the number of head-to-tail bends is counted. While simple, such manual thrashing counts are labor-intensive, of low throughput, and maybe subject to observer bias [68]. More often, the number of body bends is counted from video recordings [69], and it is possible to quantify the number of body bends from such recordings using freely available software, thus reducing operator bias. Recordings can be analyzed using ImageJ with the wrMTrck plugin (build 110,622) [70]. After normalizing the frames to that of the highest light intensity and with background removal, movies can be analyzed such that each worm is given a number for body bends per minute (BBPM). These data can then be used to determine how well the worm moves in a liquid environment. A limitation of this process is that only a few worms can be observed at any one time, and so developments have been made to observe 100's of animals in parallel [68], but again, this requires access to suitable microscopes and an understanding of the Python script.

3.5 Burrowing Assay

The swimming/thrashing assessment of *C. elegans* locomotion described in Subheading 3.4 reflects locomotion in two dimensions, which is not representative of the natural environment of *C. elegans*, where worms move in three-dimensional space consisting of soil, rotten fruit, and water droplets [71]. Consequently, assays have been developed that make use of a gel environment in which the worms are able to burrow, therefore engaging in complex neuromuscular actuation [72, 73]. Pluronic® F127 is a biocompatible thermo-reversible hydrogel, and by placing worms in this gel, a nature-like environment forms that promotes a distinct movement [72] that is likely to be more physically demanding on the neuromuscular system due to the extra vector for movement. The optically transparent hydrogel allows for diverse applications, including the evaluation of muscle-defective mutants, separation and recovery of fast/slow burrowers for differential gene expression analysis, and testing disease models of protein aggregation and locomotory impairment.

In advance of the experiment, Pluronic® gel is prepared to 26% *w/w* with distilled water, using a clean and dry injection needle to mix the contents, ensuring all Pluronic® gel flakes are introduced to water. This will cause a lump of gel, but this is expected. Cotton buds should not be used when mixing the solution as this will affect the final percentage of the gel. When preparing the gel, glass beakers should be completely dry, or the gel viscosity will be affected, which in turn will affect the results. Once mixed, the beaker is covered with parafilm and left at 4 °C for at least 12 h before being used in an experiment, and thereafter care must be taken toward the handling of the gel during the experiment (*see* **Note 4**). On the day of the experiment, in excess of 90 staged worms per condition are picked onto an unseeded NGM plate at room temperature for 30 min to remove bacterial residue. During this time, a droplet of the gel (~30 μL) is placed on the bottom of each well in a 12-well plate into which worms ($n = 30$) are picked. Care must be taken to ensure that the total time taken to pick the worms does not exceed 30 min, as this will dry out the gel droplet and consequently impact the reliability of the results. Then, the wells are filled with Pluronic® gel (*see* **Note 5**). After a short period of time, the edge of the gel is gently ripped with a worm pick; if no healing of the gel is observed, the gel is considered solid. At this time, a bacteria/isoamyl alcohol solution is added. This solution is prepared by concentrating a standard *OP50* culture by centrifuging 1.5 mL of overnight culture in 1.5 mL Eppendorf tubes at 7000 rpm for 2 min. The supernatant is removed, leaving 60 μL. To this, 1% isoamyl alcohol in M9-buffer is added and the bacterial pellet resuspended. Then, 25 μL of bacteria/isoamyl alcohol solution is added to the surface of the gel, and this is the point at which the experiment begins, and time is recorded as $T = 0$. The number of worms at the surface of the gel is observed every 15 min over a 120-min period (Fig. 4a).

After 2 h, approximately 60% of the L4 stage worms (young worms) reach the surface of the Pluronic® gel (Fig. 4b). Strikingly, worms that are 48 h older display an enhanced rate of movement so that more worms reach the surface of the gel more quickly. This is likely due to the fact that nematodes exhibit peak movement rate in the first few days of adulthood [74]. As the worm ages, they move more slowly and with less vigor, with differences clearly seen at 7 days post L4. This decline in mobility will continue until death and is linked to a loss in muscle tone and decline in muscle structure [74].

3.6 Nematode Electropharyngeograms (EPGs)

The pharynx is a neuromuscular pump at the anterior end of the alimentary tract in *C. elegans* (Fig. 1b), for which proper feeding rate and the timing of the stereotypic pharyngeal movements are required for efficient feeding and survival. Pump rate and timing are controlled by a series of 20 neurons, representing 14 different neuronal types, which contain both neuropeptides and small molecule neurotransmitters. The pharyngeal muscle generates a myogenic rhythm due to the release of acetylcholine from cholinergic excitatory motor neurons (MC neurons) that is similar to the rhythmic motion in circulatory systems, such as the mammalian heart muscle [75]. In *C. elegans*, deviations from expected pumps per minute are a powerful and sensitive indicator of toxicity [76], environmental changes such as diet [77] or due to disease, as seen in the DMD model [78].

Counting the stereotypical motion of pharyngeal pumping is usually undertaken in a stationary environment with an abundance of bacteria on an NGM plate. Traditional feeding assays rely on manual scoring of the number of grinder movements (i.e., pumps) over a defined period of time, usually 3× 20 s, using a benchtop stereoscope [79]. However, counting pump number is time-

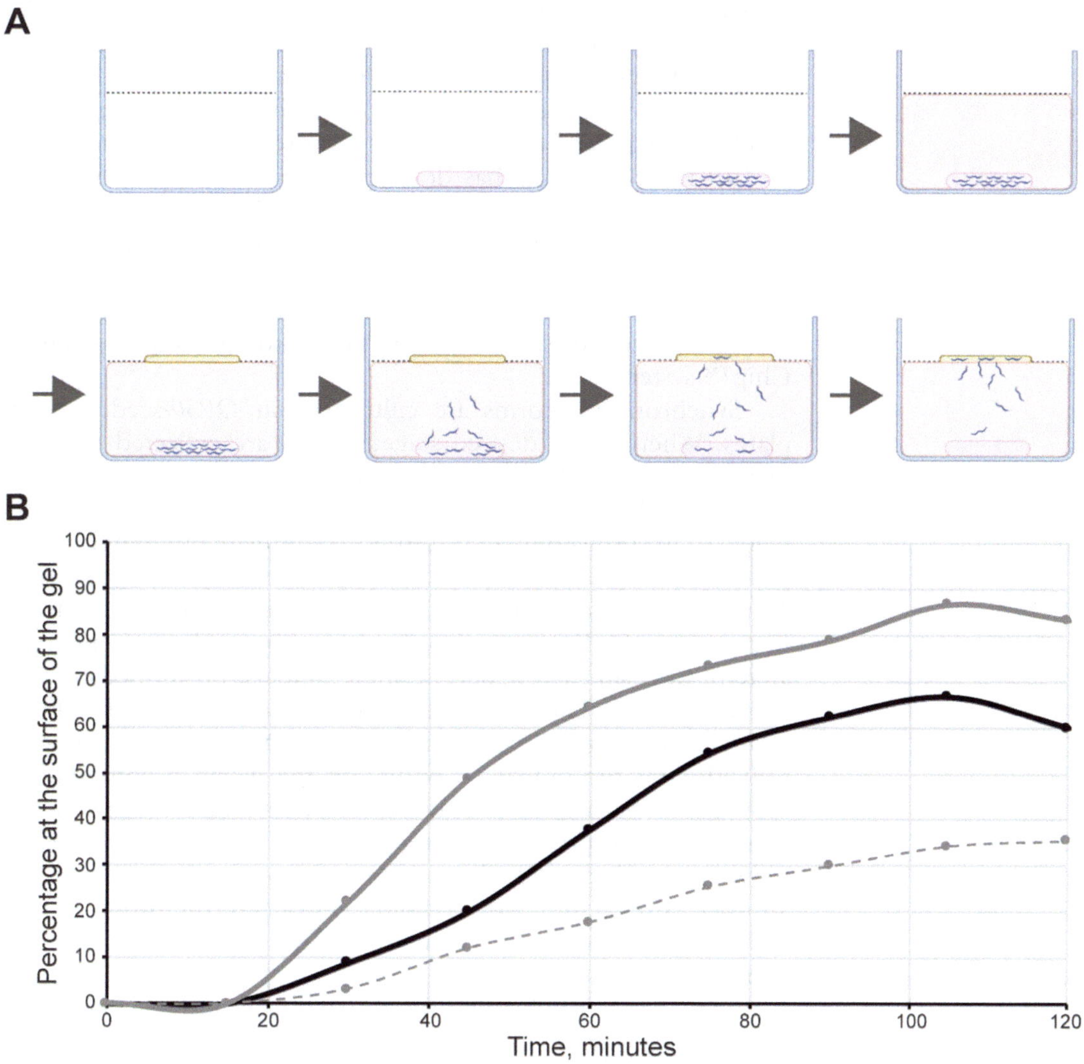

Fig. 4 Worms are able to burrow through Pluronic® gel mimicking a 3D environment. (**a**) Schematic representation of the setup of the Pluronic® gel experiment. In a 12-well plate, a line indicates the height of the gel. Worms are picked into a droplet of the Pluronic® gel at the bottom of each well, and the wells are filled with Pluronic® gel to the designated line. When a bacteria/isoamyl alcohol mixture is added to the surface, this is the start of the experiment ($T = 0$). The number of worms at the surface of the gel is counted every 15 min for a total of 120 min. (**b**) Graph to show the rate of worms migrating to the surface of the gel. Approximately 60% of young worms, L4 stage (solid black line), will reach the surface within 2 h. Worms that are 48 h older (solid grey line) are able to reach the surface of the Pluronic® gel much more quickly and in greater numbers. However, worms that are 1 week old (dashed grey line) struggle to reach the surface, with just 30% finding their way to the top after 2 h

consuming, and multiple tests must be undertaken due to variation and human error. In contrast, the use of electropharyngeograms (EPGs) allows the action potentials of pharyngeal muscles to be measured in live nematodes but is limited in the number of *C. elegans* that can be screened and is more labor intensive [75].

InVivo Biosystems has developed a ScreenChip™ system to directly measure EPGs and how they are affected by age, disease, or different compounds [77, 78, 80]. EPG recordings are taken using the ScreenChip™ system and the NemAquire™ software as described in the manufacturer's instructions. Here, we provide a general method to generate an EPG recording using the ScreenChip™ system.

Synchronized worms are cultivated on *OP50*-seeded NGM plates. When at the desired stage, worms are collected from the plates with M9-buffer into a 1.5 mL Eppendorf tube and washed twice with M9-buffer by gravity before being resuspended in freshly made 10 mM serotonin/M9-buffer to stimulate pumping. During this time, the ScreenChip™ can be prepared (*see* **Note 6**). After 10 min incubation, worms are gently resuspended and slowly transferred from the Eppendorf to the inlet tube of the ScreenChip™ system, and using air pressure, a single animal is positioned between the electrodes. Then, once the tubing has been removed, the relative noise should drop significantly from unreadable to below 600 μV, and the individual pumps of the worm become visible. Now the recording can start using the NemAquire™ software. If a worm does not produce a pumping signal or multiple worms are between the electrodes, allow that worm to pass through and load another. After 2 min, the recording is stopped, the tubing is reattached, and the worm is washed out of the capillary so that a new worm can be loaded.

Each 2-min EPG recording is analyzed using the NemAnalysis™ software from InVivo Biosystems. Individual traces can be analyzed and a representative image generated (Fig. 5a). Data provided from the system show the number of pumps occurred during the recording, their duration, frequency, and amplitude, which can be plotted and analyzed statistically. Differences can be observed between worms of the same age but on different diets or in disease models. In the example shown, a 1-week-old nematode model for Parkinson's disease (strain, *NL5901*) has significantly fewer pumps compared to the same age as a wild-type animal (Fig. 5b).

3.7 Oxygen Consumption

A steadily increasing body of evidence is associating mitochondrial dysfunction with aging and placing it central to many diseases, including neurodegenerative disease, cancer, and diabetes. Mitochondria are responsible for the production of ATP, via substrate breakdown, e.g., glycolysis, followed by the tricarboxylic acid cycle providing substrates for the electron transport chain—consisting of

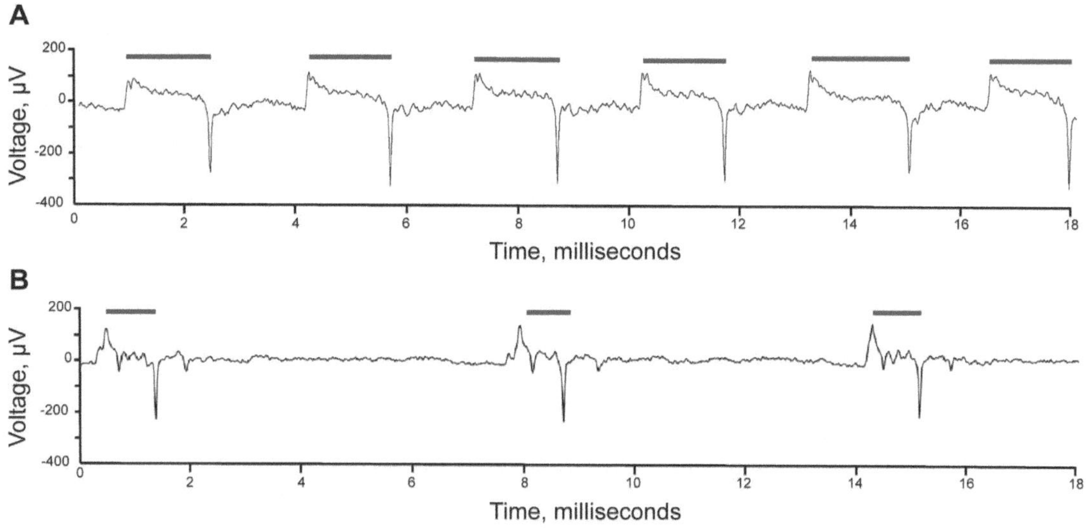

Fig. 5 Representative electropharyngeogram (EPG) from a wild-type worm and a worm with Parkinson's disease at 7 days post L4. Worms are loaded into the InVivo Biosystems ScreenChip™, and recordings are obtained using the associated NemAquire™ software and analyzed with NemAnalysis™ software. Each recording is 2 min long with a representative section of 18 ms (ms) shown. The red bars indicated a single complete pump. (**a**) A one-week-old wild-type worm, strain *N2,* which pumps six times in 18 ms, with each pump lasting for approximately 121 ms. The pumps are separated by 289 ms. (**b**) A worm expressing the human alpha-synuclein protein fused to YFP is a Parkinson's disease model (strain *NL5901*) and at 1 week of age has much fewer pumps per minute compared to wild-type worms of the same age. In 18 ms, there are just three pumps, which last for just 101 ms

the first four complexes of the oxidative phosphorylation system (OXPHOS). The proton motive force established by OXPHOS Complexes I, III, and IV subsequently drives ATP synthesis at Complex V. At Complex IV, terminal electrons react with oxygen, which results in a measurable oxygen consumption rate (OCR) using oxygraphic chambers, for example, in the seahorse extracellular flux analyser. While initially developed for cells, the seahorse platform is able to measure the oxygen consumption rate (OCR) in live *C. elegans* [81–83], thus providing a key indicator of mitochondrial function in a multicellular organism. The Seahorse XF analyser can be used to measure the OCR in nematode disease models [54, 84], in worms that have been exposed to various drugs [85, 86], or nutraceuticals [87].

Prior to working with the Seahorse XF Analyser, a series of optimization runs must be undertaken to ensure reliable estimation of OCR and to reduce variability. Additional experimental parameters require optimization, including the number of worms per well; the concentration of cyanide-*p*-trifluoromethoxyphenylhydrazone (FCCP), a protonophore that uncouples the mitochondrial membrane potential; and sodium azide, which is an inhibitor of the mitochondrial complex IV and thereby mitochondrial respiration (*see* **Note 7**).

To prepare worms for exposure to the Seahorse XF Analyser, first age-synchronized L1s are grown on NGM agar until they reach the desired stage. Worms are collected in a 15 mL Falcon tube using prechilled M9-buffer and washed three times in order to remove bacteria. The washing step is performed by centrifugation for 2 min at 500 g, removal of the supernatant, and addition of 10 mL fresh M9-buffer. After the final wash, worms are allowed to settle by gravity for 5 min, and the supernatant containing any last remaining eggs or small larvae is removed. The entire washing procedure is repeated twice.

The worms are then distributed into Seahorse XF96 (96-well) cell culture microplates, using at least six replicates per experimental condition. The pipette tips used to transfer the worms into the wells are first dipped into PBS with 0.05% (*v/v*) Triton X to prevent sticking of the worms to the plastic. Staged worms are aliquoted into each well of the plate, except for the four corner wells, which are used for background subtraction. Each well is then topped up to a total of 175 μL of M9-buffer, and the plate is now ready for analysis using the precooled Seahorse XF96 Flux analyser (*see* **Note 8**). The plate is left for approximately 1 h for the worms to calibrate, and then measurements can begin (Fig. 6a). Basal respiration is measured (four time points, mixing and measuring), followed by determination of the maximal respiration after FCCP addition (three time points) and non-mitochondrial respiration after sodium

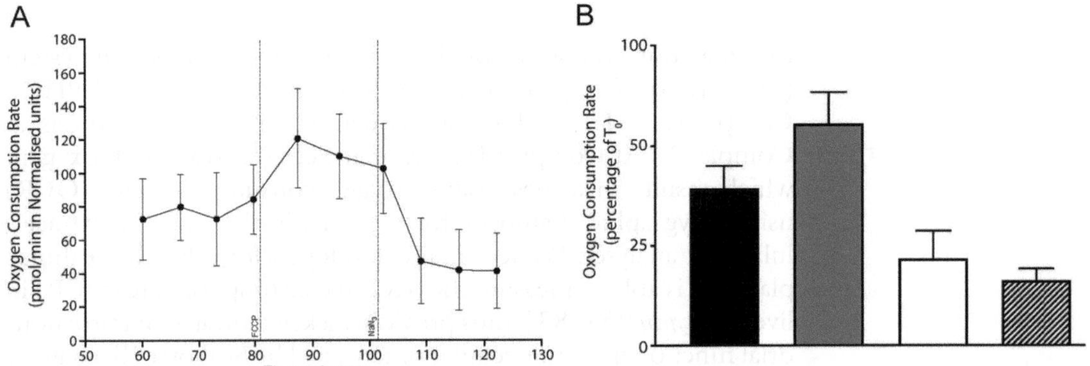

Fig. 6 Oxygen consumption rates in *C. elegans.* (**a**) Example trace of the oxygen consumption rate (OCR) in wild-type animals. The average of three replicates with standard deviation is shown. Dashed lines indicate when the cyanide-*p*-trifluoromethoxyphenylhydrazone (FCCP) and sodium azide (NaN_3) were added. Time points before 60 min are not shown, as this is the time required to allow the worms to settle in their new environment in the Seahorse machine. (**b**) Respiration states in wild-type *C. elegans.* T_0 is taken as the mean of the three measurements prior to the addition of FCCP. Routine respiration (black bar) is 78% with maximum respiration (the OCR after addition of FCCP; grey bar) 111%. Residual respiration is the non-mitochondrial oxygen consumption left after addition of sodium azide (white bar). Spare capacity is the difference between routine and maximum respiration (striped bar). Errors bars are standard error of the mean, calculated with at least three replicates

azide addition (three time points). Finally, the oxygen consumption (OCR) may be analyzed using Agilent's Wave software to calculate the routine, maximal, residual respiration (i.e., non-mitochondrial respiration) (Fig. 6b). Non-mitochondrial respiration is subtracted from all other respiratory rates, to correct for oxygen leakage during the experiment. Routine respiration is the oxygen consumption that is used to meet ATP demand under resting conditions, which in Fig. 6b is at ~79%. Maximum respiration is attained by adding the uncoupling agent FCCP, thereby causing the oxidative phosphorylation at maximum capacity. This increase is indicated by the spare capacity, representing the capacity to generate ATP via oxidative phosphorylation when energy demands increase. After the addition of sodium azide, the maximum respiration drops to the residual respiration to ~44% of T_0 as shown in Fig. 6b.

3.8 Other Considerations

The experiments described here are relatively simple ways in which physical performance or fitness can be measured in *C. elegans* as a proxy for mammals and humans. This is achieved due to the conserved muscle composition and organization between mammals and *C. elegans*. Additionally, such experiments in the nematode also provide a means to investigate how age, diet, drugs, and environment may affect muscle biology. However, there are a number of other methods that can be utilized to provide measurements of other parameters of muscle health.

Muscle strength is lost as humans age, but a decline in strength is also associated with disease and inactivity. It is important that there is a standard measure of strength in model organisms that have high relevance to human physiology. To this end, a microfluidics-based platform, NemaFlex, is able to measure strength in *C. elegans*, which can be robustly linked to maximum voluntary force in vertebrates [88]. In the NemaFlex system, the nematode crawls across the microfluidic device, deflecting micropillars as it moves [54, 85]. The pillars with maximum deflections are translated to force, and from this, the muscle strength of the nematode can be calculated.

Exercise is a powerful intervention with a variety of benefits to human health; however, how exercise promotes such benefits remains elusive. By engaging *C. elegans* in a series of swimming sessions, a so-called exercise regime, it becomes possible to dissect the molecular mechanisms behind the health benefits of exercise [89]. Worms that were given a 90-min swimming session were found to break down intramuscular lipids used to fuel high energy motion and were accompanied by a transient increase in mitochondria oxidative conditions, similar to that which occurs in mice and humans [90]. Strikingly, there was specific exercise-dependent transcriptional changes in genes associated with the oxidative stress pathways and downregulation of transcripts involved in glucose metabolism [89, 90].

The ability to maintain protein homeostasis is a key metric for health assessment, with elevated protein degradation observed during aging and in disease. Although many methods exist to do this using radiolabeled probes [91, 92], alternative methods can be just as informative. For example, by assessing the levels of cytosolic muscle proteins in transgenic worm strains [52] or using photo-convertible fluorescent proteins tagged to a human disease-causing protein such as Huntingtin [92], it is possible to observe and quantify the turnover of the protein.

In summary, given the range of neuromuscular assessment techniques and the vast array of genomic tools, the nematode worm *C. elegans* makes an excellent model system for forward and reverse translation of human neuromuscular research.

4 Notes

1. A single *OP50* colony is grown in LB medium overnight at 37 °C. Then, using a sterile 10 mL serological pipette (Stripette), the bacterial suspension can be gently spread over the NGM plate. The surface of the agar should not be broken, nor should the bacteria reach the edges of plates. The seeded plate should be left overnight at room temperature before being used.
2. When bleaching, always use glass pipettes to transfer worms to prevent them from sticking to plastic tips. Ideally leave the worms to hatch overnight in 15 mL Falcon tubes filled to 10 mL with M9-buffer on a rotating wheel. If this is not possible, place the eggs in a non-vented petri dish with enough M9-buffer to fully cover the bottom of the plate in a box with moist tissue at 15 °C.
3. MitoTracker is dissolved in Dimethylsulfoxide (DMSO) to a stock concentration of 940 μM and then further diluted to a working stock of 4.7 μM in M9-buffer. Age-synchronized worms (L4 or young adults) are then picked directly into 20 μL of the MitoTracker/M9-buffer solution and incubated at 20 °C [52]. After 1 h, the worms are washed twice with M9-buffer and then transferred to a microscope slide for imaging.
4. The properties of Pluronic® gel are thermodynamic, meaning the gel is liquid at 16 °C and solid at 21 °C, which makes the gel very susceptible to heat and conduction of heat. Therefore, a preexperiment check is required to ensure proper gelation. This is achieved by monitoring the gelation of a single drop of the gel on a plastic surface. Retrieve the Pluronic® gel from the

4 °C refrigerator by holding it at the top edge of the glass beaker, and place it on the bench on top of an empty upturned petri dish. Using a clean glass pipette, homogenize the gel by stirring for a few seconds, then dip a clean p200 pipette tip into the gel, and place a droplet on a petri dish lid. Return the Pluronic® gel to the refrigerator. After 1 min, the gel droplet should have solidified and be malleable. If the droplet remains liquid, re-stir the Pluronic® gel to ensure full homogenization. If the top of the droplet remains liquid but the bottom is resistant, the Pluronic® gel is not suited for use and has probably shifted in % *w/w*.

5. The amount of Pluronic® gel needed depends on the size of the well plate being used. For assays using Corning® 12-well plates, 2.73 mL Pluronic® gel per well is required so that the gel has a final height of 70 mm.
6. To prepare the ScreenChip™, gently fill the entire inlet tube with serotonin/M9-buffer (without worms). Insert the inlet tube into the ScreenChip™, and flush the ScreenChip™ with the solution, ensuring no debris, worms, or air bubbles are present between the electrodes. A relative noise recording (of approximately 10 s) should be performed before an experiment to ensure noise ratios remain below 20 μV and no spontaneous spikes (caused by surrounding activity of electronics or vibration) is seen. The amplitude should not exceed 20 μV in both the positive and negative voltage. When the relative noise ratio is below 20 μV, the device is ready to run with worms.
7. Before the experimental run, the concentrations of FCCP and sodium azide must be optimized. FCCP is dissolved in ethanol to 20 mM and then further diluted x125 in M9-buffer to 160 μM. Sodium azide is dissolved in MilliQ water to 10 M and then further diluted to a final concentration of 90 mM. For our experiments, we have found that the most optimal final concentrations were 20 μM for FCCP and 10 mM for sodium azide. The number of worms per well must also be optimized, and we found 20 worms/well to be sufficient.
8. The probes must be calibrated and equilibrated before each experiment. Twenty-four hours before each run, a sensor plate should be calibrated in Seahorse calibrant (200 μL per well) at 37 °C without CO_2. The Seahorse XF respirometers are designed to function at 37 °C, but this temperature is too high for *C. elegans*. Therefore, the machine must be precooled to 20–25 °C so that the oxygen consumption measurements are all performed at temperatures ranging from 22 to 25 °C.

Acknowledgments

We would like to thank David van de Klashorst, Ambra Miagostovich, Stefan van Asten, and Jinke Baan Hofman for contributing to the worm images and Tom Schirris at the Radboud University Medical Center for supporting us with the seahorse experiments.

References

1. Brenner S (1974) The genetics of *Caenorhabditis elegans*. Genetics 77:71–94
2. Tissenbaum HA (2015) Using *C. elegans* for aging research. Invertebr Reprod Dev 59: 59063
3. Son HG et al (2019) Age-dependent changes and biomakers of aging in *Caenorhabditis elegans*. Aging Cell 18(2):e12853
4. Ermolaeva MA, Schumacher B (2014) Insights from the worm: the *C. elegans* model for innate immunity. Semin Immunol 26(4):303–309
5. Hunt PR (2017) The *C. elegans* model in toxicity testing. J Appl Toxicol 37(1):50–59
6. Giunti S et al (2021) Drug discovery: insights from the invertebrate *Caenorhabitis elegans*. Pharmacol Res Perspect 9(2):e00721
7. Caldwell KA et al (2020) Modeling neurodegeneration in *Caenorhabditis elegans*. Dis Model Mech 13(10):dmm046110
8. Markaki M, Tavernarakis N (2010) Modelling human diseases in *Caenorhabditis elegans*. Biotechnol J 5(12):1261–1276
9. Kaletta T, Hengartner MO (2006) Finding function in novel targets: *C. elegans* as a model organism. Nat Rev Drug Discov 5(5): 387–398
10. Koch K et al (2014) *Caenorhabditis elegans* as model system in pharmacology and toxicology: effects of flavonoids on redox-sensitive signalling pathways and ageing. Sci World J 2014: 920398
11. Kwok TCY et al (2006) A small-molecule screen in *C. elegans* yields a new calcium channel antagonist. Nature 441(7089):91–95
12. Lucanic M et al (2018) A simple method for high throughput chemical screening in *Caenorhabditis elegans*. J Exp Biol 133:e56892
13. Teo E et al (2020) A high throughput drug screening paradigm using transgenic *Caenorhabditis elegans* model of Alzheimer's disease. Transl Med Aging 4:11–21
14. Sohrabi S et al (2021) High-throughput behavioral screen in *C. elegans* reveals Parkinson's disease drug candidates. Commun Biol 4: 203
15. Corsi AK et al (2015) A transparent window into biology: a primer on *Caenorhabditis elegans*. Genetics 200(2):387–407
16. Byerly L et al (1976) The life cycle of the nematode *Caenorhanditis elegans*.I: wild type growth and reproduction. Dev Biol 51(1): 23–33
17. Shaye DD, Greenwald I (2011) OrthoList: a compendium of *C. elegans* genes with human orthologs. PLoS One 9(1):10
18. Lehner B et al (2006) Systematic mapping of genetic interactions in *Caenrohabditis elegans* identified common modifiers of diverse signalling pathways. Nat Genet 38(8):896–903
19. Sulston JE, Horvitz HR (1977) Post-embryonic cell lineages of the nematode, *Caenorhabditis elegans*. Dev Biol 56(1):110–156
20. Consortium TCeS (1998) Genome sequence of the nematode *C. elegans*: a platform for investigating biology. Science 282(5396): 2021–2018
21. Hillier LW et al (2005) Genomics in *C. elegans*: so many genes, such a little worm. Genome Res 15(12):1651–1660
22. Lai CH et al (2000) Identification of novel human genes evolutionarily conserved in *Caenorhabditis elegans* by comparative proteomics. Genome Res 10(5):703–713
23. Fire A et al (1998) Ingestion of bacterially expressed dsRNAs can produce specific and potent genetic interference in *Caenorhabditis elegans*. Nature 391:806–811
24. Timmons L et al (2001) Ingestion of bacterially expressed dsRNAs can produce specific and potent genetic interferance in *Caenorhabditis elegans*. Gene 263(1–2):102–112
25. Kamath RS, Ahringer J (2003) Genome-wide screening in *C. elegans*. Methods 30:313–321
26. Chalfie M et al (1994) Green fluorescent protien as a marker for gene expression. Science 263:802–805
27. White JG et al (1986) The structure of the nervous system of the nematode. Philos Trans R Soc Lond Ser B Biol Sci 314(1165):1–340

28. Cook SJ et al (2019) Whole-animal connectomes of both *Caenorhabditis elegans* sexes. Nature 571(7763):63–71
29. Pirri JK et al (2009) A tyramine-gated chloride channel coordinates distinct motor programs of a *Caenorhabditis elegans* escape response. Neuron 62:526–538
30. Collins KM et al (2016) Activity of the *C. elegans* egg-laying behaviour circuit is controlled by competing activation and feedback inhibition. eLife 5:e211126
31. de Bono M, Maricq AV (2005) Neuronal substrates of complex behaviours in *C. elegans.* Annu Rev Neurosci 28:451–501
32. Stephens GJ et al (2008) Dimensionality and dynamics in the behaviour of *C. elegans.* PLoS Comput Biol 4(4):e1000028
33. Hughes S, Celikel T (2019) Prominent inhibitory projections guide sensorimotor computation: an invertebrate perspective. BioEssays 41(10):e1900088
34. Lecroisey C et al (2007) The *C. elegans* dense body: anchoring and signalling structure of the muscle. J Muscle Res Cell Motil 28(1): 79–87
35. Francis R, Waterston RH (1991) Muscle cell attachment in *Caenorhabditis elegans.* J Cell Biol 114(3):465–479
36. Gieseler K et al (2017) Development, structure and maintenance of *C. elegans* body wall muscle. WormBook
37. Ono S (2014) Regulation of structure and function of sarcomeric actin filaments in striated muscle of the nematode *Caenorhabditis elegans.* Anat Rec 297(9):1548–1559
38. Buckingham SD, Sattelle DB (2009) Fast, automated measurement of nematode swimming (thrashing) without morphometry. BMC Neurosci 10:85
39. Benian GM, Epstein HF (2011) *Caenorhabditis elegans* muscle: a genetic and molecular model for protein interactions in the heart. Circ Res 109(9):1082–1095
40. Stiernagle T (2006) Maintenance of *C. elegans.* WormBook
41. Meneely PM et al (2019) Working with worms: *Caenorhabditis elegans* as a model organism. Curr Protoc Essent Lab Tech 19:e35
42. Riley SP et al (2017) Culture of *Escherichia coli* and related bacteria. Curr Protoc Microbiol 15: 15
43. Breimann L et al (2019) Light-microscopy methods in *C. elegans* research. Curr Opin Syst Biol 13:82–92
44. Mackenzie JM Jr et al (1978) Muscle development in *Caenorhabditis elegans*: mutants exhibiting retarded sarcomere construction. Cell 15(3):751–762
45. Waterston RH et al (1980) Mutants with altered muscle structure in *Caenorhabditis elegans.* Dev Biol 77(2):271–302
46. Kim H et al (2017) Coherent raman imaging of live muscle sarcomeres assisted by SFG microscopy. Sci Rep 7:9211
47. Qadota H et al (2018) Protein phosphatase 2A is crucial for sarcomere organization in *Caenorhabditis elegans* striated muscle. Mol Biol Cell 29(17):2084–2097
48. Schultz J et al (2017) The secreted MSP domain of *C. elegans* VAPB homolog VPR-1 patterns the adult striated muscle mitochondrial reticulum via SMN-1. Development 144: 2175–2186
49. Saraskija S, Norman KR (2018) Analysis of mitochondrial structure in the body wall muscle of *Caenorhabditis elegans.* Bio Protoc 8(7): e2801
50. Kim E et al (2013) Long-term imaging of *Caenorhabditis elegans* using nanoparticle-mediated immobilization. PLoS One 8(1): e53419
51. Koopman WJH et al (2005) Mitochondrial network complexity and pathological decrease in complex I activity are tightly correlated in isolated human complex I deficiency. Am J Phys Cell Phys 289(4):C881–C890
52. Gaffney CJ et al (2014) Methods to assess subcellular compartments of muscle in *C. elegans.* J Vis Exp 2014(93):52043
53. Dilberger B et al (2019) Mitochondrial oxidative stress impairs energy metabolsim and reduces stress resistance and longevity of *C. elegans.* Oxid Med Cell Longev 2019: 6840540
54. Hewitt JE et al (2018) Muscle strength deficiency and mitochondrial dysfunction in a muscular dystrophy model of *Caenorhabditis elegans* and its functional response to drugs. Dis Model Mech 11(12):dmm036137
55. Regmi SG et al (2014) Age-dependent changes in mitochondrial morpholgy and volume are not predictors of lifespan. Aging 6(2):118–130
56. Glenn CF et al (2004) Behavioral deficits during early stages of aging in *Caenorhabditis elegans* result from locomotory deficits possibly linked to muscle frailty. J Gerontol A 59(12): 1251–1260
57. Ryu D et al (2016) Urolithin A induces mitophagy and prolongs lifespan in *C. elegans* and increases muscle function in rodents. Nat Med 22(8):879–888
58. Byrne JJ et al (2019) Disruption of mitochondrial dynamics affects behaviour and lifespan in

Caenorhabditis elegans. Cell Mol Life Sci 76(10):1967–1985
59. Baek J-H et al (2002) Using machine vision to analyze and classify *Caenorhabditis elegans* behavioral phenotypes quantitatively. J Neurosci Methods 118(1):9–21
60. Anderson GL et al (2004) Assessing behavioral toxicity with *Caenorhabditis elegans*. Environ Toxicol Chem 23(5):1245–1240
61. Yemini E et al (2013) A database of *C. elegans* behavioral phenotypes. Nat Methods 10(9): 877–879
62. Swierczek NA et al (2011) High-throughput behavioral analysis in *C. elegans*. Nat Methods 8(7):592–598
63. Tsibidis GD, Tavernarakis N (2007) Nemo: a computational tool for analyzing nematode locomotion. BMC Neurosci 8:86
64. Ramot D et al (2008) The parallel worm tracker: a platform for measuring average speed and drug-induced paralysis in nematodes. PLoS One 3(5):e2208
65. Leifer AM et al (2011) Optogenetic manipulation of neural activity in freely moving *Caenorhabditis elegans*. Nat Methods 8:147–152
66. Matthew MD et al (2012) WormScan: a technique for high-throughput phenotypic analysis of *Caenorhabditis elegans*. PLoS One 7(3): e33483
67. Nawa M et al (2012) Reduced expression of BTBD10, an Akt activator, leads to motor neuron death. Cell Death Differ 19(8):1398–1407
68. Koopman M et al (2020) Assessing motor-related phenotypes of *Caenorhabditis elegans* with the wide field-of-view nematode tracking platform. Nat Protoc 15:2071–2106
69. Luersen K et al (2014) Gait-specific adaptation of locomotor activity in response to dietary restriction in *Caenorhabditis elegans*. J Exp Biol 217(14):2480–2488
70. Schindelin J et al (2015) The ImageJ ecosystem: an open platform for biomedical image analysis. Mol Reprod Dev 82(708):518–529
71. Frezal L, Felix M-A (2015) The natural history of model organisms: *C. elegans* outside the petri dish. eLife 4:e05849
72. Lesanpezeshki L et al (2019) Pluronic gel-based burrowing assay for rapid assessment of neuromuscular health in *C. elegans*. Sci Rep 9(1):15246
73. Bilbao A et al (2018) Roll manuvers are essential for active reorientation of *Caenorhabditis elegans* in 3D media. PNAS 115(16):e3616–e3625
74. Herndon LA et al (2002) Stochastic and genetic factors influence tissue-specific decline in ageing *C. elegans*. Nature 419:808–814
75. Trojanowski NF et al (2016) Pharyngeal pumping in *Caenorhabditis elegans* depends on tonic and phasic signalling from the nervous system. Sci Rep 6:22940
76. Wellenberg A et al (2021) Use of *C. elegans* as a 3R-compliant *in vivo* model for the chemoprevention of cisplatin-induced neurotoxicity. Exp Neurol 341:113705
77. van de Klashorst D et al (2020) Montmorency tart cherry (*Prunus cerasus* L.) acts as a calorie restriction mimetic that increases intestinal fat and lifespan in *Caenorhabditis elegans*. J Funct Foods 68:103890
78. Ellwood RA et al (2021) Mitochondrial hydrogen sulphide supplementation improves health in the *C. elegans* Duchenne muscular dystrophy model. PNAS 118(9):e2019343118
79. Hao L, Buttner EA (2014) Methods for studying the mechanisms of action of antipsychotic drugs in *Caenorhabditis elegans*. J Exp Biol 84: 50864
80. Weeks JC et al (2016) Microfluidic EPG recordings show striking pharyngeal pumping phenotype in a *C. elegans* Alzheimer's disease model. MicroPubl Biol. https://doi.org/10.17912/W2WC7M
81. Koopman M et al (2016) A screening-based platform for the assesssment of cellular respirtaion in *Caenorhabditis elegans*. Nat Protoc 11(10):1798–1816
82. Sarasija S, Norman KR (2019) Measurement of oxygen consumption rates in intact *Caenorhabditis elegans*. J Vis Exp 144:e59277
83. Ng LF, Gruber J (2019) Measurement of respiration rate in live *Caenorhabditis elegans*. Bio Protoc 9(10):e3243
84. Lopes AFC et al (2020) A *C. elegans* model for neurodegenertaion in Cockayne syndrome. Nucleic Acids Res 48(19):10973–10985
85. Du Preez G et al (2020) Oxygen consumption rate of *Caenorhabditis elegans* as a high-throughput endpoint of toxicity testing using the Seahorse XFe96 Extracellular Flux Analyzer. Sci Rep 10:4239
86. Andreux PA et al (2014) A method to identify and validate mitochondrial modulators using mammalian cells and the worm *C. elegans*. Sci Rep 4:5285
87. Fang EF et al (2017) Tomatidine enhances lifespan and healthspan in *C. elegans* through mitophagy induction via the SKN-1/Nrf2 pathway. Sci Rep 7:47208

88. Rahman M et al (2018) NemaFlex: a microfluidics-based technology for standardized measurement of muscular strength of *C. elegans*. Lab Chip 18(15):2187–2201
89. Laranjeiro R et al (2019) Swim exercise in *Caenorhabditis elegans* extends neuromuscular and gut healthspan, enhances learning ability, and protects against neurodegeneration. PNAS 116(47):23829–23839
90. Laranjeiro R et al (2017) Single swim sessions in *C. elegans* induce key features of mammalian exercise. BMC Biol 15:30
91. Visscher M et al (2016) Proteome-wide changes in protein turnover rates in *C. elegans* models of longevity and age-related disease. Cell Rep 16(11):3041–3051
92. Pigazzini ML, Kirstein J (2020) In vivo quantification of protein turnover in aging *C. elegans* using photoconvertable Dendra2. J Vis Exp 160:e61196
93. Fox RM et al (2007) The embryonic muscle transcriptome of *Caenorhabditis elegans*. Genome Biol 8(9):R188

Chapter 15

Methodologies to Quantify Skeletal Muscle Blood Flow/Perfusion

Eleanor J. Jones and Bethan E. Phillips

Abstract

Blood flow to skeletal muscles is an important determinant for delivery of oxygen and nutrients and removal of metabolic by-products, thereby facilitating muscle health and contractile function. Reflecting this, impairments in muscle blood supply occurring with advancing age or disease can restrict functional ability and be detrimental to overall health. Multiple methods have been used to measure muscle blood flow in both clinical and research environments, each with advantages and limitations to their use. Extensively used, contrast-enhanced ultrasound is a relatively simple, minimally invasive method using intravenously injected microbubbles to directly record perfusion activity in the skeletal muscle microvasculature. However, despite the relative ease of this technique and the clear utility of information on skeletal muscle blood flow when studying the neuromuscular system, research in this space is currently limited. The development of methodologies for measuring skeletal muscle blood flow has future potential for many applications in neuromuscular research.

Key words Skeletal muscle, Contrast-enhanced ultrasound, Perfusion, Blood flow

1 Introduction

The systemic circulation functions in transporting blood around the body to the many organ systems. Oxygenated blood is pumped by the heart through the arteries and progressively smaller vessels on the arterial side of the systemic circulation until it reaches capilliarized tissue. In relation to skeletal muscle, perfusion is important for the delivery of oxygen and nutrients, each of which are required for aerobic respiration to generate energy for muscle contraction and also for the removal of by-products—most notably carbon dioxide. In addition, other factors such as hormones, growth factors, and immune cells are also transported to skeletal muscle via the blood, enabling numerous other physiological functions such as protein synthesis, cell signaling, and inflammation [1].

Skeletal muscle is highly vascularized, which is functionally important for its contractile and metabolic actions. During

Philip J. Atherton and Daniel J. Wilkinson (eds.), *Neuromuscular Assessments of Form and Function*, Neuromethods, vol. 204,
https://doi.org/10.1007/978-1-0716-3315-1_15,

exercise, for example, there is a substantial increase in blood flow to the muscles to enable the delivery of sufficient oxygen. These increases are proportional to metabolic demand, and blood flow during exercise can be up to 100-times higher than resting values in large muscle groups such as the knee extensors [2]. It has been suggested that nutritive and non-nutritive blood flow distribution routes to supply skeletal muscle exist, with distribution between the two routes having significant effects on muscle metabolism [3]. For example, blood flow through the non-nutritive route acts as a vascular shunt, whereas so-called nutritive blood flow is through direct contact with muscle cells. The distribution of the circulation through these routes is controlled by precapillary sphincters located where arterioles end to form the microvasculature and influenced by feedback from active muscle fibers [4]. Further, in addition to muscle activity having a role in venous return by increasing pressure through compression of the veins, repeat bouts of exercise (i.e., in the form of exercise training) can enhance muscle blood flow by enhancing vasodilatory responses likely via nitric oxide produced by the endothelium, and/or by increasing blood flow capacity through capillary angiogenesis [5].

Contrary to the potential for exercise training to enhance skeletal muscle perfusion, reductions in skeletal muscle perfusion represent impaired oxygen and nutrient delivery, and these reductions are seen with advancing age and a myriad of clinical conditions such as type 2 diabetes and peripheral artery disease [6, 7]. Consequently, poor perfusion and resultant ischemia potentially underlie a number of metabolic situations observed in these conditions, including impaired glucose uptake and reduced muscle protein synthesis [8]. In addition, reduced muscle perfusion has also been implicated in impaired muscle function, a parameter independently associated with morbidity and mortality and directly associated with reduced independence and increased fall risk, particularly in older adults [9].

The ability to record blood flow and tissue perfusion in skeletal muscle enables further understanding of neuromuscular control in health, in response to disease and following environmental stimuli (i.e., exercise, nutrition, and drugs). Therefore, the aim of this chapter is to review the different techniques used to measure skeletal muscle blood flow and perfusion. It will then specifically focus on the use of contrast-enhanced ultrasound (CEUS) and its utility in neuromuscular research.

2 Methods

2.1 Doppler Ultrasound

Doppler ultrasound is a noninvasive method for continuous recording of blood velocity in conducting vessels that are the main supply route to a specified region. Most commonly, this technique is used to measure flow in the brachial or femoral arteries

[10]. However, although this method records blood flow in the major vessels supplying, for example, a limb, it is unknown what proportion will be delivered to the muscle and as such only provides an indirect estimate of skeletal muscle blood flow. This method calculates the velocity of blood flow through a large artery from the Doppler frequency shift that occurs as transmitted ultrasound waves of a specific frequency are reflected by the red blood cells passing through the vessel [11]. Doppler ultrasound allows for simultaneous imaging of the artery and a pulsed Doppler velocity signal, which provides the potential for determining the artery diameter as well as an estimation of blood flow and shear stress. The blood velocity multiplied by the cross-sectional area of the artery is used to determine blood flow. However, to obtain accurate flow estimates using Doppler, the limb should be in a fixed position, preferably around 60 degrees to the angle of the ultrasound beam, as the probe angle can influence velocity measurements, engendering this method highly operator-dependent [12]. In addition, anatomical features such as the bifurcation of the femoral artery cause turbulent blood flow, affecting the accuracy of results [13], and as such, recording distance from such features must be considered and maintained consistent. Beyond measuring blood flow velocity through a large vessel, Doppler ultrasound can also be used to assess flow-mediated dilation responses, e.g., when studying reactive hyperemia (a commonly used and accepted method of determining endothelial function [10]).

2.2 Xenon-133

The inert radioactive isotope Xenon-133 is a highly diffusible gas, which when combined with saline can be injected intravenously and used to measure regional blood flow. This technique works by externally detecting clearance of Xenon-133 within the blood vessels using a crystal ratemeter and plotting clearance curves to assess blood flow in tissues with small arterial supplies such as skeletal muscle [14]. This method has been previously used to successfully quantify regional muscle blood flow changes in response to exercise [15, 16]. However, this method is not only invasive (requiring a venous cannula for the delivery of the isotope), but it also uses a radioactive agent and as such requires costly detection and shielding equipment (likely the reason for this becoming a less favored method in recent years). It also has limited reliability for measurements over time due to the position of the injection site (which is difficult to precisely replicate), having implications for Xenon-133 distribution and subsequent perfusion [17].

2.3 Near-Infrared Spectroscopy

Near-infrared spectroscopy (commonly known as NIRS) is a method primarily used to measure deoxygenation/oxygenation in skeletal muscle microvasculature and calculates muscle blood flow from equations for hemoglobin and oxyhemoglobin kinetics [18]. To obtain the required information for this calculation, a detector probe is placed on the muscle of interest, which detects the absorption of near-infrared waves that vary in wavelength depending on hemoglobin oxygenation [19]. This can then be used in combination with venous occlusion by cuff inflation to obtain muscle blood flow by measuring the rate of oxygen saturation during reperfusion on release of cuff occlusion [20]. As this method has more sensitivity for smaller vessels, it facilitates regional measurements of blood flow, making it more appropriate than Doppler for studying skeletal muscle. The advantages of this method are that it is noninvasive and can be performed during dynamic activities, which is common when studying skeletal muscle blood flow (i.e., during exercise). This method can also be used in combination with dye dilution to record muscle blood flow by detecting changes in absorption as blood replaces the dye bolus [21]. However, this method does not directly measure muscle blood flow, and when used with occlusion, the timing of cuff inflation can affect the reliability of results [22].

2.4 Thermodilution

Thermodilution with iced saline is used to measure regional blood flow and is often considered an alternative to indicator dye dilution combined with NIRS, as described above. Whereas the NIRS dye dilution method measures changing levels of infrared absorption from a dye (often indocyanine) injected intravenously [23], the thermodilution method involves a continuous intravenous infusion of cooled saline at a rate to cause a ~1 °C drop in blood temperature. Blood temperature is recorded using a thermistor catheter inserted into the vein, and muscle blood flow is proportional to the change in blood temperature following thermodilution [24]. This technique has been used successfully to measure muscle blood flow during exercise and in clinical populations [25, 26], with the benefits that it does not require advanced equipment for spectrophotometry nor multiple blood sampling, and is not complicated by recirculation of dye [10]. However, this technique is relatively invasive, and the temperature of the saline can cause discomfort. Further, it must not be performed for too long or too often to prevent tissue cooling.

2.5 Contrast-Enhanced Ultrasound

CEUS is now the most used real-time method for assessing tissue blood flow. Already used in clinical settings to determine kidney [27], cardiac [28], or tumor [29] perfusion, it can assess real-time microvascular blood volume (MBV) and blood flow velocity (MFV), facilitating determination of total microvascular blood flow (MBF) [30]. CEUS works by measuring the acoustic backscatter of intravascular gas-filled microbubbles, which are stabilized by an outer lipid capsule, using ultrasound. Tissue perfusion can be assessed through destruction of these microbubbles using high mechanical index (MI) ultrasound energy, often known as a flash, and the assessment of microbubble reappearance in a region of tissue captured by an ultrasound probe. A range of microbubble preparations are available, but choice is often restricted by legislation and international availability of the contrast agent [31]. Although this method is easy to set up and quick to measure, it is minimally invasive due to the requirement of a venous cannula to infuse the microbubbles, and the contrast agents required are relatively expensive.

CEUS has been used to investigate skeletal muscle microvascular function in response to exercise [32, 33] and disease [34]. However, one of the main limitations of CEUS for the assessment of muscle perfusion is the lack of validation for its application over time [35]. As such, the majority of skeletal muscle CEUS assessments are performed as a single measure or used to assess responses to an acute stimulus (e.g., nutrition or an acute bout of exercise). Nonetheless, this does not preclude carefully made measurements, e.g., using detailed anatomical landmarking, to compare across health vs. disease.

The remainder of this chapter will focus on CEUS due to its simplicity to perform and ability to directly measure skeletal muscle perfusion, making it a favorable method for use in both research and clinical settings. A comparison of the aforementioned methods is summarized in Table 1, and although this is not exhaustive, with more advanced techniques such as arterial spin labeling magnetic resonance imaging (ASL-MRI) and positron emission tomography (PET) also being used to measure skeletal muscle blood flow, these methods require expensive technical equipment and have limited capability for dynamic use [36].

Table 1
Overview of methods to assess skeletal muscle blood flow

Method	Measures	Strengths	Limitations
Doppler US	Blood flow in large vessels	Noninvasive	Operator- and technique-dependent
Xenon-133	Systemic clearance of radioactive agent		Expensive equipment Radioactive agent
NIRS	Deoxygenation/oxygenation of microvasculature	Noninvasive Used in dynamic conditions	Not a direct measure of blood flow
Thermodilution	Systemic reduction in temperature	Expensive equipment not needed	Tissue cooling
CEUS	Reperfusion of microbubbles in specific region	Direct measure in muscle	No validation for repeated measures over time

Abbreviations: *CEUS* contrast-enhanced ultrasound, *NIRS* near-infrared spectroscopy, *US* ultrasound

3 A Focus on CEUS

3.1 Principles of Contrast Agents

For real-time imaging of skeletal muscle microvascular blood flow using ultrasound, a contrast agent consisting of "microbubbles" is continuously infused intravenously. The microbubbles are formed of a stabilizing outer lipid layer encasing gas (most commonly air or a fluorocarbon), enabling visualization via ultrasound and distinguishing them from blood cells [37]. This microbubble structure also allows for the measurement of MBV and MFV by quantifying microbubble replenishment following destruction using a high-MI ultrasound "flash," which uses energy to disrupt the lipid outer shell causing the gas to diffuse away, resulting in the loss of the contrast ability. Although these principles are common to all microbubbles, the chemical components forming the microbubbles can vary depending on the agent used. For example, Sonovue™ (Bracco, Milan, Italy) uses sulfur hexafluoride (SF_6) microbubbles, whereas Optison™ (GE HealthCare, USA) and Definity® (Lantheus Medical Imaging Inc., USA) use perflutren (C_3F_8). These contrast agents are required to be reconstituted from their storage state to an active form before use. For example, Sonovue™ is provided as 25 mg of lyophilized powder in an SF_6 atmosphere, which needs to be combined with 5 mL 0.9% saline to form a dispersion with 8 μL of SF_6 microbubbles mL^{-1}, following vigorous shaking [31].

Using Sonovue™ as an example, the mean diameter of contrast microbubbles is 2.5 μm with 90% <6 μm; therefore, they are slightly smaller than red blood cells, which have a mean diameter of 6–8 μm [31]. Sonovue™ microbubbles have been demonstrated to be safe for use by having minimal impact on blood flow properties because of their size and as such do not present problems with blocking small exchange vessels [38]. In order for signals from the microbubbles to be detected by the ultrasound scanner, the microbubbles resonate at harmonics of the fundamental frequency of the ultrasound, therefore separating it from the signal reflected by surrounding tissue. By high-pass filtering of the signal, the reflected wave from the tissue is suppressed, making it appear dark, whereas detection of the harmonics, predominantly reflected by intravascular microbubbles, makes blood vessels appear white, independent of their size [39] (Fig. 1).

In addition to differences in composition, contrast agents can also vary in microbubble concentration at a given infusion rate. Given the linear relationship between the acoustic intensity (AI) generated by resonating microbubbles and their concentration [30], a difference in concentration will affect calculations of microvascular blood flow parameters, and this must be considered when interpreting CEUS-derived data. For example, Sonovue™ has a

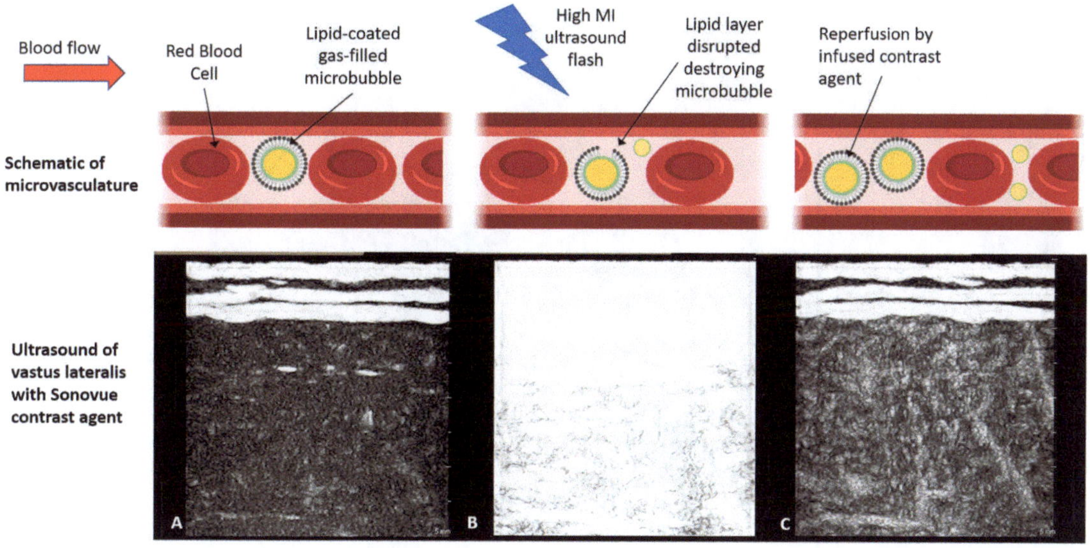

Fig. 1 Use of contrast agents to image muscle blood flow. (**a**) Muscle at rest. (**b**) High mechanical index (MI) flash to destroy microbubbles. (**c**) Reperfusion of muscle following bubble destruction after a short contraction

concentration of 93,000 bubbles/mL of plasma, whereas Definity® achieves a microbubble concentration more than eight times higher [40].

3.2 Setup and Operation of CEUS for Recording from Skeletal Muscle

Without rationale for assessment of a specific region of skeletal muscle (i.e., for the assessment of distal or proximal regions, which are known to have distinct physiological features [41]), the optimal position for ultrasound probe placement is suggested to be over the mid-belly of the chosen muscle and needs to be located prior to a CEUS protocol commencing. For the assessment of limb muscles, the ultrasound probe can be secured onto the muscle of choice using a housing unit and is appropriately secured (e.g., with Velcro™ circumferential straps) (Fig. 2). This setup is important as there needs to be minimal movement of the probe to maximize consistency of the muscle area across images for later analysis. When securing the probe into the housing unit, sufficient ultrasound gel should be used to obtain a clear image throughout the recording period, and before microbubble infusion begins, adequate time should be allowed for the temperature and pH of the probe to equilibrate with the skin.

Microbubbles are reconstituted according to manufacturer instructions and can then be delivered via an infusion pump and a low volume peripheral venous cannula connection line to an 18 GA (1.3 mmØ) cannula sited in the medial cubital vein of the preferred arm. The rate of microbubble infusion depends on the protocol requirements and amount of contrast agent available. For example,

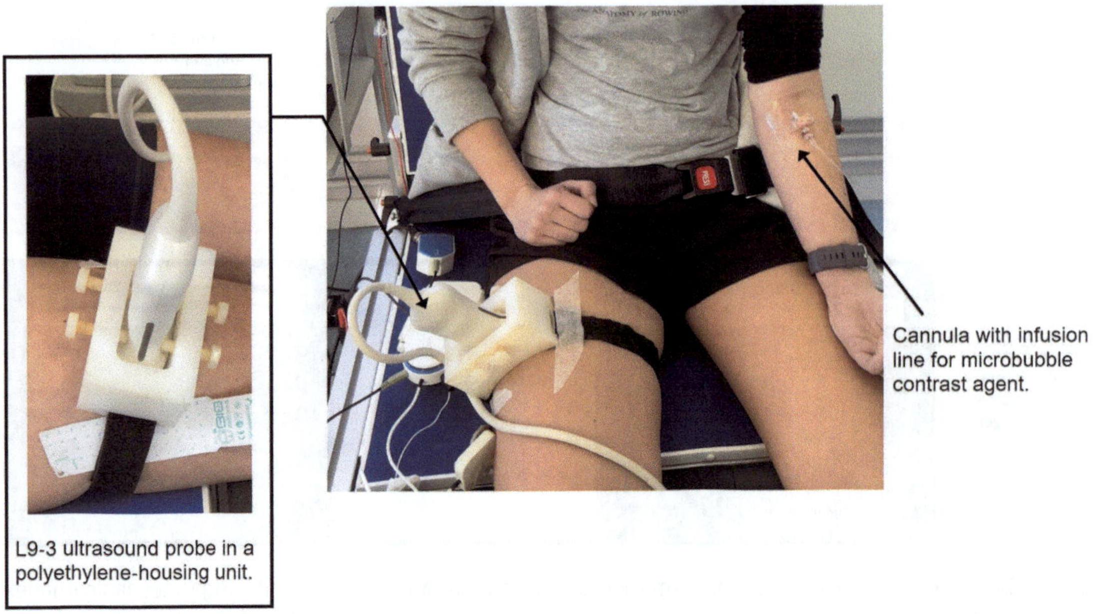

Fig. 2 Setup for contrast-enhanced ultrasound (CEUS) imaging of the vastus lateralis muscle performed using a Philips iU22 scanner and L9-3 probe (Philips Healthcare, USA)

Sonovue™ infused at a priming rate of 2 mL/min^{-1} for 1 min followed by 1 mL/min^{-1} for 3 min allows one vial of Sonovue™ to provide adequate volume for a 4-min recording period [31]. Achieving a steady-state concentration of microbubbles in the skeletal muscle vasculature before recording, whereby the muscle microvascular concentration of microbubbles is constant, is necessary in order to derive parameters of blood flow using this technique. Steady-state concentration of microbubbles in muscle vasculature can be achieved within ~2 min [31] using Sonovue™ at 1–2 mL/min^{-1}, with significantly longer time needed for alternative contrast agents (i.e., ~6 min for Definity® infused at ~1.35 mL/min^{-1} [40]).

As the CEUS signal can be influenced by the ultrasound settings (i.e., depth) and acoustic parameters (i.e., frequency), it is important that these remain constant throughout [36]. For example, recordings made in "Contrast Mode" (which allows use of a low mechanical index for visualization and prevents destruction of the contrast agent) minimize tissue-dependent signal while maximizing the proportion of recorded signal that is reflected off microbubbles by selective register of harmonics [39]. Depending on the requirements, other settings such as depth, frequency, and resolution should be adjusted to maximize the size and focus of the area of interest and to optimize temporal resolution. It is important that settings remain constant not only throughout a measurement but also across sessions or conditions if making multiple assessments to allow comparison across recordings and calculated parameters.

One setting that is key to CEUS is the mechanical index (MI). The MI measures the potential for mechanical damage to occur to tissue by ultrasound pulses and is a unitless index. A low MI can be achieved using higher frequency ultrasound [39]. For many clinical and research purposes, an MI of ~0.08 enables contrast agent microbubbles to resonate without being destroyed, which allows visualization. Previous protocols to assess MBV in skeletal muscle have used a combination of short, intermittent high-MI flashes (MI > 1.0) followed by continuous, real-time, low MI recordings to measure microbubble replenishment after disruption (caused by the high-MI flash), thereby facilitating measures of MFV and MBV [31, 40] (Fig. 1).

3.3 Analysis of CEUS Data

CEUS video recordings can be exported to quantification software (e.g., Q-Lab (Philips) or VueBox™ (Bracco)), which permit the selection of regions of interest (ROI) from the recorded ultrasound image to (if selected appropriately) minimize signal contribution from connective tissue and large rapid-filling vessels (Fig. 3). From selected ROI, the mean background AI from tissue echogenicity (where all microbubbles will have been destroyed [31, 40]) can be calculated from the period directly following a flash (~0.5 s) and subtracted from all subsequent values from recordings occurring during microbubble replenishment. These values can then be

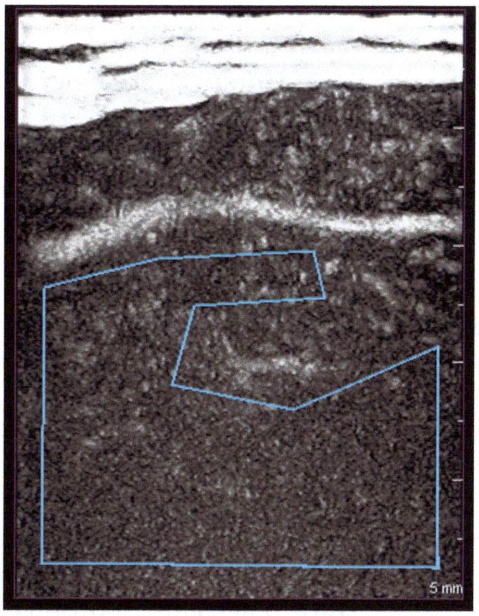

Fig. 3 Selected region of interest (ROI) for analysis of echogenicity from captured images, excluding rapid-filling vessels

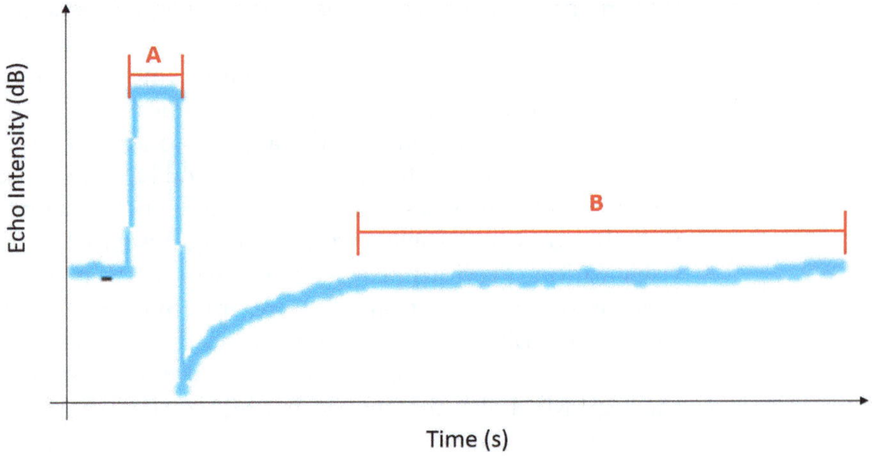

Fig. 4 A reperfusion curve showing the echo intensity during the flash (A) and the microvascular blood volume (plateau of the curve, B)

plotted as a replenishment curve and used to calculate microvascular flow parameters of MFV, MBV, and MBF (Fig. 4).

MFV can be calculated by measuring the reperfusion rate of the microbubbles in the muscle microvasculature following destruction with a high-MI flash, given the infusion of contrast is continuous. The measurement of the concentration of microbubbles at steady state will provide an estimation of microvascular cross-sectional area or MBV, and this can be determined independently or as the point of plateau from a replenishment curve after a high-MI flash. MBF can then be calculated by multiplying MFV by MBV [30]. Alternatively, this can be expressed in equation form by expressing mean AI against time and fitting it to the exponential function of one phase association:

$$y = y_0 + A\,(1 - e - \beta t),$$

where A is the plateau in AI (MBV) and β is the rate constant of microvascular flow and is proportional to MFV, with their product $A\beta$, which is proportional to MBF [30, 31, 42].

3.4 Step-By-Step Guide for the Assessment of Microvascular Blood Flow with CEUS

The following guide is for the measurement of microvascular blood flow in the vastus lateralis muscle using a Philips iU22 ultrasound machine and L9-3 probe (Philips Healthcare, USA) with Sonovue™ contrast agent (Bracco, Milan, Italy).

3.4.1 Ultrasound Machine Setup

After switching on the machine, the identification fields should be completed before making any measurements to facilitate data capture. Thereafter, individual participant-based settings (i.e., depth) can be adjusted for specific requirements. The low mechanical index (MI) must be set at 0.08, with a high MI above 0.8 and

1.32 recommended. The machine should be in contrast mode with 2D gain set at 95%, contrast resolution should be C30, and the frame rate should be set at 21 Hz. In order to overcome ultrasound attenuation and achieve a uniform ultrasound signal throughout the tissue, a graduated gain with depth is recommended (with no gain cursor later than 62), e.g., by having the time gain compensation (TGC) sliders set at 62, 64, 66, 68, 70, 71, 71, and 71 for a 4 cm depth recording. To allow video capture as detailed below, in "setup," "print/network" should be selected, followed by "prospective," and the capture time is set to 30 s. Ensure that "contrast timer" and "manual flash" are both available options on the touch screen at this time.

3.4.2 Participant Preparation

After applying sufficient ultrasound gel between the skin and the ultrasound probe to prevent wave reflection at the boundary of air, the ultrasound probe can be secured over the mid-belly of the muscle (or alternative location of choice) using a housing unit and appropriately secured (e.g., with Velcro™ circumferential straps) to ensure minimal probe movement. In addition, the participant should remain still and relaxed throughout all CEUS assessments. Before microbubble infusion, adequate time should be allowed for the temperature and pH of the probe to equilibrate with the skin. A cannula should be inserted into a vein of the preferred arm for infusion of the contrast agent (Fig. 2).

3.4.3 Sonovue™ Preparation

Preparation and administration of the contrast agent should be done by a medically qualified person in accordance with manufacturer's instruction. Sonovue™ must be prepared and reconstituted before use, and this is done by injecting 5 mL of sodium chloride 9 mg/mL (0.9%) solution into the contents of the Sonovue™ vial. The vial is then shaken vigorously for 20 s to obtain a white milky homogeneous liquid, after which the desired volume of the dispersion can be drawn into a syringe. Sonovue™ should be shaken until administered and infused using an oscillating infusion pump (e.g., Bracco VueJect®, to prevent bubble destruction or separation), and PVC connection line is attached to the venous cannula.

3.4.4 CEUS Measurement Protocol

Before starting CEUS measurements, instruct the participants to remain very still and relaxed throughout; this is vital for accurate analysis. Participants should be instructed not to talk, unless to raise awareness of discomfort or unease at any time. The Sonovue™ should be infused at 2 mL/min for 1 min and 1 mL/min thereafter. To first record the rise of microbubbles to systemic steady state, as the infusion begins, select "contrast timer" on the touch screen, and immediately press "capture" to start a 30 s cine recording. At the end of the first cine, press "capture" immediately, and then repeat once more to capture 90 s of rise to steady state. To next record microvascular replenishment, select "capture" immediately followed by "manual flash" (on touch screen). This will record a 30 s replenishment curve (cine). At the end of the first cine, repeat the "capture-flash" for as many as the protocol requires (often three). At the end of the assessment session, the data can be saved by using the "end exam" button, and then export for analysis.

3.4.5 Data Handling and Analysis

CEUS cine recordings can be exported to Q-Lab (Philips) quantification software. In Q-Lab, a freehand region of interest (ROI, Fig. 3) can be selected from the first "capture-flash" cine, which should include as much muscle tissue as possible while excluding any visible artifact (e.g., shadow or connective tissue) and large rapid-filling vessels that are apparent directly after the flash. From this selected ROI, the acoustic index values occurring during microbubble replenishment can then be plotted into graphical software (e.g., GraphPad Prism) as a replenishment curve (Fig. 4) and used to calculate microvascular flow parameters of MFV, MBV, and MBF from the A and β values obtained from a one phase association nonlinear regression of mean echogenicity values.

4 Applications in Neuromuscular Research

With blood flow an important factor for muscle health, simple noninvasive methods for recording muscle blood flow such as CEUS are needed for further understanding of skeletal muscle maladaptation/adaptation, including in situations of disease and/or deconditioning. Early research using dilution techniques had limited ability to measure skeletal muscle blood flow, particularly during muscle contraction. CEUS has the potential to be used to study muscle perfusion in a range of physiological scenarios, including during acute contraction and in response to chronic intervention periods (e.g., exercise training or immobilization), but confirmation of its repeatability over time, including in situations where the underlying tissue composition may alter (i.e., with hypertrophy or atrophy), would enable more widespread use.

Considering the application of CEUS in neuromuscular research, changes in muscle blood flow have frequently been linked to muscle fatigue and task failure [43, 44], with associated mechanisms responsible for this suggested to include accumulation of metabolites such as lactate and potassium ions, affecting afferent feedback, hemoglobin desaturation, and reduced substrate delivery [45, 46]. Additionally, during sustained isometric contractions, blood flow can be temporarily limited and even occluded at high force levels due to increases in intramuscular pressure and anatomical obstruction [47], resulting in profound hyperemia on relaxation [48]. Related to these alterations in muscle blood flow during exercise, a body of research has demonstrated that resistance exercise training (RET) with occlusion can increase muscle strength and hypertrophy to a greater extent than standard (non-occluded) RET alone. This research not only highlights the key role of blood flow in muscle adaptation [49] but also offers a potential exercise training format for those unable to perform high-intensity RET, as low-load RET with occlusion has been shown to induce equivalent gains in muscle mass and function to higher intensity RET performed without occlusion [50]. The mechanisms behind the favorable outcomes associated with RET with occlusion are postulated to include metabolic stress from ischemia and mechanical tension [51], which impact neuromuscular function and muscle contractile properties during exercise [52].

At the muscle fiber level, type 1 fibers require a greater blood supply than type 2 fibers because of their greater oxidative capacity [53]. However, beyond this, there is currently limited research investigating muscle blood flow and its relationship with specific properties at the muscle fiber level, including those related to the motor unit (MU) such as MU size, number, and firing rates; all of which are known to impact muscle function [54]. Indeed, it is known that blood flow is important for the delivery of neurotrophins required for plasticity, repair, and regeneration of the motor nerves for MU remodeling to prevent muscle atrophy and fiber loss [55, 56]. However, to date, this relationship has been primarily based on measures of limb or large vessel blood flow, and as such, the true relationship between muscle blood flow and MU properties is unclear.

CEUS can be used to measure muscle blood flow and determine the relationship with neuromuscular function in a wide range of populations, including older adults, those with chronic disease and athletes. With age, research has demonstrated reductions in both muscle blood flow and control of blood flow [57–59], with these declines linked to reduced muscle strength and function [60]. Further, reductions in muscle blood flow have also been demonstrated to have a negative impact on muscle function and metabolism in individuals with type 2 diabetes [61].

5 Conclusion

Skeletal muscle blood flow is an important physiological parameter to study in neuromuscular research due to its role in muscle function per se and in maladaptation/adaptation with disease and exercise training. There are many methods that can be used to measure blood flow; however, CEUS offers the benefits of directly measuring muscle blood flow using a technique that is easy to use, cost-effective, and reliable. CEUS is currently underutilized in the research environment, and assessment of CEUS repeatability over chronic time frames would enable more widespread use in research, with scope for more work to be done particularly on the relationship between neuromuscular parameters such as MU firing patterns and muscle blood flow.

6 Notes

Although CEUS presents an effective methodology to directly measure skeletal muscle microvascular blood flow it is not without limitations. Of particular relevance to the study of neuromuscular physiology, this method is not suitable for use during dynamic movement, as accurate measurement is reliant upon maintaining the same ROI across multiple images. Although probe movement can be minimized through use of a custom-built housing unit secured to the participant and this has been shown to facilitate CEUS assessments before and after some forms of activity (e.g., knee extensions) [62], assessments in temporal proximity to more dynamic forms of activity (e.g., running or cycling) remain challenging. In addition, the repeatability of CEUS over time has not yet been validated. To achieve this, the probe position would need to be replicated at each measurement session to obtain the same ROI across sessions. This can be achieved by determining probe placement against structural anatomic landmarks (e.g., bone features) but cannot account for physiological changes in the limb that may occur over the study period, e.g., muscle hypertrophy from exercise training or adipose losses from dietary intervention. Finally, recording time is somewhat limited by the volume of contrast agent available, largely owing to the financial cost per vial (~ £50). Despite these limitations of CEUS and the relatively wide variety of alternative methods available to record blood flow, most are not able to directly assess muscle microvascular blood flow and may not be suitable for specific protocol requirements, especially those aiming to explore the relationship between muscle microvascular and neuromuscular function.

References

1. Kusters YHAM, Barrett EJ (2016) Muscle microvasculature's structural and functional specializations facilitate muscle metabolism. Am J Physiol Metab 310:E379–E387. https://doi.org/10.1152/ajpendo.00443.2015
2. Saltin B, Rådegran G, Koskolou MD, Roach RC (1998) Skeletal muscle blood flow in humans and its regulation during exercise. Acta Physiol Scand 162:421–436. https://doi.org/10.1046/j.1365-201X.1998.0293e.x
3. Clark MG, Rattigan S, Clerk LH, Vincent MA, Clark ADH, Youd JM et al (2000) Nutritive and non-nutritive blood flow: rest and exercise. Acta Physiol Scand 168:519–530. https://doi.org/10.1046/j.1365-201x.2000.00704.x
4. Segal SS (2005) Regulation of blood flow in the microcirculation. Microcirculation:33–45. https://doi.org/10.1080/10739680590895028
5. Laughlin MH, Roseguini B (2008) Mechanisms for exercise training-induced increases in skeletal muscle blood flow capacity: differences with interval sprint training versus aerobic endurance training. J Physiol Pharmacol. NIH Public Access:71–88. Available: /pmc/articles/PMC2654584/

6. Groen BBL, Hamer HM, Snijders T, van Kranenburg J, Frijns D, Vink H et al (2014) Skeletal muscle capillary density and microvascular function are compromised with aging and type 2 diabetes. J Appl Physiol 116:998–1005. https://doi.org/10.1152/japplphysiol.00919.2013
7. Duerschmied D, Zhou Q, Rink E, Harder D, Freund G, Olschewski M et al (2009) Simplified contrast ultrasound accurately reveals muscle perfusion deficits and reflects collateralization in PAD. Atherosclerosis 202: 505–512. https://doi.org/10.1016/j.atherosclerosis.2008.05.046
8. Adelnia F, Cameron D, Bergeron CM, Fishbein KW, Spencer RG, Reiter DA et al (2019) The role of muscle perfusion in the age-associated decline of mitochondrial function in healthy individuals. Front Physiol 10: 427. https://doi.org/10.3389/fphys.2019.00427
9. Landi F, Cruz-Jentoft AJ, Liperoti R, Russo A, Giovannini S, Tosato M et al (2013) Sarcopenia and mortality risk in frail olderpersons aged 80 years and older: results from iLSIRENTE study. Age Ageing 42:203–209. https://doi.org/10.1093/ageing/afs194
10. Casey DP, Curry TB, Joyner MJ (2008) Measuring muscle blood flow: a key link between systemic and regional metabolism. Curr Opin Clin Nutr Metabol Care. NIH Public Access:580–586. https://doi.org/10.1097/MCO.0b013e32830b5b34
11. Hoskins R (1990) Measurement of arterial blood flow by Doppler ultrasound. Clin Phys Physiol Meas 11:1–26
12. Limb RG, and skeletal muscle blood flow measurements at rest and during exercise in human subjects (1999) Proceedings of the Nutrition Society. CAB International, pp 887–898. https://doi.org/10.1017/S0029665199001196
13. Rådegran G (1997) Ultrasound doppler estimates of femoral artery blood flow during dynamic knee extensor exercise in humans. J Appl Physiol 83:1383–1388. https://doi.org/10.1152/jappl.1997.83.4.1383
14. Lassen NA, Lindbjerg J, Munck O (1964) Measurement of blood-flow through skeletal muscle by intramuscular injection of Xenon-133. Lancet 283:686–689. https://doi.org/10.1016/S0140-6736(64)91518-1
15. Styf J, Korner L, Suurkula M (1987) Intramuscular pressure and muscle blood flow during exercise in chronic compartment syndrome. J Bone Jt Surg - Ser B 69:301–305. https://doi.org/10.1302/0301-620x.69b2.3818765
16. Grimby G, Häggendal E, Saltin B (1967) Local xenon 133 clearance from the quadriceps muscle during exercise in man. J Appl Physiol 22: 305–310. https://doi.org/10.1152/jappl.1967.22.2.305
17. Marcus ML, Bischof CJ, Heistad DD (1981) Comparison of microsphere and Xenon-133 clearance method in measuring skeletal muscle and cerebral blood flow. Circ Res 48:748–761. https://doi.org/10.1161/01.RES.48.5.748
18. Nioka S, Kime R, Sunar U, Im J, Izzetoglu M, Zhang J et al (2006) A novel method to measure regional muscle blood flow continuously using NIRS kinetics information. Dyn Med 5: 5. https://doi.org/10.1186/1476-5918-5-5
19. Mancini DM, Bolinger L, Li H, Kendrick K, Chance B, Wilson JR (1994) Validation of near-infrared spectroscopy in humans. https://doi.org/101152/jappl19947762740. 1994 [cited 27 Jan 2021]. https://doi.org/10.1152/JAPPL.1994.77.6.2740
20. Van Beekvelt MCP, Colier WNJM, Wevers RA, Van Engelen BGM (2001) Performance of near-infrared spectroscopy in measuring local O2 consumption and blood flow in skeletal muscle. J Appl Physiol 90:511–519. https://doi.org/10.1152/jappl.2001.90.2.511
21. Louvaris Z, Habazettl H, Wagner H, Zakynthinos S, Wagner P, Vogiatzis I (2018) Near-infrared spectroscopy using indocyanine green dye for minimally invasive measurement of respiratory and leg muscle blood flow in patients with COPD. J Appl Physiol 125:947–959. https://doi.org/10.1152/japplphysiol.00959.2017
22. Lucero AA, Addae G, Lawrence W, Neway B, Credeur DP, Faulkner J et al (2018) Reliability of muscle blood flow and oxygen consumption response from exercise using near-infrared spectroscopy. Exp Physiol 103:90–100. https://doi.org/10.1113/EP086537
23. Boushel R, Langberg H, Olesen J, Nowak M, Simonsen L, Bulow J et al (2000) Regional blood flow during exercise in humans measured by near-infrared spectroscopy and indocyanine green. J Appl Physiol 89:1868–1878. https://doi.org/10.1152/jappl.2000.89.5.1868
24. Andersen P, Saltin B (1985) Maximal perfusion of skeletal muscle in man. J Physiol 366:233–249. https://doi.org/10.1113/jphysiol.1985.sp015794
25. Hillig T, Krustrup P, Fleming I, Osada T, Saltin B, Hellsten Y (2003) Cytochrome P450 2C9 plays an important role in the regulation of exercise-induced skeletal muscle blood flow and oxygen uptake in humans. J Physiol 546:

307–314. https://doi.org/10.1113/jphysiol.2002.030833

26. Richardson RS, Leek BT, Gavin TP, Haseler LJ, Mudaliar SRD, Henry R et al (2004) Reduced mechanical efficiency in chronic obstructive pulmonary disease but normal peak Vo2 with small muscle mass exercise. Am J Respir Crit Care Med 169:89–96. https://doi.org/10.1164/rccm.200305-627oc
27. Schneider A, Johnson L, Goodwin M, Schelleman A, Bellomo R (2011) Bench-to-bedside review: contrast enhanced ultrasonography—a promising technique to assess renal perfusion in the ICU. Crit Care BioMed Central:1–8. https://doi.org/10.1186/cc10058
28. Porter TR, Xie F (2010) Myocardial perfusion imaging with contrast ultrasound. JACC Cardiovasc Imaging:176–187. https://doi.org/10.1016/j.jcmg.2009.09.024
29. Kuenen MPJ, Mischi M, Wijkstra H (2011) Contrast-ultrasound diffusion imaging for localization of prostate cancer. IEEE Trans Med Imaging 30:1493–1502. https://doi.org/10.1109/TMI.2011.2125981
30. Wei K, Jayaweera AR, Firoozan S, Linka A, Skyba DM, Kaul S (1998) Quantification of myocardial blood flow with ultrasound-induced destruction of microbubbles administered as a constant venous infusion. Circulation 97:473–483. https://doi.org/10.1161/01.CIR.97.5.473
31. Mitchell WK, Phillips BE, Williams JP, Rankin D, Smith K, Lund JN et al (2013) Development of a new Sonovue™ contrast-enhanced ultrasound approach reveals temporal and age-related features of muscle microvascular responses to feeding. Physiol Rep:1. https://doi.org/10.1002/PHY2.119@10.1002/(ISSN)2051-817X(CAT)WOMENINPHYSIOLOGY(VI)GENDERAGE
32. Krix M, Weber MA, Kauczor HU, Delorme S, Krakowski-Roosen H (2010) Changes in the micro-circulation of skeletal muscle due to varied isometric exercise assessed by contrast-enhanced ultrasound. Eur J Radiol 76:110–116. https://doi.org/10.1016/j.ejrad.2009.05.007
33. Hildebrandt W, Schwarzbach H, Pardun A, Hannemann L, Bogs B, König AM et al (2017) Age-related differences in skeletal muscle microvascular response to exercise as detected by contrast-enhanced ultrasound (CEUS). McNeil P, editor. PLoS One 12:e0172771. https://doi.org/10.1371/journal.pone.0172771
34. Amarteifio E, Weber M-A, Wormsbecher S, Demirel S, Krakowski-Roosen H, Jöres A et al (2011) Dynamic contrast-enhanced ultrasound for assessment of skeletal muscle microcirculation in peripheral arterial disease. Investig Radiol 46:504–508. https://doi.org/10.1097/RLI.0b013e3182183a77
35. Thomas KN, Cotter JD, Lucas SJE, Hill BG, van Rij AM (2015) Reliability of contrast-enhanced ultrasound for the assessment of muscle perfusion in health and peripheral arterial disease. Ultrasound Med Biol 41:26–34. https://doi.org/10.1016/j.ultrasmedbio.2014.06.012
36. Gliemann L, Mortensen SP, Hellsten Y (2018) Methods for the determination of skeletal muscle blood flow: development, strengths and limitations. Eur J Appl Physiol. Springer Verlag:1081–1094. https://doi.org/10.1007/s00421-018-3880-5
37. Quaia E (2005) Classification and safety of microbubble-based contrast agents. Contrast Media in Ultrasonography. Springer-Verlag, Berlin/Heidelberg, pp 3–14. https://doi.org/10.1007/3-540-27214-3_1
38. Schneider M, Broillet A, Tardy I, Pochon S, Bussat P, Bettinger T et al (2012) Use of intravital microscopy to study the microvascular behavior of microbubble-based ultrasound contrast agents. Microcirculation 19:245–259. https://doi.org/10.1111/j.1549-8719.2011.00152.x
39. Quaia E (2005) Physical basis and principles of action of microbubble-based contrast agents. Contrast Media in Ultrasonography. Springer-Verlag, Berlin/Heidelberg, pp 15–30. https://doi.org/10.1007/3-540-27214-3_2
40. Sjøberg KA, Rattigan S, Hiscock N, Richter EA, Kiens B (2011) A new method to study changes in microvascular blood volume in muscle and adipose tissue: real-time imaging in humans and rat. Am J Physiol Circ Physiol 301:H450–H458. https://doi.org/10.1152/ajpheart.01174.2010
41. Blazevich AJ, Gill ND, Zhou S (2006) Intra- and intermuscular variation in human quadriceps femoris architecture assessed in vivo. J Anat 209:289–310. https://doi.org/10.1111/j.1469-7580.2006.00619.x
42. Krix M, Weber M-A, Krakowski-Roosen H, Huttner HB, Delorme S, Kauczor H-U et al (2005) Assessment of skeletal muscle perfusion using contrast-enhanced ultrasonography. J Ultrasound Med 24:431–441. https://doi.org/10.7863/jum.2005.24.4.431
43. Sjøgaard G, Savard G, Juel C (1988) Muscle blood flow during isometric activity and its relation to muscle fatigue. Eur J Appl Physiol

Occup Physiol 57:327–335. https://doi.org/10.1007/BF00635992

44. Clark BC, Collier SR, Manini TM, Ploutz-Snyder LL (2005) Sex differences in muscle fatigability and activation patterns of the human quadriceps femoris. Eur J Appl Physiol 94:196–206. https://doi.org/10.1007/s00421-004-1293-0
45. Enoka RM, Duchateau J (2008) Muscle fatigue: what, why and how it influences muscle function. J Physiol. Wiley:11–23. https://doi.org/10.1113/jphysiol.2007.139477
46. Lanza IR, Wigmore DM, Befroy DE, Kent-Braun JA (2006) In vivo ATP production during free-flow and ischaemic muscle contractions in humans. J Physiol 577:353–367. https://doi.org/10.1113/jphysiol.2006.114249
47. McNeil CJ, Allen MD, Olympico E, Shoemaker JK, Rice CL (2015) Blood flow and muscle oxygenation during low, moderate, and maximal sustained isometric contractions. Am J Physiol Integr Comp Physiol 309:R475–R481. https://doi.org/10.1152/ajpregu.00387.2014
48. Wesche J (1986) The time course and magnitude of blood flow changes in the human quadriceps muscles following isometric contraction. J Physiol 377:445–462. https://doi.org/10.1113/jphysiol.1986.sp016197
49. Hill EC, Housh TJ, Keller JL, Smith CM, Schmidt RJ, Johnson GO (2018) Early phase adaptations in muscle strength and hypertrophy as a result of low-intensity blood flow restriction resistance training. Eur J Appl Physiol 118:1831–1843. https://doi.org/10.1007/s00421-018-3918-8
50. Takarada Y, Takazawa H, Sato Y, Takebayashi S, Tanaka Y, Ishii N (2000) Effects of resistance exercise combined with moderate vascular occlusion on muscular function in humans. J Appl Physiol 88:2097–2106. https://doi.org/10.1152/jappl.2000.88.6.2097
51. Pearson SJ, Hussain SR (2015) A review on the mechanisms of blood-flow restriction resistance training-induced muscle hypertrophy. Sports Med. Springer International Publishing, pp 187–200. https://doi.org/10.1007/s40279-014-0264-9
52. Husmann F, Mittlmeier T, Bruhn S, Zschorlich V, Behrens M (2017) Published ahead of Print. Impact of blood flow restriction exercise on muscle fatigue development and recovery. Med Sci Sports Exerc 50:436
53. Mackie BG, Terjung RL (1983) Blood flow to different skeletal muscle fiber types during contraction. Am J Physiol Heart Circ Physiol 14:265–275. https://doi.org/10.1152/ajpheart.1983.245.2.h265
54. Piasecki M, Ireland A, Piasecki J, Stashuk DW, Swiecicka A, Rutter MK et al (2018) Failure to expand the motor unit size to compensate for declining motor unit numbers distinguishes sarcopenic from non-sarcopenic older men. J Physiol 596:1627–1637. https://doi.org/10.1113/JP275520
55. Cheng A, Wan R, Yang JL, Kamimura N, Son TG, Ouyang X et al (2012) Involvement of PGC-1α in the formation and maintenance of neuronal dendritic spines. Nat Commun 3:1–12. https://doi.org/10.1038/ncomms2238
56. English AW, Wilhelm JC, Ward PJ (2014) Exercise, neurotrophins, and axon regeneration in the PNS. Physiology 29:437–445. https://doi.org/10.1152/physiol.00028.2014
57. Holloszy JO, McCully KK, Posner JD (1995) The application of blood flow measurements to the study of aging muscle. J Gerontol Ser A Biol Sci Med Sci 50A:130–136. https://doi.org/10.1093/gerona/50A.Special_Issue.130
58. Dinenno FA, Jones PP, Seals DR, Tanaka H (1999) Limb blood flow and vascular conductance are reduced with age in healthy humans: relation to elevations in sympathetic nerve activity and declines in oxygen demand. Circulation 100:164–170. https://doi.org/10.1161/01.CIR.100.2.164
59. Payne GW (2006) Effect of inflammation on the aging microcirculation: impact on skeletal muscle blood flow control. Microcirculation 13:343–352. https://doi.org/10.1080/10739680600618918
60. Wigmore DM, Damon BM, Pober DM, Kent-Braun JA (2004) MRI measures of perfusion-related changes in human skeletal muscle during progressive contractions. J Appl Physiol 97:2385–2394. https://doi.org/10.1152/japplphysiol.01390.2003
61. Lalande S, Gusso S, Hofman PL, Baldi JC (2008) Reduced leg blood flow during submaximal exercise in type 2 diabetes. Med Sci Sport Exerc 40:612–617. https://doi.org/10.1249/MSS.0b013e318161aa99
62. Herrod PJJ, Atherton PJ, Smith K, Williams JP, Lund JN, Phillips BE (2021) Six weeks of high-intensity interval training enhances contractile activity induced vascular reactivity and skeletal muscle perfusion in older adults. GeroScience. https://doi.org/10.1007/s11357-021-00463-6

INDEX

Philip J. Atherton and Daniel J. Wilkinson (eds.), *Neuromuscular Assessments of Form and Function*, Neuromethods, vol. 204, https://doi.org/10.1007/978-1-0716-3315-1, © Springer Science+Business Media, LLC, part of Springer Nature 2023

N

O

P

S

T